AF726200

Complete Plain-English Walkthrough From Zero to FHIR Playbook

Guide to FHIR, HL7, and Modern Health Data Interoperability Deployment

C. Louis-Charles, PhD

Cybersoft Publishing LLC

Cybersoft Publishing LLC
Fort Washington, MD 20744

First Edition 2026

Disclaimer: Riverside General Medical Center and all characters described in this book are illustrative composites used to explain healthcare interoperability concepts. They do not depict any real institution, person, or situation. This book is for general education only. It does not provide medical, legal, or compliance advice. Consult appropriate licensed professionals for organizational decisions.

Contents

Introduction: What You Are Holding

What You Are Holding

There is a moment, familiar to nearly everyone who has worked in healthcare administration or health information technology, when the full weight of the system's fragmentation becomes impossible to ignore. It might come during a frantic phone call between a discharging nurse and a receiving clinic, both parties searching for a medication list that exists somewhere but cannot be found in the time available. It might come during a compliance review when someone traces a missing lab result across three different systems and discovers that each acknowledged receipt, but none triggered the workflow that should have followed. It might come during a boardroom conversation when a vendor demonstrates seamless, real-time data exchange, and a quiet voice at the table asks, "Why don't we do it that way?" And the honest answer is: "Because the technology we have and the technology they are demonstrating were built thirty years apart, speak different languages, and have never been properly introduced."

OVERCOMING HEALTHCARE SYSTEM FRAGMENTATION:
Bridging the Legacy-Modern Gap
CURRENT CHAOS: MOMENTS OF REALIZATION
Scenario A: Med List Lost (Time Ran Out)
MEDICATION LIST
Discharging Nurse
MEDICATION LIST
Receiving Clinic
Scenario B: Dead-End Labs (Workflow Failed)
ACKNOWLEDGE RECEIPT
Compliance Officer
Database
Database
Database
Workflow Stopped
THE TRUE BARRIER: THE ANSWER - TECHNOLOGY GAP
Seamless Data Exchange
Why don't we do it that way?
M7v2 Exchange
DATA HUB
BUILT 30 YEARS APART
1994
30
30 years apart
2024
SPEAK DIFFERENT LANGUAGES
Legacy HL7v2 text based code)
Collision/ Error
FHIR JSON
NEVER PROPERLY INTRODUCED
silo 2
Disconnected Systems
CONCERNED FRUSTRATION
CONCERNED PATIENT
PAPER RECORDS
LIMITED VISIBILITY
EMPOWERED PROVIDERS
ENGAGED PATIENTS
BETTER DECISIONS
There is a moment, familiar to everyone in healthcare... searching for medication lists, tracing lost lab workflows, and a simple boardroom question. Built 30 years apart, different languages, never properly introduced.

What you are holding is the book that explains that gap, not in abstractions, not in vendor marketing language, and not in the specialized shorthand that health IT professionals use among themselves when they have forgotten that most of the people making decisions about healthcare data are not health IT professionals. What you are holding is an accessible, story-driven guide to two of the most consequential and least understood acronyms in modern healthcare: FHIR and HL7. It is not a vendor implementation manual for a specific electronic health record platform, though it will give you the conceptual vocabulary to read one intelligently. It is not a hands-on laboratory guide with server installations and command-line walkthroughs, though it will give you the mental model to follow one without getting lost. It is not a legal compliance document, and nothing in these pages should be taken as formal legal or regulatory advice. What it is, precisely, is the book that the authors of vendor manuals assume you have already read.

Diagram Intro.1: The Gap Between Legacy and Modern Health Data Exchange

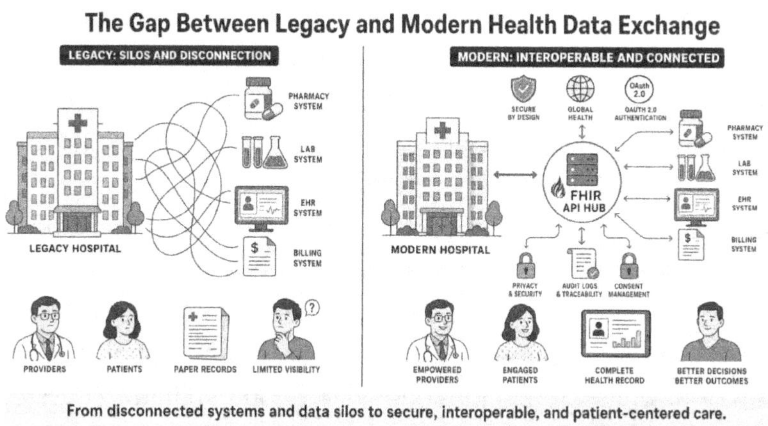

That assumption (that readers arrive already equipped with a working understanding of healthcare data standards) is one of the quiet structural failures of health IT education. It explains why talented hospital administrators sit through technical meetings they cannot fully parse and nod at decisions they cannot fully evaluate. It explains why junior developers assigned to a health IT project spend their first three months in a state of low-grade confusion, piecing together a conceptual framework from documentation that was written for people who already have it. It explains why compliance officers can quote regulations fluently but struggle to describe the technical mechanism by which their organization would demonstrate compliance if a federal auditor asked a specific question. The goal of this book is to close that gap. Not by simplifying healthcare data into something it is not, but by explaining what it actually is, building from first principles, and using the experiences of four fictional but thoroughly plausible people to make the concepts stick.

You may read this book linearly if you are a learner who wants to build knowledge progressively, from the historical origins of HL7 to the architectural decisions of an FHIR migration. The book is deliberately sequenced to reward that approach. Each chapter adds vocabulary and context that makes the next chapter easier, and each chapter's characters carry forward the consequences of the decisions they made in the previous one. Alternatively, if you are a reference reader who needs to understand a specific concept quickly -- clinical terminologies before a vendor meeting, the information blocking rule before a compliance review, data exchange patterns before an architecture decision -- you may enter at any chapter. Each chapter is written to stand alone, with enough context provided that you will not be lost even if you skipped the ones before it.

The two uses of this book are not in tension; they are the same book read at different speeds.

The North Star for every page of this book is the same. Healthcare interoperability has long been treated as a specialist domain: dense, jargon-heavy, and effectively gatekept by years of technical experience that most healthcare professionals do not have and should not need to do their jobs well. This book challenges that assumption. It demonstrates that the core concepts of healthcare data exchange are learnable, logical, and genuinely fascinating once the acronyms are stripped away and the human stakes are made visible. The goal is not to produce experts overnight. The goal is to produce informed participants who can ask better questions, evaluate vendor claims more critically, make decisions with genuine understanding of what they are deciding, and advocate for their patients and their institutions with the kind of confidence that comes only from actually knowing what you are talking about. That is a more modest goal than it sounds, and it is the only goal worth pursuing.

Three plain-English definitions are worth carrying into every chapter of this book, because they appear on nearly every page and they mean something more specific than casual usage suggests.

FHIR (pronounced "fire") stands for Fast Healthcare Interoperability Resources. It is a modern, web-based standard for exchanging healthcare data that uses the same foundational technologies as the websites and applications you use every day: REST APIs, JSON data structures, and OAuth authorization. FHIR was developed by HL7 International and published as a formal standard beginning in 2014. It defines a library of data structures called Resources -- Patient, Observation, MedicationRequest, DiagnosticReport, and dozens more -- each representing a discrete piece of clinical or administrative information in a format that any modern software system can read, write, and exchange over the internet. When someone says their organization is "moving to FHIR," they mean they are adopting this standard as the primary language for data exchange, replacing or supplementing older formats that do not use web-native technologies.

HL7 stands for Health Level Seven International. It is simultaneously the name of the standards organization that created FHIR and of a much older family of messaging standards that still powers the majority of healthcare data exchange in the United States today. The "Level Seven" in HL7's name refers to the seventh layer of the ISO networking model -- the application layer -- which is where data meaning lives, as opposed to the lower layers that govern how data travels across a network. The older HL7 standard most relevant to this book is HL7 version 2, typically written HL7 v2 or simply v2, which was first published in 1987 and has been transmitting admission notifications, lab results, medication orders, and discharge summaries between hospital systems ever since. HL7 v2 and FHIR are both HL7 standards; they are not competitors in the sense that one is right and the other is wrong. They are successive generations of the same mission, separated by thirty years of technology evolution.

Interoperability is the ability of different health information systems to exchange data and, critically, to use that data meaningfully once it arrives. The second half of that definition is the part most often omitted in casual usage. A system that can receive a lab result but cannot display it in the relevant clinical workflow is not interoperable in any useful sense; it is merely connected. A system that can transmit a medication list but does so in a format the receiving system cannot parse is not interoperable; it is noisy. True interoperability requires agreement at multiple levels: agreement on the transport mechanism (how data travels), agreement on the syntax (how data is structured), agreement on the semantics (what the data means), and agreement on the workflow (what happens after the data arrives). FHIR addresses all four of these levels, which is why it represents a genuine advance over its predecessors and why so much of this book is devoted to explaining how it does so.

Diagram Intro.2: Three Definitions Side by Side

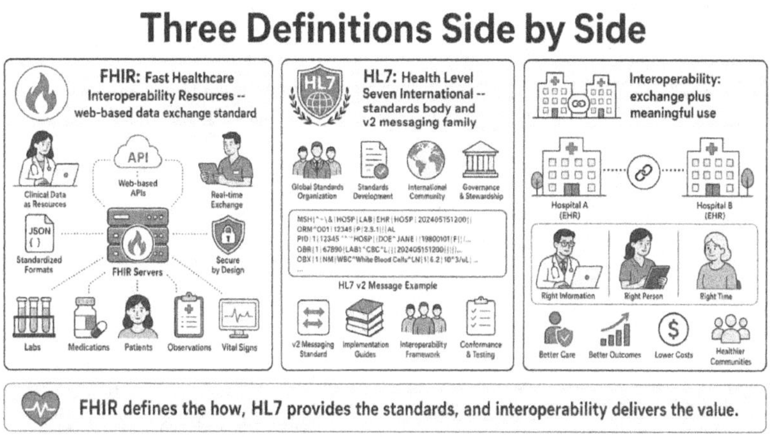

These three definitions are not trivia. They are the load-bearing vocabulary of every conversation this book contains, and investing sixty seconds in understanding them precisely now will prevent hours of confusion later. The reader who enters Chapter 1 with these three definitions in clear focus will find that the rest of the book assembles around them naturally, the way a jigsaw puzzle assembles more quickly once the corner pieces are placed.

Meet the Cast

The concepts in this book are taught through people. That choice is deliberate. Healthcare data standards are, at their core, a response to human problems: the problem of a patient whose record does not follow them from one provider to another, the problem of a nurse who cannot see a medication list at the moment of administration, the problem of an administrator who cannot prove to a federal auditor that their organization's systems are not blocking patient access to data. Abstract explanations of those problems can be accurate without being memorable. Narrative explanations -- explanations that put a named person in a specific situation and force that person to confront the problem in real time -- produce the kind of understanding that persists long after the book is closed.

Riverside General Medical Center is a 280-bed community hospital in the mid-Atlantic region — in every way that matters for this book, a typical American community hospital. It runs a legacy EHR customized over nine years and 217 active HL7 v2 integration routes connecting it to its laboratory, pharmacy, radiology, and a dozen other ancillary systems. It carries a compliance obligation under the ONC Information Blocking Rule it does not yet fully understand, and a board that has approved a "digital transformation" budget without a clear definition of what transformation means in this context. And it has four people who will spend the entire book figuring that out.

Diagram Intro.3: Riverside General's System Map at the Start of the Book

Riverside General's System Map at the Start of the Book

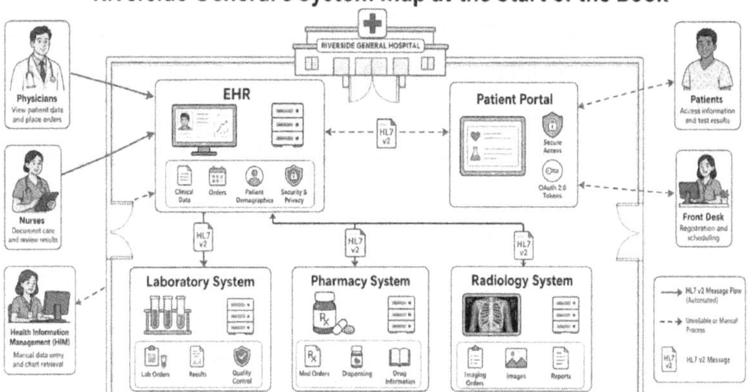

Riverside General relies on HL7 v2 messages to connect systems—creating a mix of automated and manual processes.

Patrick Osei

Patrick Osei is the administrator of Riverside General Medical Center. He holds an MBA from a state university he worked his way through on evenings and weekends while managing a medical records department, and he has spent fifteen years climbing from coordinator to director to administrator in the same health system. That trajectory matters because it means he has seen the inside of nearly every operational failure the hospital has produced. He was the coordinator who fielded the angry calls when a discharged patient's records did not arrive at the receiving facility. He was the director who sat through the post-incident review when a missing interface acknowledgment caused a lab result to sit in a queue for six hours. He was the administrator who signed the vendor contract for the integration engine upgrade that was supposed to fix both of those problems and, three years later, is still managing the exceptions that the upgrade did not address.

Patrick is practical, responsible, and deeply accountable: accountable to a board that wants results, to a medical staff that wants tools that do not break during rounds, to a compliance officer who sends him documents with alarming regularity, and to his own conscience about the patients whose records pass through every system he manages. He is not a technologist. He can navigate an EHR dashboard and read a system uptime report, but he cannot write a query or parse an HL7 message. What he can do is translate: he is the person who sits between a boardroom conversation about "digital transformation" and a whiteboard conversation about integration routes, and he has spent fifteen years learning how to make both sides feel heard.

Patrick's defining fear at the start of this book is specific and visceral. Not the abstract fear that something will go wrong, but the concrete fear, grounded in a specific morning eighteen months ago, that one more system failure will cost a patient their health and cost him his job. That morning -- a double-dose medication event caught only because a pharmacist happened to glance at a paper chart that the electronic record should have superseded -- is the event that opens Chapter 1, and it is the event that Patrick carries into every subsequent decision he makes. His defining growth by the end of the book is not that he has learned to write code or configure an integration engine. His defining growth is a genuine understanding of the infrastructure that either prevents or causes that kind of failure, and the administrative and organizational judgment to lead his institution through changing it.

Patrick is the reader's primary guide throughout this book. When a concept is introduced through Patrick's questions, those are the questions the reader should be asking. When Patrick finally understands something, the reader should feel that same click of comprehension. He is not a stand-in for the reader's ignorance. He is a stand-in for the reader's intelligence applied to a domain it has not yet mastered. The distinction matters because Patrick is not passive. He pushes back, asks follow-up questions, and occasionally rejects explanations that do not satisfy him -- which means the reader gets not just the first explanation but the better one that follows it.

Dr. Maya Chen

Dr. Maya Chen is Riverside General's Chief Medical Officer, a board-certified internist who practiced clinical medicine for twenty years before stepping into administration. She has seen three failed EHR implementations from the physician side: watching her colleagues adapt to new interfaces mid-shift, absorbing the cognitive load of systems designed for billing efficiency rather than clinical flow, and adjusting order sets that changed without warning and not always for the better. She has also seen one catastrophic interface failure in which a delayed lab result led to a patient being transferred to the intensive care unit with a condition that could have been caught four hours earlier. That event is the fact she carries into every technology discussion she enters, stated or unstated.

Maya's filter for every technology decision is simple enough to fit on a sticky note: "Will this hurt someone?" It is a question that sounds like caution but is actually a design requirement. She is not opposed to new technology; she is opposed to technology that has not been tested against the most important question in medicine. Her skepticism is not pessimism. It is clinical rigor applied to an administrative problem. She has been promised better systems more times than she can count, and she has watched most of those promises dissolve into six-month delays, scope creep, and the quiet reinstatement of paper-based backup procedures that were supposed to be temporary but became permanent because no one had the political will to remove them.

The specific professional experience that makes Maya indispensable to this book's teaching apparatus is her ability to translate between clinical meaning and technical specification. When Dev explains that a FHIR Observation resource can carry a lab result with a coded value, units of measure, a reference range, and an interpretation flag, Maya explains what happens in a clinical workflow when the interpretation flag is missing. When Sandra describes an information-blocking exception, Maya explains what it looks like from the patient's perspective. Her presence in technical chapters is not ornamental. She is the mechanism by which abstract standards acquire clinical consequence.

Dr. Chen is the book's conscience. She is the reminder -- appearing in every chapter where a technical decision carries a clinical consequence -- that healthcare data is not abstract. It is attached to a human being in a hospital bed, and the standard governing how that data flows is, ultimately, as much a clinical instrument as a technical one. When she approves a clinical decision support rule in the closing chapter, it is the narrative signal that the technology has passed the only test that ultimately matters.

Diagram Intro.4: The Four Characters and Their Domains of Authority

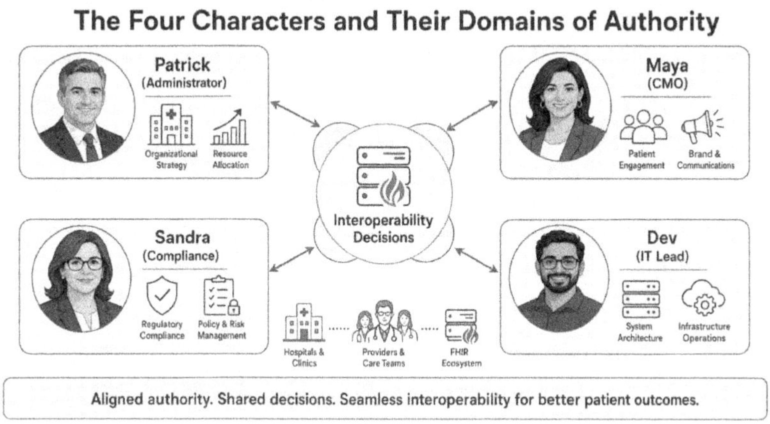

Dev Okafor

Dev Okafor is Riverside General's IT Lead and the person who built, maintains, and has essentially memorized every one of the hospital's 217 active HL7 v2 integration routes in Mirth Connect. He built the first of those routes twelve years ago as a junior analyst assigned to connect the new laboratory information system to the legacy EHR, working from HL7 v2.3 documentation he found in a binder on a shelf that had not been touched since the previous analyst left. He had no formal training in healthcare data standards. He had a working system in three weeks. He has been the single consistent technical mind behind the integration layer ever since.

Dev is brilliant and methodical. He documents everything, tests before deploying, and has a personal rule never to push a change to a production integration route on a Friday afternoon. His documentation is so thorough that the integration engine he maintains has 99.97% uptime over the past three years. His routes have not dropped a critical lab result in over a thousand days. He knows this not because he checked recently, but because checking is automatic: he reads the monitoring dashboard the way most people read email, and deviations from baseline register with the immediacy of a physical sensation.

Dev is deeply proud of what he built, and that pride is not vanity. It is professional satisfaction grounded in the knowledge that what he built works and that the people in the hospital depend on it to work every single day, including at three in the morning on a holiday weekend. His resistance to FHIR adoption at the start of the book is not stubbornness. It is the entirely rational position of a person who has watched technology migrations fail -- has watched, specifically, a middleware upgrade in 2019 that was supposed to take four hours and took eleven weeks, with seventeen production incidents and two unplanned maintenance windows -- and who is responsible for the system that cannot fail, even during the migration to the next system.

The arc of Dev's engagement with FHIR across this book is the technical heart of the narrative. He begins as a skeptic: fluent in HL7 v2, competent with the systems he manages, and unconvinced that any benefit FHIR provides is worth the risk of migrating. He ends as a practitioner: someone who has built FHIR facades over legacy systems, written SMART on FHIR application configurations, designed an audit logging architecture that satisfies Sandra's compliance requirements, and presented a 24-month migration roadmap to a board that approved it unanimously. That conversion is not a personality change. It is an education, and the reader undergoes it alongside him.

Dev is the reader's technical anchor. The pseudocode blocks in this book are written the way Dev would write them: clear, defensively structured, with error handling that assumes something will eventually go wrong because something always does. When Dev finally sits down, builds the first FHIR facade over Riverside General's legacy EHR, and watches it return a

Patient resource on the first try, the reader should feel that moment as a genuine narrative event, not merely a technical milestone.

Sandra Torres

Sandra Torres is Riverside General's Compliance and Privacy Officer. She has memorized the ONC Information Blocking Rule's eight exceptions the way other professionals memorize their children's birthdays: not because she was required to, but because forgetting one at the wrong moment has real consequences, and real consequences are the professional reality she chose when she left eight years of healthcare law practice to take a job on the provider side. That transition from outside counsel to inside compliance is the biographical fact that defines her professional posture. She understands both how the regulations are written and how they are enforced. She reads every ONC guidance document as soon as it is published and annotates it in the margins.

Sandra is risk-focused, regulatory-minded, and the first person Patrick calls when he is not sure whether something is legal, which in a hospital operating in the current regulatory environment is often. She is also the person who, from her years in litigation, understands that "we've always done it this way" is not a compliance defense, and that the gap between what a system was designed to do and what it actually does in practice is precisely where enforcement actions are born. Her relationship with technology is transactional: she wants to understand it well enough to assess its compliance risk, not to build it. She and Dev have a productive but sometimes tense professional relationship grounded in the fact that they are both rigorous and neither of them is wrong.

Sandra's presence in this book's teaching apparatus serves a function that is easy to underestimate. Compliance and privacy are not afterthoughts in healthcare data exchange; they are foundational design constraints. The information blocking rule, the HIPAA security rule, the FHIR consent resource, and the OAuth scope requirements for SMART on FHIR applications—these are not boxes to check after the architecture is designed. They are inputs to the architectural decisions themselves, and Sandra is the character who makes that clear in real time, often by asking a pointed question in a meeting that causes Dev to pause and reconsider a design choice he thought was already settled.

Sandra's growth across the book is the most complete character arc of the four. She begins by seeing FHIR primarily as a compliance obligation: another standard to certify against, another checklist to complete, another audit risk to document and manage. She ends by seeing it as a compliance tool -- a framework that, when properly implemented, makes audit logging cleaner, patient access more demonstrable, information-blocking defenses more defensible, and the relationship between the organization's technical practices and its legal obligations more transparent than ever. When she joins a federal advisory committee on health data privacy in the closing chapter, it is the narrative payoff of a character who stopped managing around the technology and started shaping it.

How Stakes Escalate Across the Book

The four characters at Riverside General do not face their challenges in isolation or in random order. The book is structured as a deliberate escalation. Each crisis is slightly larger than the last; each resolution requires applying a concept more sophisticated than the one before it; and each chapter builds on the knowledge established in the previous chapters in ways the reader will feel, even when they are not explicitly stated. Understanding why that escalation is structured as it is will help the reader approach each section with the right frame of reference.

Diagram Intro.5: The Arc of Escalating Stakes Across 14 Chapters

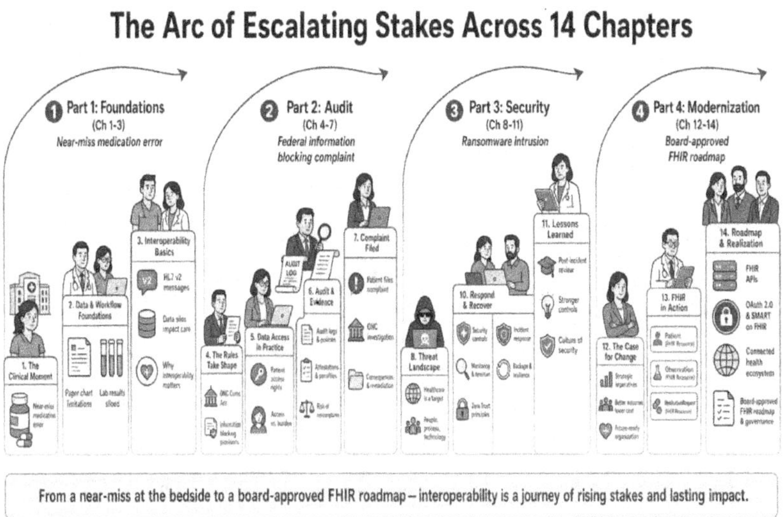

Chapters 1 through 3 open with a near miss. A double-dose medication event, caught only because a pharmacist happened to consult a paper chart that the electronic record should have superseded, forces Patrick and his team to confront the most basic question in healthcare data: why do the systems not talk to each other reliably, and what would it actually take to make them do so? These three chapters establish the vocabulary, the history, and the fundamental architecture of both HL7 v2 and FHIR. Chapter 1 situates healthcare data as a distinctive domain with distinctive problems -- privacy obligations, clinical stakes, legacy infrastructure -- that explain why health IT does not look like other IT. Chapter 2 tells the history of how HL7 v2 was created and why it became ubiquitous, giving Dev's deep familiarity with it an intellectual grounding that the reader shares by the end of the chapter. Chapter 3 introduces FHIR from first principles: what a Resource is, what an API is, what SMART on FHIR adds to the picture, and why Patrick -- who entered the chapter prepared to dismiss FHIR as another vendor initiative -- leaves it persuaded that something genuinely new is being proposed.

These three chapters do not yet require Riverside General to build anything. They require Riverside General to understand something, and the gap between the system they have and the system those concepts describe is the productive tension that carries the narrative forward. Patrick leaves Chapter 3 knowing what FHIR is. The next four chapters will show him what it costs.

Chapters 4 through 7 are organized around an escalation that begins with a letter. An anonymous information blocking complaint has been filed with the ONC against Riverside General, alleging that the hospital's patient portal makes it effectively impossible for patients to download their own records in a machine-readable format. The complaint is not frivolous. The portal was configured three years ago and has not been updated since a regulatory requirement that postdates the configuration was published. Sandra receives the notification on a Tuesday morning and forwards it to Patrick within the hour with a subject line that reads: "We need to talk today." This arc drives four of the book's most technically demanding chapters.

Chapter 4 takes Dev deep into HL7 v2 with a specific purpose: to understand the existing message flows well enough to assess which of them need to be exposed through a FHIR layer to satisfy the portal requirement. Chapter 5 examines the integration engine -- Mirth Connect in Riverside General's case -- as both a technology and an architectural pattern, and begins to describe what an FHIR facade would look like layered on top of it. Chapter 6 introduces clinical terminologies -- SNOMED CT, LOINC, RxNorm, ICD-10 -- which turn out to be the missing piece in why the portal's data cannot be made machine-readable without a substantial mapping effort. Chapter 7 introduces FHIR Profiles and the US Core Implementation Guide, which define the specific data elements and coding systems that a FHIR server must support to satisfy the regulatory requirement. By the end of Chapter 7, the team has a compliance response drafted and a preliminary technical plan. The stakes in this arc are financial, reputational, and legal, and that context shadows every technical decision.

The escalation in Chapters 8 through 11 arrives without warning on a Wednesday night, which is when escalations that matter tend to arrive. A ransomware intrusion penetrates the hospital's network perimeter, encrypts three file servers, and triggers a network isolation protocol that takes the integration engine offline for ninety minutes. The attackers do not successfully exfiltrate patient data. But the forensic investigation that follows reveals three architectural security gaps in the FHIR layer that Dev built during the compliance response: insufficient audit logging that would have made it impossible to prove, in the event of a breach, exactly which data was accessed and by whom; overly broad OAuth scopes in the SMART on FHIR application configuration that granted two vendor applications access to patient data they did not need; and a missing FHIR Consent resource implementation that would have prevented certain data sharing operations from proceeding without documented patient authorization.

None of these gaps caused the ransomware incident. But any one of them, in a different scenario, could have been catastrophic -- and the forensic team's report lands on Sandra's desk with three findings highlighted in red. This arc gives Dev and Sandra their most technically demanding challenge of the book: rebuilding the security and privacy architecture of the FHIR layer not as a ground-up design project but as a post-incident remediation, working under time pressure, with a board that is asking for weekly updates and a medical staff that is asking whether their patients' data was compromised. Chapter 8 covers the regulatory dimension of the incident, anchored in Sandra's expertise. Chapter 9 explores data exchange patterns in depth, specifically the patterns that need to be redesigned to close the first security gap. Chapter 10 takes Maya and Patrick into the patient access dimension, addressing what patients need to be told, what they have the right to access, and how to fill the Consent resource gap. Chapter 11, driven by Sandra and Dev together, completes the security and privacy architecture and closes all three findings.

Diagram Intro.6: The Three Security Gaps Found in Post-Incident Review

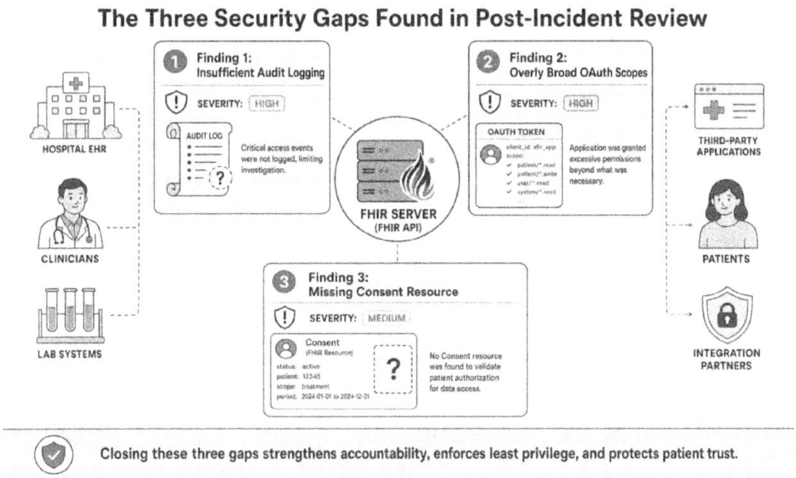

Between Chapter 11 and Chapter 12, the narrative pauses for a short interlude describing the 90 days of remediation work that followed the forensic report. The characters do not explain concepts in the interlude; they implement them. It is a compressed view of what it actually looks like to rebuild a system's security posture while keeping it running, and it serves as both a narrative breath and a practical illustration of how the concepts from Chapters 8 through 11 translate into operational reality.

Chapters 12 through 14 move from remediation to architecture. The board, having been briefed on the ransomware incident and the three security findings, and having watched the team respond with demonstrable competence, has approved a 24-month FHIR modernization roadmap. This is not a reactive project; it is a designed one, and it represents the culmination of everything the team has learned. Chapter 12 covers the architectural patterns available for FHIR modernization: facade patterns, parallel-run patterns, sunset patterns, and the considerations that govern which is appropriate for which part of the legacy system. Chapter 13 covers migration planning in detail: how to scope a migration, how to sequence it, how to test it, and how to communicate it to clinical staff in language they will actually understand and act on. Chapter 14 looks forward: FHIR R5, AI applications built on FHIR data, genomic data standards, and the questions that Riverside General has not yet answered but will need to soon.

The escalation across these fourteen chapters is not just narrative; it is pedagogical. A reader who enters Chapter 12 without having read the preceding eleven chapters will find it comprehensible but somewhat flat: a series of well-explained technical concepts without much emotional texture. A reader who has followed the characters from the near-miss through the compliance response through the ransomware incident will find Chapter 12 resonant differently, because they will understand why every architectural decision is difficult, what is at stake if it is wrong, and what it means to the specific people making it. The cumulative weight of the narrative is, precisely, the mechanism by which this book teaches. The closing section, which returns to Riverside General eighteen months after the board approved the roadmap, does not summarize what happened. It shows what changed, and asks the reader to evaluate that change in light of everything they now know.

The reader who finishes this book will not merely have read about healthcare interoperability. They will have followed four real people -- an administrator, a chief medical officer, an integration engineer, and a compliance officer -- as they built it, broke it, fixed it, and finally understood why it matters. That understanding is the point. It is the only thing this book is trying to give you, and it will be worth considerably more than the time it takes to acquire it.

1 Why Healthcare Data Is Different

Scenario

Patrick Osei did not expect the week to begin with a crisis. He arrived at Riverside General on a Tuesday morning with a mental agenda organized around a budget review, a vendor presentation to reschedule, and a follow-up with facilities about the new wing. The message from Dr. Maya Chen was waiting before he finished unlocking his office door.

The subject line read: "Incident Report -- Need You This Morning." A patient in the cardiology unit had received a double dose of a blood thinner. The hospital's electronic health record had not synchronized with the pharmacy dispensing system during the overnight window. A nurse had checked the EHR, seen no record of a morning administration, and reordered. The patient was found in time. The word "near-miss" appeared three times in two paragraphs.

Patrick read the message standing in his doorway, coat still on, coffee going cold. He had been at Riverside General for six years, had shepherded two major EHR upgrades, and had sat through more vendor demonstrations promising "seamless interoperability" than he could count. Six years in healthcare administration had taught him one thing with certainty: when a system fails in a hospital, someone gets hurt. Sometimes the harm is financial. Sometimes reputational. This time, it had almost been physical, and the failure mechanism was not a power outage or a hardware crash. It was a gap between two software systems that were supposed to communicate but didn't.

The emergency board meeting was convened by nine o'clock. Patrick sat at the far end of the conference table and listened as the chief information officer walked through the sequence of events, the pharmacy director described her team's corrective protocols, and the risk management officer outlined the reporting obligations triggered. The explanations were detailed and earnest. But a different question kept forming in his mind, one the technical presentation was not quite answering.

He had spent the previous weekend at his bank's ATM, which had processed a transaction involving his checking account, his bank's network, a national payments clearinghouse, and a merchant's point-of-sale system across two state lines, returning the correct balance within three seconds. Four separate systems, at least two belonging to organizations with no particular reason to cooperate, all exchanging relevant data in real time. Why can a bank tell you your account balance from any ATM in the world, but this hospital cannot reliably synchronize a medication record between its own pharmacy and its own patient floors?

The board meeting ended with an action list. Patrick drove home that evening with the quieter version of the same question still open: what exactly is it about healthcare data that makes this so hard? The answer to that question is this chapter.

Why It Matters

Healthcare data is not merely complicated. It is categorically different from almost every other kind of data that organizations manage. That difference has practical consequences for every decision about how to collect, store, move, and protect it.

The category distinction begins with the stakes attached to accuracy. In commercial data management, a single incorrect record is typically a nuisance. A mislabeled customer order generates a return shipment and a refund. In clinical data management, a single incorrect record can injure or kill a patient. A mislabeled medication dose, an allergy record that failed to transfer between systems, a lab result that arrived 12 hours late because the sending and receiving systems disagreed on the timestamp format: each is a potential patient safety event. The asymmetry is not a matter of degree. It is a matter of kind, and it shapes every design choice in healthcare IT, from redundancy requirements on message delivery systems to audit logging obligations on every access event.

The stakes are further complicated by the number of data categories a hospital must manage simultaneously. A single patient encounter at Riverside General generates clinical data (the diagnosis, ordered medications, lab results, vital signs recorded every four hours), administrative data (the admission record, bed assignment, insurance verification, discharge planning notes), and financial data (billing codes, insurance authorization, the claim submitted to the payer, the remittance advice received in return). All three data streams flow simultaneously, for the same patient, often accessed by different people using different systems for different purposes. The physician needs clinical data to make care decisions. The case manager needs administrative data to coordinate the discharge. The billing department needs financial data to submit the claim before the payer's filing deadline.

The complexity multiplies when the encounter expands beyond the hospital's walls. A single patient's care at Riverside General might involve the hospital's EHR, the primary care physician's separate practice management system, a reference laboratory, a pharmacy benefits manager, a commercial insurer, a state immunization registry, and a federal database managed by the

Centers for Medicare and Medicaid Services. Not one of these systems was designed to work with the others. Most were built by different vendors in different decades. All have different ideas about what a "patient identifier" means, what format a date should take, and what code should represent a diagnosis. This is not a technology failure in the conventional sense. It is the predictable outcome of an industry that built its information systems department by department, vendor by vendor, and decade by decade, without a mandate to share and without a common language to share in. Understanding why that situation developed is the foundation for understanding what FHIR is, why it matters, and what it can and cannot solve.

The Hospital as a City of Strangers

There is an analogy that Patrick would have found useful in that board meeting. Imagine a city that grew up over forty years without central planning, where each neighborhood was built by a different developer using different construction standards and a different local language. The pharmacy neighborhood speaks one proprietary dialect. The laboratory speaks another. The radiology suite has its own vocabulary and filing conventions. The administrator's office manages a records system that can read none of the above with complete accuracy. A patient walking through this city carries a medical history that no single neighborhood can fully read, because each neighborhood has only ever learned its own dialect.

Healthcare data silos are exactly this kind of city. They are not the product of negligence or malice. They are the cumulative result of systems designed to operate within a department or product line, purchased and configured without any expectation that they would one day need to exchange information with systems they were never designed to address. The laboratory information system that Riverside General's pathology department purchased in the late 1990s was an excellent system for its purpose. It was not designed to push results into an EHR that did not yet exist, and when that EHR arrived a decade later, the interface between them required a custom software project that took eight months and cost more than either system cost to purchase.

The layering of systems over time makes the situation more complex with each passing year. Each new system arrives with its own data model, its own API assumptions, and its own understanding of fundamental concepts like patient identity, encounter, and order. When two systems need to exchange data, the hospital typically builds a point-to-point interface: a custom piece of software that translates the output of one system into something the other can accept. A hospital with a dozen clinical systems can have dozens of such interfaces, each custom-built, each requiring its own maintenance, each a potential point of failure. When Patrick's pharmacy system and EHR failed to synchronize overnight, the interface between them relied on a timing assumption that no longer held after a recent software update. Nobody had noticed, because the interface had worked for three years without incident, and monitoring resources were focused on systems that were actively causing problems.

Diagram 1.1: The Hospital as a City of Strangers

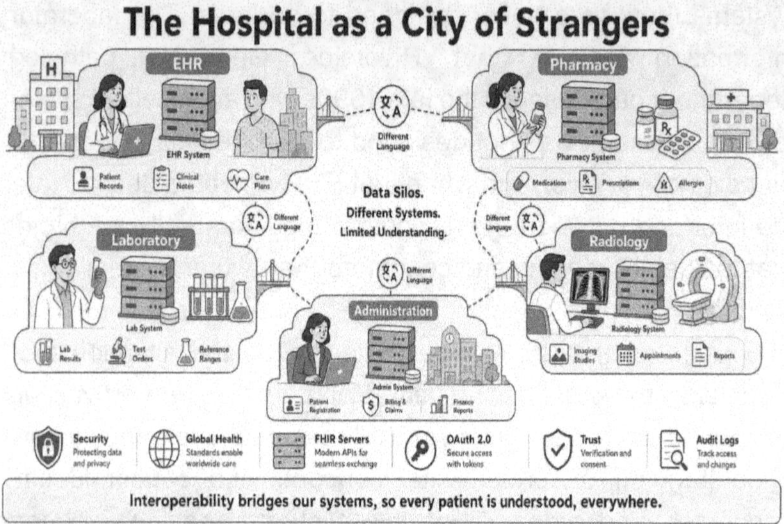

The city-of-strangers analogy clarifies something important: the problem is not that the neighborhoods are too far apart. The pharmacy and the EHR are in the same building, running on the same network, managed by the same IT department. The problem is that they speak different languages, and the translator between them was imperfect. Fixing that translation problem, permanently and at scale, is the mission for which healthcare interoperability standards exist.

Clinical, Administrative, and Financial Data

A hospital is, among other things, a data factory operating three simultaneous production lines, and the products of those three lines must be kept synchronized even though they are managed by different people, stored in different systems, and measured by different quality standards.

The first production line generates clinical data. Clinical data is the record of what happened to the patient medically: the chief complaint, the admitting physician's history, vital signs recorded at regular intervals, medications ordered and administered, laboratory tests ordered and results, imaging studies performed and interpreted, and diagnoses assigned at each stage of the encounter. Clinical data must be precise enough to drive care decisions. When a physician orders a medication, the order must specify the drug, dose, route, and frequency with enough precision that a pharmacist can detect a dose outside the normal range for this patient's weight, or a drug interaction with another medication the patient is already receiving.

The second production line generates administrative data: the registration event that created the patient's record, the insurance verification, the bed assignment, the transfer events that moved the patient between units, and the discharge event that closed the encounter and triggered billing and post-discharge follow-up. Administrative data must be timely enough to coordinate care. If the bed assignment system does not communicate in real time with the nursing unit, patients wait in the emergency department for beds that are ready but have not been released in the system. If the transfer event does not propagate to the pharmacy, medication orders are delivered to the wrong floor.

Diagram 1.2: Three Data Streams in One Encounter

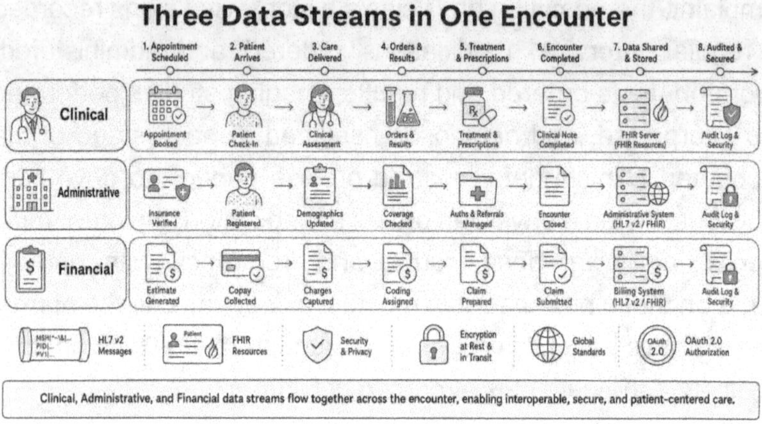

The third production line generates financial data: the procedure codes assigned to each service rendered, the diagnosis codes that justify medical necessity, the insurance authorization obtained before elective procedures, the claim submitted to the payer, the explanation of benefits returned, the payment posted to the account, and the balance billed to the patient. Financial data must be accurate enough to ensure the hospital gets paid and timely enough to meet payers' filing deadlines. A claim submitted after a payer's filing window is denied regardless of its clinical accuracy.

What makes the three-stream architecture genuinely difficult is not that any single stream is unmanageable in isolation. It is that all three must be derived from the same underlying encounter and must remain consistent with each other. The diagnosis in the clinical record must match the diagnosis on the insurance claim. The medication administered at 8 a.m. must appear in both the clinical administration record and the pharmacy dispensing record. When these streams fall out of synchronization, the consequences spread across all three simultaneously: a clinical documentation error becomes a billing error, which becomes a compliance risk, and a pharmacy synchronization failure becomes a patient safety event, which becomes an incident report, which becomes a regulatory notification.

Patrick understood the three-stream problem in the abstract before the near-miss. What the incident made concrete was that "must agree" is not a philosophical requirement. It is a technical one, and meeting it requires every system that touches a patient encounter to speak a language every other system understands.

The Multi-Stakeholder Ownership Problem

Healthcare data is unusual in the number of parties who have a legitimate claim to it. In most industries, data ownership is relatively simple: the organization that generated the data owns and controls it. Healthcare data does not work this way, and the departure from simple ownership is not a design choice. It is a legal and ethical requirement rooted in the data itself.

A patient's health record is simultaneously subject to the interests of at least seven categories of stakeholder. The patient has a legal and ethical right to access their own health information and to direct how those records are shared. The treating physician has a professional interest in the accuracy and completeness of the clinical documentation they created. The hospital bears custodial responsibility for the records and is the legal entity accountable to regulators for their management. The insurer has a contractual right to review clinical records in support of claims adjudication and prior authorization decisions. The reference laboratory may be required to retain copies for regulatory purposes independent of the hospital. The state public health department has a statutory right to receive certain reportable data. Federal agencies, including the Centers for Medicare and Medicaid Services, impose data retention and access requirements on any organization participating in Medicare or Medicaid.

Each of these stakeholders has a different relationship to the record, a different level of authorized access, and a different regulatory framework governing what they may do with it. A healthcare information system serving all of these stakeholders must enforce access controls that are simultaneously context-dependent, role-dependent, and purpose-dependent. This is not a problem that access control lists designed for corporate IT infrastructure were built to solve.

Diagram 1.3: The Multi-Stakeholder Data Ownership Web

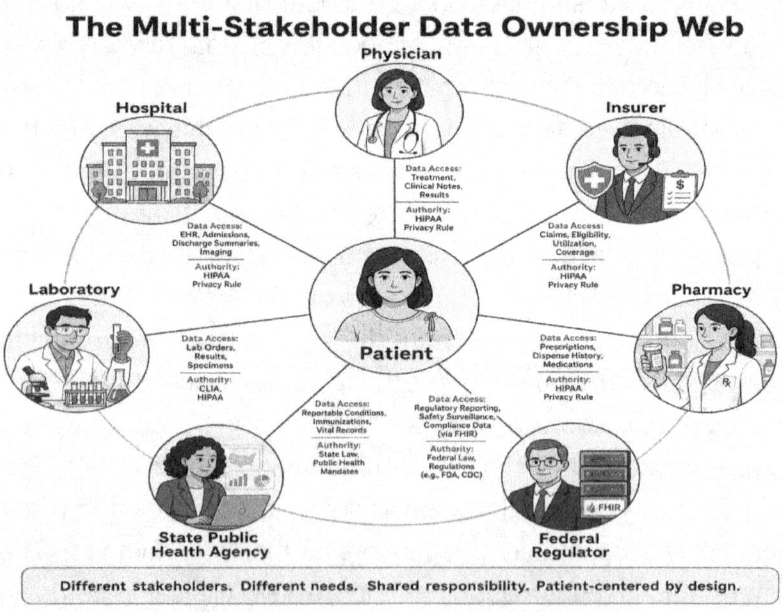

The multi-stakeholder problem also has a temporal dimension. Patients change insurance plans, change primary care physicians, move to different states, and age into Medicare eligibility. Each transition raises a data portability question: who is responsible for ensuring that the new stakeholder can access records created by the previous stakeholder? The 21st Century Cures Act of 2016 and its implementing regulations represent the federal government's most ambitious attempt to resolve this by establishing a legal right to data portability and a technical standard (FHIR) through which that portability can be implemented. How well that attempt is succeeding is a question that later chapters are better positioned to answer.

The multi-stakeholder problem is genuinely hard in a way that neither good intentions nor good technology can fully resolve. The interests of different stakeholders are not always aligned. The patient who wants their complete record shared broadly for continuity of care may have strong reasons to keep mental health records separate from records accessible to an employer-sponsored health plan. The insurer that wants complete clinical documentation for prior authorization review is the same entity whose cost-containment interests are sometimes at odds with the treating physician's clinical judgment. Technical standards create the infrastructure for navigating these tensions. They do not resolve them.

Why Accuracy Means Something Different Here

To understand why healthcare data systems are built with delivery confirmation at every step, audit logs on every access, redundant transmission pathways, and manual review processes that persist long after automation could theoretically replace them, it helps to be precise about what accuracy means in a clinical context.

In commercial data management, accuracy is measured statistically. A fulfillment center that ships the correct item 99.5% of the time is operating at a high level of quality. The 0.5% error rate generates customer service contacts and replacement shipments: undesirable but manageable. In clinical data management, the same statistical framework can yield different conclusions. A medication administration system that is correct 99.5% of the time means 0.5% of administrations are incorrect. At a hospital administering thousands of medication doses per day, that rate is dozens of patients per week receiving the wrong drug, dose, or route. Downstream safety checks will catch some of those errors. Many will not, because safety systems are only as good as the data flowing into them.

The medication near-miss that opened this chapter is a specific illustration of this principle. The pharmacy system and the EHR both functioned correctly in a narrow technical sense: the pharmacy system dispensed the medication as ordered, and the EHR recorded the order as placed. The failure was at the interface between them. The synchronization that was supposed to propagate the first-dose administration event from the pharmacy system to the EHR did not occur within the time window in which the second-dose decision was made. The result was that the EHR, which is the system a nurse consults before ordering a medication, had an incomplete picture of the patient's medication history. The nurse acted correctly on the information available. The information available was wrong.

Diagram 1.4: The Accuracy Asymmetry Between Healthcare and Commerce

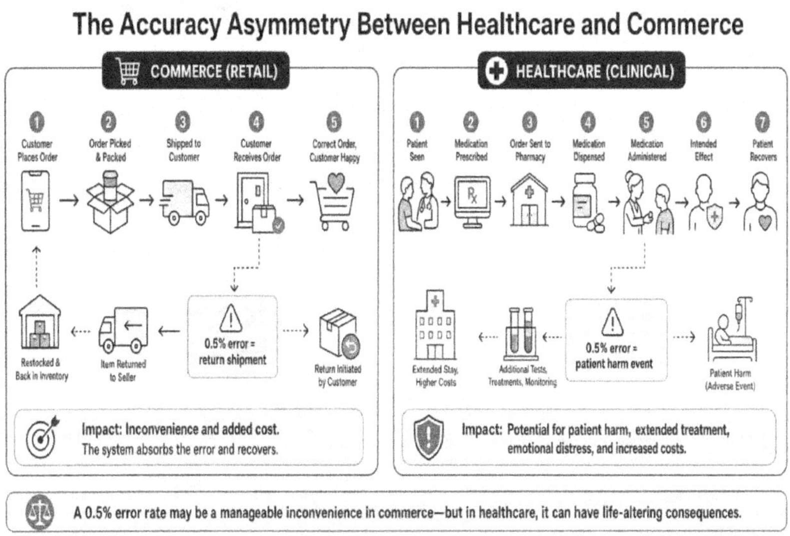

This accuracy requirement explains design choices that seem excessive by commercial IT standards. Healthcare messaging standards require delivery confirmation: when a system sends a message, it expects an acknowledgment confirming receipt and processing. If the acknowledgment does not arrive within a defined window, the sending system must retransmit or generate an alert for human review. Most commercial APIs operate on a fire-and-forget model, assuming success and handling errors when they surface. Healthcare cannot afford fire-and-forget because an undetected delivery failure may mean a patient does not receive a critical medication or that their allergy record does not propagate to the prescribing system.

Healthcare data systems also require audit logs on every access event. Every time a record is viewed, modified, transmitted, or deleted, the system must record who acted, when, from which system, and for what stated purpose. These logs are a mechanism for reconstructing the information environment at any given moment in a patient's care. The Riverside General pharmacy incident generated exactly this kind of reconstruction: investigators traced the exact time the first-dose event was recorded in the pharmacy system, the exact time synchronization was scheduled to run, and the exact time the nurse consulted the EHR -- and determined that the gap between those timestamps was the proximate cause of the near-miss. Patrick read the report three weeks later and, for the first time, understood exactly what each system had and had not known at each decision point. The audit trail was the difference between "we had a near-miss, we don't know why" and "we had a near-miss, here is the precise failure, and here is the remediation."

The Paper Legacy

There is a persistent misconception about the current state of healthcare digitization: that the problem of paper-based data is essentially solved. EHR adoption, driven by the HITECH Act's meaningful use incentive programs after 2009, moved enormous amounts of documentation into digital form. It did not eliminate paper, and the gap between what professionals assume about healthcare digitization and the reality they encounter on arrival is one of the most reliable sources of surprise for those entering the field from other industries.

Faxed referral letters, handwritten orders, paper intake forms, printed discharge summaries, and handwritten medication reconciliation lists are not historical artifacts in most health systems. They are current practice. The reasons are structural. Digital systems designed to replace paper workflows often failed to account for the complexity of clinical work. EHR systems deployed during the wave of meaningful use-driven adoption were frequently designed by engineers who understood databases but had limited experience with clinical workflow, resulting in systems that required more clicks, more structured data entry, and more time than the paper forms they replaced.

The cost of digitizing legacy paper records has consistently exceeded the budgets allocated to accomplish it. Decades of paper charts represent a data resource of immense clinical value, but converting them to digital format suitable for clinical use requires human review and structured data entry that current optical character recognition technology cannot fully replace. Organizations that have attempted comprehensive chart digitization projects typically find the effort larger, slower, and more expensive than initial estimates suggested, and the resulting digital records require additional remediation before they can be safely used in clinical decision support systems.

Diagram 1.5: The Paper Legacy in Healthcare Data Entry

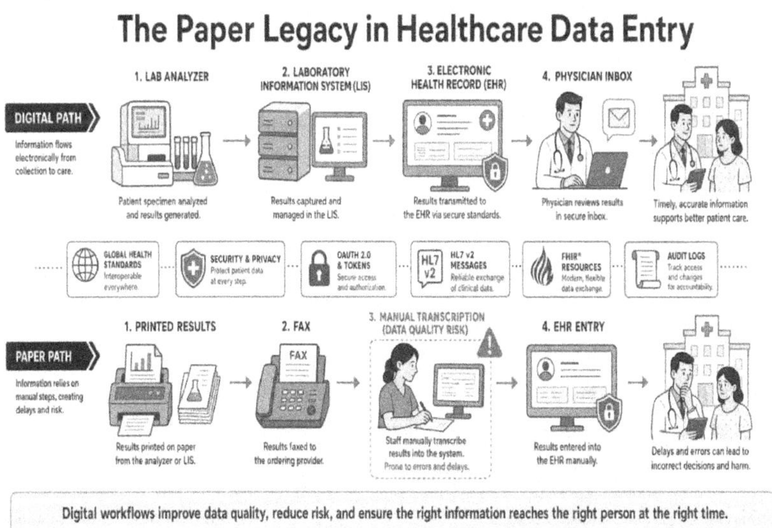

Every paper-originated data point that enters a digital system is a transcription event, and every transcription event is a potential source of error. A medical assistant transcribing a patient's current medications from a handwritten list is doing cognitive work under time pressure, in a noisy environment, with interruptions. A printed discharge summary that a patient brings to a new physician's office is three transcription steps removed from the original documentation, each of which is a potential source of divergence.

The paper legacy is not a problem that FHIR solves on its own. FHIR is a standard for exchanging data between digital systems that agree to use it. Data on paper cannot be exchanged via FHIR until it has been digitized, and digitization at the quality required for clinical use is labor-intensive and error-prone. Any realistic plan for healthcare interoperability must account for the paper-to-digital transition as a prerequisite. Implementers who learn this lesson late typically discover it when data quality problems created by unaccounted-for paper sources begin to corrupt the outputs of otherwise well-designed FHIR integrations.

Sensitivity, Consent, and the Need-to-Know Standard

Healthcare data occupies a distinct category of sensitivity that creates design requirements with no close analog in commercial data management, and that are frequently underestimated by technologists entering healthcare from other industries.

Health records can reveal psychiatric diagnoses, substance use treatment histories, reproductive health procedures, genetic variants associated with serious diseases, or infectious disease testing and treatment. These categories have consequences for patients that extend far beyond the clinical encounter: they affect employment, insurance, relationships, and social standing in ways that can be severe and lasting. A commercial data breach that exposes a customer's purchase history is a privacy violation. A healthcare data breach that exposes a patient's psychiatric diagnosis or HIV status may be a life-altering event.

Federal law recognizes this distinction. HIPAA's minimum necessary standard requires that health information be used or disclosed only to the extent necessary to accomplish the purpose of the disclosure. This is a standard of deliberate minimization: before sharing health data, the disclosing party must consider what is actually needed for the specific purpose and limit the disclosure accordingly.

Diagram 1.6: The Minimum Necessary Access Principle

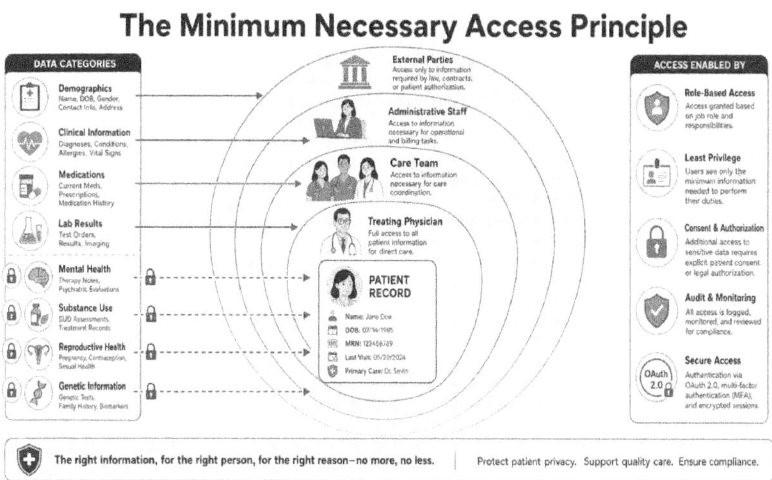

The minimum necessary standard creates a design requirement that goes beyond conventional role-based access control. It is not sufficient to define that a nurse has access to a patient's record and a billing clerk does not. A nurse caring for an orthopedic patient should have access to orthopedic records, but may not have a legitimate need for the patient's psychiatric history or substance use treatment records. A billing coordinator processing a cardiology claim has a legitimate need for the procedure and diagnosis codes relevant to that claim, but not for the patient's full medication list. Implementing this granularity requires a data model that understands the semantic categories of clinical information and an access control layer that enforces purpose-based restrictions at the record level, not just the user level.

Most legacy healthcare systems were not built with this level of granularity, and the gap between what HIPAA requires and what legacy systems can technically enforce is a persistent source of compliance risk. Sandra Torres, Riverside General's compliance and privacy officer, maintains a list of systems the hospital's legal counsel has flagged as potentially noncompliant with the minimum necessary standard, and the list has grown longer, not shorter, as the hospital has added systems. As FHIR-based patient access APIs enable external applications to request health data with granularity and scale that legacy systems were never designed to support, managing those gaps will become a larger, not smaller, part of any organization's data governance program.

Why Bigger Infrastructure Misses the Point

Patrick Osei came to his role with a mental model of technology problems shaped by his experience elsewhere. Technology problems, in that model, were capacity problems: not enough storage, not enough processing speed, not enough bandwidth. The solution to a capacity problem is more capacity. This mental model is not wrong in general. It is wrong specifically for healthcare interoperability.

Consider what happens if Patrick authorizes a significant infrastructure investment: new servers, doubled network bandwidth, and a monitoring platform that alerts the IT team to any communication delay within 30 seconds. The pharmacy system and the EHR are now connected by the fastest, most monitored network Riverside General has ever operated. The near-miss medication event happens again anyway.

The reason is that the interface between the pharmacy system and the EHR translates data by mapping medication codes using a lookup table built when the interface was first implemented. In the years since, the pharmacy system added a new formulation of a blood thinner with an internal code that does not appear in the lookup table. When the pharmacy system sends an administration event for that medication, the interface does not know what to do with the code, silently drops the message, and returns a success acknowledgment anyway, because it was programmed to be forgiving about unmapped codes so it would not block all messages every time a new medication was added.

Diagram 1.7: Semantic Mismatch vs. Infrastructure Failure

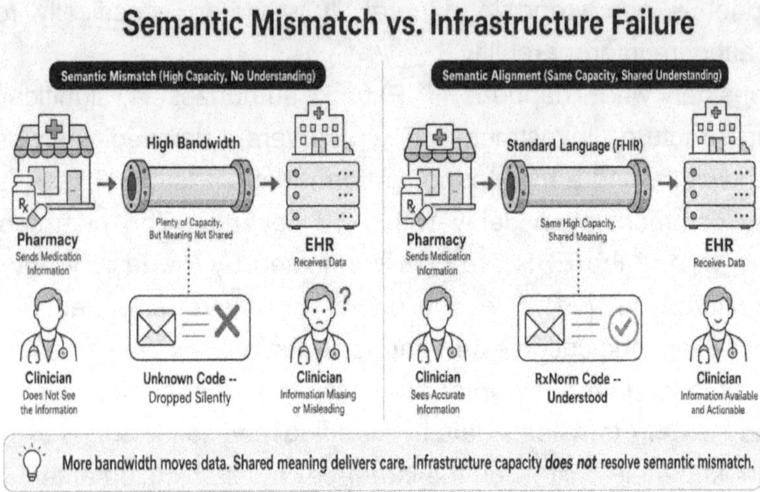

The failure in this scenario is not a capacity failure. It is a semantic failure. The two systems communicate in a shared format but do not share a vocabulary. The concept of "this specific medication formulation" is represented differently in the two systems, and the interface was not designed to handle cases where those representations diverge. More bandwidth does not solve a vocabulary mismatch. A monitoring platform cannot detect a message that was silently dropped due to an unrecognized code, because, from the platform's perspective, the message was sent and acknowledged successfully.

This is the core distinction that motivates the existence of healthcare interoperability standards. Standards like FHIR, HL7, SNOMED CT, LOINC, and RxNorm do not primarily exist to move data faster. They exist to establish shared vocabularies: agreements about how concepts should be represented so that a receiving system can be confident it understands what a message means. A medication administration event sent as an FHIR MedicationAdministration resource, with the medication coded in RxNorm and the dose in standardized units, can be received and interpreted by any system that implements FHIR and uses RxNorm coding, without custom interface logic and without a lookup table that requires maintenance every time a new formulation is added. The semantic agreement is built into the standard.

Dev Okafor, Riverside General's IT lead, described it this way in a planning meeting several months after the incident: imagine a large international conference where every delegate speaks a different language. The conference organizer, frustrated by the communication breakdown, purchases the most powerful audio system available. The microphones are perfect, the amplifiers are perfect, and no one understands anything any better than before, because the problem was never the volume. The problem was the language. FHIR is the common language. The infrastructure is the microphone. Without the language, the microphone does not help.

The Regulatory Ecosystem

No account of why healthcare data is different would be complete without addressing the regulatory environment in which it lives. Healthcare data is subject to one of the most complex and overlapping sets of legal obligations of any data category in the United States, and those obligations explain many of the design choices that characterize healthcare information systems.

HIPAA is the framework most people associate with healthcare data regulation, and it deserves that prominence. The HIPAA Privacy Rule establishes who may use and disclose protected health information, for what purposes, and under what conditions. The HIPAA Security Rule establishes the administrative, physical, and technical safeguards that covered entities must implement to protect electronic protected health information. The HITECH Act of 2009 strengthened HIPAA's enforcement provisions, extended many requirements to business associates, and created the incentive programs that drove EHR adoption. Together, HIPAA and HITECH establish the baseline compliance framework for any organization that handles health data in the United States.

HIPAA is not the only regulatory layer, and in many cases not the most restrictive. State privacy laws in many states impose requirements stricter than HIPAA's federal floor, particularly for sensitive categories such as mental health records, substance use treatment records, HIV status, and genetic information. An organization operating in multiple states must navigate a patchwork of requirements that may differ materially from each other and from the federal baseline.

Diagram 1.8: The Regulatory Stack for Healthcare Data

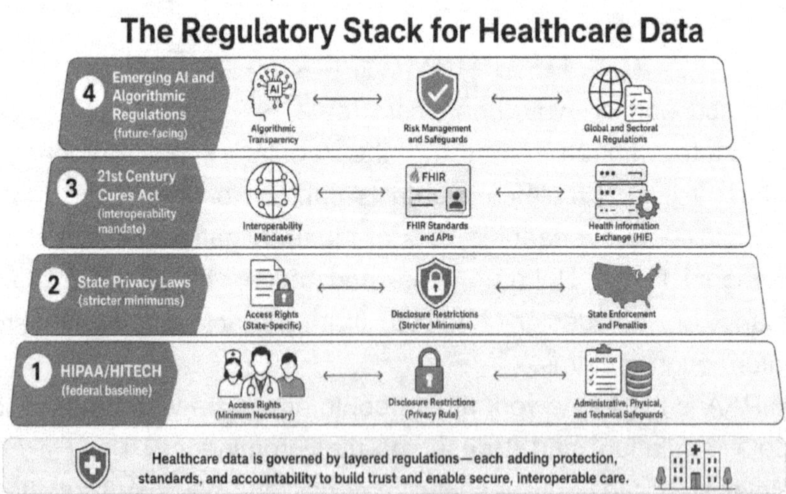

Beyond privacy and security, healthcare organizations are subject to a regulatory layer focused specifically on interoperability. The 21st Century Cures Act, enacted in 2016, introduced the concept of information blocking: taking actions that interfere with the access, exchange, or use of electronic health information. The Cures Act made information blocking a violation of federal law for covered health IT developers, health information exchanges, and healthcare providers. It also required HHS to adopt a common interoperability standard: FHIR. The ONC Cures Act Final Rule, published in 2020, requires certified EHR developers to implement FHIR-based patient access APIs, clinician access APIs, and payer-to-provider data exchange APIs on a defined compliance timeline. For Patrick, sitting in the planning meetings that followed the medication near-miss, this regulatory context is more than background. It is the business case for the technical investments described in this book's later chapters. The question is not whether Riverside General will eventually need to implement FHIR. That answer is already determined. The question is whether the implementation will be proactive and planned, or reactive and rushed. Sandra Torres would add: regulatory obligations around healthcare data are not static. They are evolving, driven by legislation, rulemaking, state law, and litigation about the boundaries of existing rules. The architecture decisions made today -- the data models adopted, the access control frameworks implemented, the audit logging configurations deployed -- will either accommodate future regulatory requirements or require expensive retrofitting when those requirements arrive. Building compliance into the architecture is not a luxury. It is a risk management requirement. Sandra raised one of those evolving obligations in the same planning meeting. She had been reviewing the hospital's annual

HIPAA Security Risk Analysis the previous week, she said, and the audit logging configuration on the integration engine did not capture which downstream system actually consumed a given message. The hospital could prove that a message was left in the EHR. It could not, in a way an auditor would accept, prove that the receiving system had read it. That gap had not yet caused a problem. Under the Cures Act information blocking rule, it would eventually become one. Patrick wrote it down in the column of his notebook he used for items he intended to revisit and turned the meeting back to the medication-near-miss agenda. Sandra noticed the column. She had seen items move into and out of it for two years.

What This Chapter Has Established

To understand FHIR, you need to understand why it exists. FHIR does not exist because healthcare IT professionals decided to modernize their messaging standards for the pleasure of modernization. It exists because the characteristics of healthcare data -- its multi-category complexity, its life-or-death accuracy requirement, its multi-stakeholder ownership structure, its deep sensitivity, its paper legacy, and the regulatory environment that governs it -- created a set of problems that neither market forces nor institutional goodwill could solve, and that required a standard, mandated, technically precise solution. Patrick's question from the board meeting has a real answer, and it is not "healthcare is simply behind." Healthcare data is doing something fundamentally more complex than a bank transaction, subject to constraints with no commercial parallel, in an environment shaped by decades of independently built systems that were never designed to communicate. The banking system did not solve its interoperability problem through the heroic efforts of individual banks. It solved the problem by adopting shared standards that all participants agreed to implement. Healthcare is attempting the same solution, forty years later, with FHIR as the standard and the 21st Century Cures Act as the mandate.

Before the technical ground of the chapters that follow can be productively covered, it is worth sitting for one more moment with the insight Patrick carried home from the incident review: the systems in that hospital were not broken. They were functioning exactly as designed. The pharmacy system dispensed medications. The EHR recorded orders. The interface between them synchronized data on a schedule that had been adequate for three years but was no longer. The failure was not a malfunction. It was a gap between two systems that spoke different internal languages, connected by a translator that was working correctly but had never been designed to handle the specific situation that arose. Closing gaps like that one, permanently and at scale, is the work the rest of this book describes.

Diagram 1.9: The Road from Data Complexity to FHIR Adoption

The Road from Data Complexity to FHIR Adoption

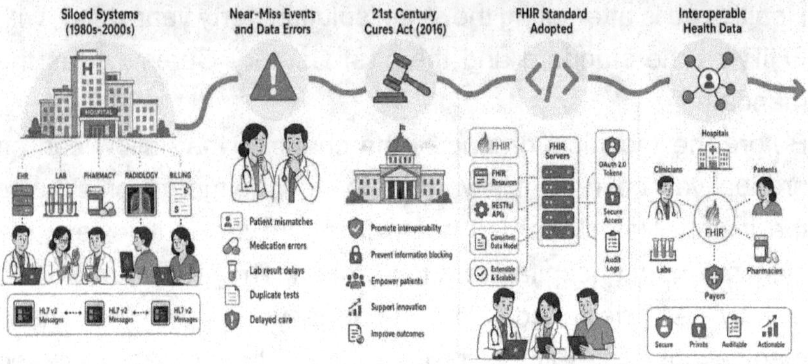

From fragmented systemns to a connected future—FHIR enables secure, trusted, and actionable health data for better care.

Takeaways

Healthcare data is categorically different from commercial data, not merely more complex. The life-or-death accuracy requirement, the multi-stakeholder ownership structure, and the sensitivity of clinical information create a set of characteristics with no close parallel in retail, financial, or logistics data management. Every assumption about data systems that transfers from another industry must be re-examined before it is applied to healthcare.

Data silos in healthcare are a structural outcome, not a technical failure. They formed because the regulatory, financial, and competitive environment of the past four decades did not require interoperability, and because the cost of connecting systems was consistently deferred in favor of the cost of running them. Solving the silo problem requires changing the incentive structure, which is precisely what the 21st Century Cures Act and its implementing regulations attempt to do.

Clinical, administrative, and financial data streams must coexist and remain synchronized within every patient encounter. A FHIR implementation that handles clinical data elegantly but cannot exchange administrative or financial information has solved only a fraction of the problem. Real-world health systems require all three data streams to flow reliably and simultaneously, and the systems that manage them must be designed to communicate with one another.

Accuracy in healthcare is a safety requirement, not a quality preference. Every design choice in a healthcare data system -- the delivery confirmation architecture, the redundancy level of the database, the granularity of the audit log -- ultimately traces to the recognition that an error in a medication record or a lab result can injure a patient. Systems that treat accuracy as a quality metric optimized for cost efficiency will eventually produce patient safety events.

The paper legacy cannot be ignored in any realistic interoperability plan. A substantial fraction of health data still originates on paper, and every FHIR implementation project that does not account for paper-originated data sources will encounter data quality problems that no amount of technical architecture could have anticipated. The data normalization challenge that precedes most FHIR projects is larger than vendors typically acknowledge, because the legacy of paper-based workflows is larger than most organizations have measured.

Sensitivity and consent are architectural requirements, not compliance afterthoughts. The minimum necessary access principle must be reflected in the data model, the access control design, and the audit logging configuration of every system that handles protected health information, from the first day of development. Systems that defer sensitivity-aware access control to a post-deployment compliance review will face remediation costs that exceed the cost of building it correctly in the first place.

Interoperability requires semantic agreement, not just technical connectivity. Two systems can be physically connected with maximum bandwidth and minimum latency and still fail to communicate if they do not share a vocabulary for representing clinical concepts. The standards covered in this book -- FHIR, HL7, SNOMED CT, LOINC, RxNorm -- exist to create that shared vocabulary. Infrastructure investments are necessary but not sufficient; the semantic layer is what makes interoperability real, and the chapters ahead begin building that layer from the ground up.

2 A Brief History of Healthcare Messaging

The Document on the Conference Table

Dev Okafor arrived at the Monday morning briefing with a manila folder tucked under one arm and a cup of coffee going cold in his other hand. He set the folder on the table first because it mattered more. Inside was a printout he had made the previous Friday after Patrick Osei, Riverside General's administrator, had asked the question Dev had been waiting years to hear: "Why does our system look the way it does?" Dev had gone back to his desk, pulled up a scan of the original HL7 version 2.1 specification from 1988, printed forty pages of it, and slid it into that folder.

When Patrick arrived and took his seat, Dev pushed the folder across the table without a word. Patrick opened it, scanned the first page, and looked up. "This is a fax format," he said. It was not entirely wrong. Dev said, "This is still running your hospital." Patrick stared at the page again, then at Dev, then back at the page. It was the particular expression of a man recalibrating something fundamental about the system he thought he understood.

That exchange captures something important about healthcare interoperability that no architectural diagram can fully convey: the history is not background material; it is the operating environment. The systems Dev maintains at Riverside General; the 217 Mirth Connect routes he monitors from a dashboard that has more red indicators than he would like, the integration patterns that confuse his junior engineers- all of it makes sense when you understand why the decisions were made the way they were, and when they were made. History in healthcare technology is not nostalgia. It is context.

This chapter is the story of how healthcare data exchange became what it is: the founding of HL7 in 1987, the anatomy of the message format that has outlasted every technology prediction made about it, the ambitious failure of HL7 version 3, the rise of the web and the gap it revealed between healthcare and every other industry, the creation of FHIR and the design philosophy that made it different from everything that came before, the regulatory moment that turned FHIR from an interesting standard into a compliance requirement, and the version map every practitioner needs to navigate the current landscape.

Diagram 2.1: The Whiteboard Where the Question Began

Patrick leaned back in his chair and said the other thing Dev had been waiting to hear: "How did it get this complicated?" Dev had a standard answer: nobody planned for it to be complicated.

Every layer of complexity was a reasonable response to a real problem at the moment it was introduced. The pipe-delimited messages came first because they were the most practical solution available in 1987. The transformation engine came because different systems spoke different dialects of the same standard. The custom translation logic came because the dialects diverged faster than the standard could constrain them. The complexity is not a collection of mistakes. It is a stratigraphic record, and reading it correctly requires knowing when each layer was deposited and why.

1987: The Year Seven Vendors Agreed

In the spring of 1987, representatives from seven healthcare technology companies gathered in Philadelphia to confront a problem that had been bleeding money from every one of their organizations. Each hospital deployment required a custom interface between the clinical systems involved because there was no agreed-upon format for exchanging clinical data across disparate applications. An ADT system from one vendor could not communicate with a laboratory information system from another without someone writing bespoke code to translate between them. Because each vendor had made different assumptions about data structure, each translation was different. The integrators were reinventing the same wheel in slightly different shapes for every contract they signed, and the hospitals were paying for the privilege.

The meeting in Philadelphia produced the first version of Health Level Seven, a name that carried a specific meaning for anyone familiar with the OSI networking model. The OSI model divided network communications into seven layers, from the physical wiring at layer one up through the application layer at layer seven, where user-facing protocols and the meaning of exchanged data were defined. The founders of HL7 deliberately named their standard for that seventh layer: they were not solving a networking problem; they were solving the problem of what the data meant once it arrived, how it was structured, and how a receiving system could be expected to parse it.

The first specification described a structured text message format for a small set of clinical transactions: patient admissions, discharges, and transfers; laboratory orders and results; financial billing events. It was not comprehensive or elegant. It left enormous room for interpretation, which meant two organizations could both claim to be "HL7 compliant" and still be unable to exchange data without a translation layer between them. But it accomplished something no previous effort had managed: seven competing vendors agreed on it, and that agreement, however imperfect, created the first critical mass of interoperability the industry had ever seen.

Dev had explained this to Patrick over coffee months earlier. "In 1987, if you wanted to connect your lab system to your ADT system, you called both vendors to write custom interfaces, paid them to test whether the two custom interfaces worked together, and usually they did not. HL7 v2 meant both vendors could agree to use the same message format, at least in principle, so the interface work was connected to a known standard rather than to each other's proprietary format." Patrick had nodded. "In principle," he repeated, with the tone of a man who had sat through enough system implementations to know what that phrase actually meant.

Diagram 2.2: HL7 Founding Problem: Spaghetti to Standard

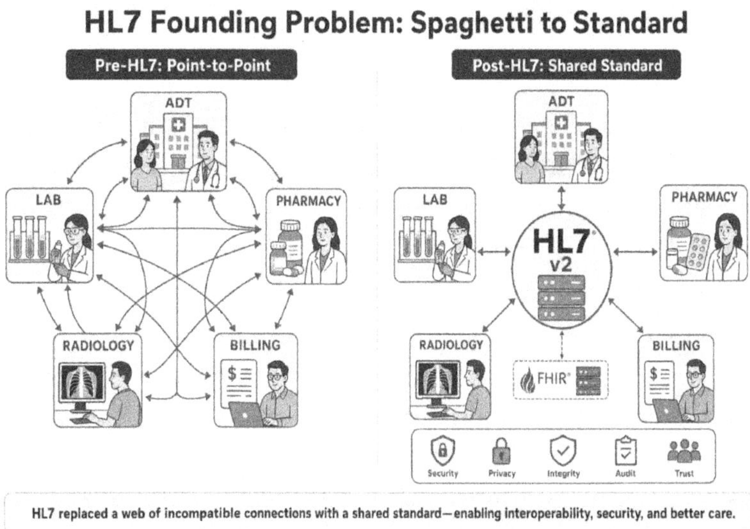

The first version of HL7 was followed by version 2.1 in 1988, and the standard entered a period of iterative refinement spanning the next three decades. Versions 2.2 through 2.9 added new message types, refined existing segments, and expanded the range of clinical transactions the standard could carry. The versioning history matters for a specific practical reason: when Riverside's laboratory information system sends an HL7 message to the ADT system, both systems need to agree on which version they are using, because the same field in the same segment can mean subtly different things across versions, and those differences compound when messages pass through transformation engines that were configured in different years by different engineers. Dev's 217 Mirth routes are, in part, a map of every time Riverside added a new system or updated an old one without fully reconciling the version assumptions embedded in the existing routes.

The Pipe-Organ Syntax: How HL7 v2 Works

An HL7 v2 message is a sequence of named segments, each on its own line, where fields within each segment are separated by the pipe character and subfields within those fields are separated by the caret. The message begins with an MSH segment (the message header), which identifies the sending and receiving applications, the message type, the date and time, and the message control identifier. A PID segment carries the patient's name, date of birth, identifiers, and contact information. An EVN segment records the event that triggered the message. Every segment has a three-letter code, and the standard specifies the order of segments within a message type.

Dev used the pipe organ metaphor to explain HL7 to people without a background in health IT. A pipe organ works because each rank of pipes is a distinct register producing a specific tonal character, and the organist controls which ranks are active by pulling or pushing a set of stops. An HL7 v2 message operates on the same principle: each segment is a register, and the combination of segments determines what the receiving system understands about the transaction. A message containing MSH, EVN, PID, and PV1 is almost certainly an ADT event. Add OBR and OBX segments, and the message carries observation data. Remove the EVN and add ORC and you have an order message. The syntax is composable, and that composability is the key to its longevity.

The compact, pipe-delimited format had practical advantages in the hospital infrastructure of the late 1980s and 1990s that are easy to underestimate. Hospital networks then operated over serial connections and early Ethernet with bandwidth measured in kilobits per second. An HL7 v2 message carrying a complete patient admission event might weigh thirty kilobytes or less as plain text. The same event expressed in XML could be ten or twenty times larger, because XML's verbosity, its explicit element tags opening and closing around every field value, trades compactness for self-description. In a 1990s hospital network environment, that tradeoff was not theoretical. It was the difference between an interface that worked and one that could not keep up with message volume during peak admission hours.

Diagram 2.3: How a Single Admission Becomes a String of Pipe-Delimited Segments — and Why One Missing Field Breaks the Downstream System

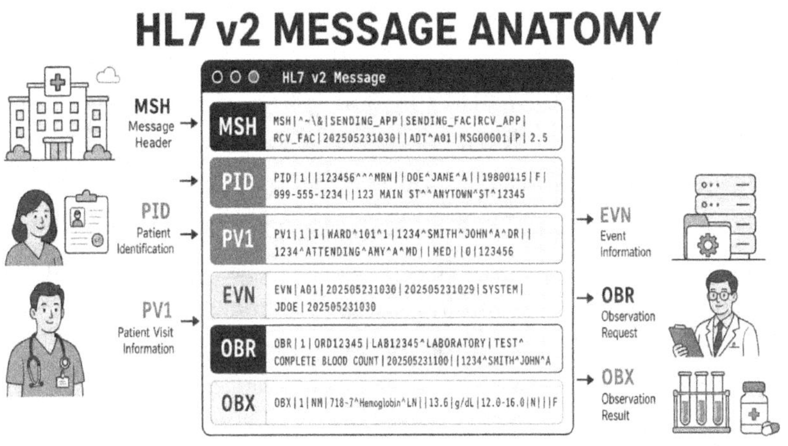

ⓘ An HL7 v2 message is built from segments. Each segment has a specific purpose and works together to share accurate healthcare information.

What made HL7 v2 genuinely readable, in a way no subsequent healthcare data standard has fully replicated, was that a skilled developer could look at a raw message in a terminal window and understand it without a specialized parser. The three-letter segment codes were mnemonic. The field order within each segment was consistent and documented. When Dev encountered a misbehaving interface at Riverside, his first diagnostic step was frequently to pull a sample message from the Mirth Connect log and read it, the way a fluent reader reads a sentence. That legibility was a consequence of the fact that the people who wrote the specification were also the people who would be debugging implementations.

The flip side of that flexibility was the proliferation of local extensions and custom Z-segments that Dev had cataloged in a spreadsheet, not because anyone had asked him to, but because he had learned from experience that the alternative was worse. HL7 v2 allows implementations to add proprietary segments by prefixing them with Z outside the standard segment namespace. In practice, Z-segments became the primary mechanism by which every vendor differentiated their implementation from every other vendor's, and by which organizations embedded local business logic so deeply into their integration infrastructure that removing it later became an archaeological project rather than an engineering task. Dev's spreadsheet listed forty-seven distinct Z-segment patterns across Riverside's current interfaces, each a decision someone made at some point for a reason that was probably good at the time.

Growing Pains: The HL7 v2 Dialect Problem

By the mid-1990s, the problem that HL7 v2's flexibility had created was visible to everyone who worked in healthcare IT. The standard specified the message format but left significant latitude in how fields were used, which optional segments were included, and how event trigger codes were assigned to specific clinical workflows. The result was what practitioners described as "the dialect problem". Every vendor's HL7 implementation was technically conformant with the standard and practically incompatible with every other vendor's implementation without a translation layer. "HL7 v2 compliant" was a statement about syntax, not about semantics, and the semantic incompatibilities were where the actual integration work lived.

Dev had inherited this problem from every predecessor in the Riverside IT department who had made pragmatic decisions about interface configuration. When a new cardiology information system was deployed six years before Dev arrived, the integration engineer had configured the ADT feed to use a specific field in the PV1 segment to carry a location code that Riverside's billing department needed. Three years later, a new radiology system was deployed, and another integration engineer configured the ADT feed using the same field to carry a different code because the radiology system had a different location structure. Both configurations were HL7 v2 compliant. Neither engineer had done anything wrong. Dev had spent two days figuring out why the billing department was getting incorrect location codes on radiology orders before he found the discrepancy, and the fix required modifying three Mirth routes and coordinating with two vendors across a six-week change management process.

Diagram 2.4: The v2 Dialect Problem at Riverside General

The v2 Dialect Problem at Riverside General

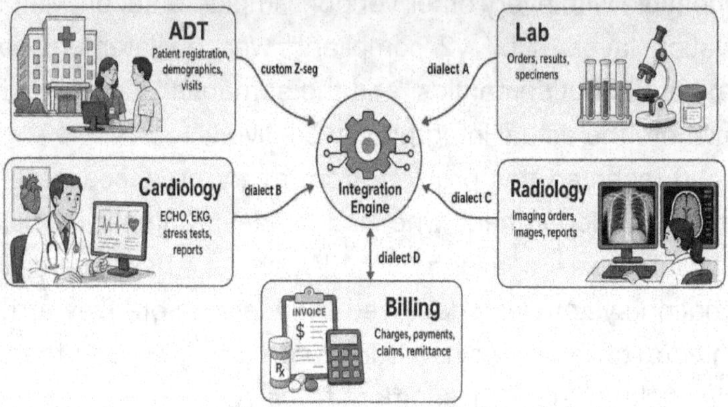

Five systems. Five dialects. Endless custom mapping. Higher costs. More risk. Slower innovation.

This pattern repeated across every major HL7 v2 installation in the country, and by 1999 it was clear to the leaders of HL7 International that the standard needed a more rigorous foundation. A standard that is strict enough to guarantee semantic interoperability will impose a compliance cost that many organizations cannot sustain; they will either not adopt it or adopt it only partially. A standard permissive enough to allow widespread adoption will generate a population of technically compliant but practically incompatible implementations. The history of HL7 is, in large part, the history of the field trying to navigate that tension, and each major standard generation represents a different point on that tradeoff curve. Understanding where each standard fell on that curve, and why, is the foundation for understanding why each successive standard was designed the way it was.

Patrick asked Dev during one of their standing meetings why the hospital could not simply upgrade all its systems to the latest version of the standard and thereby solve the dialect problem.

Dev took a moment before answering. "The dialect problem is not a version problem," he said. "It is a configuration problem. Two systems running HL7 v2.7 can still be completely incompatible if the teams that configured them made different assumptions about field usage. The version tells you which segments and fields are available. It does not tell you what the sending system is going to put in those fields, or whether the receiving system is going to interpret it the way the sender intended." Patrick wrote something in his notebook. Dev noticed, and felt, for the first time in a while, that the conversation was going somewhere useful.

HL7 v3 and the CDA: Ambition and Adoption

In the late 1990s, HL7 International undertook the most ambitious standards revision in the organization's history. The goal was to replace the flexible, loosely specified v2 framework with a new generation of standards built on a formal information model, a comprehensive, precisely specified representation of every healthcare concept that an HL7 message might need to convey. The result was HL7 version 3, and its foundation was the Reference Information Model, a unified ontology of healthcare data that had taken years of work by hundreds of experts to construct.

The Reference Information Model was genuinely impressive as an intellectual achievement. It defined patient, provider, encounter, order, result, and medication concepts with a rigor that v2's pragmatic segment definitions had never attempted. Every element in every HL7 v3 message was typed, constrained, and related to other elements through formally specified associations. The XML encoding v3 used was verbose but self-describing: a receiving system encountering a v3 message could, in principle, understand its structure without out-of-band configuration or custom parsing logic.

The problem was implementation. Conforming to HL7 v3 required not just writing code that produced well-formed XML, but also understanding the Reference Information Model deeply enough to map local data structures onto v3's formally specified concepts. The RIM was large and complex enough that this mapping exercise represented a major organizational investment. Vendors that had spent a decade building HL7 v2 implementations had little incentive to undertake a complete rewrite for a standard their customers were not yet demanding. And customers were not yet demanding it because the vendors had not implemented it. The adoption cycle never achieved the critical mass that would have driven both sides of it forward.

Diagram 2.5: HL7 v3 Architecture and the RIM

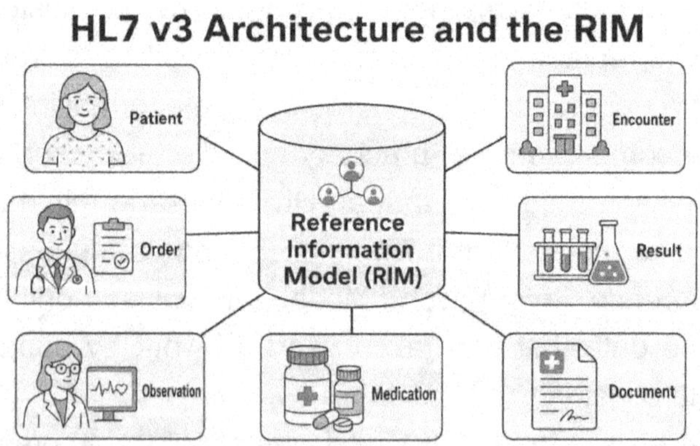

The RIM provides a shared, stable foundation for consistent healthcare information exchange.

The one component of the HL7 v3 framework that achieved meaningful adoption was the Clinical Document Architecture, or CDA. The CDA was a v3-derived standard for representing clinical documents specifically: structured clinical notes, discharge summaries, referral letters, and similar documents that had historically existed only as free-text narratives or scanned images. A receiving system could reliably parse a clinical document encoded as CDA, extract its header metadata for indexing and routing, and display its body to a clinician in a consistent format regardless of which system generated it.

The CDA achieved its widest adoption through the Consolidated CDA, a constrained profile developed under the ONC's Standards and Interoperability Framework and subsequently mandated by the Meaningful Use program. When a patient was discharged from Riverside General and referred to a specialist, the hospital was required to send a Consolidated CDA document that summarized the encounter, the patient's current medications, active problems, and allergies. Dev had configured Riverside's Consolidated CDA generation when the Meaningful Use requirements came into effect, and it had worked reliably enough that he rarely thought about it. But the CDA success was, in an important sense, an exception that proved the rule: a constrained, tightly specified profile of a v3-derived standard achieved adoption when a regulatory program required it and a government body defined exactly what the profile had to contain. The broader v3 messaging framework never benefited from those conditions, and by the mid-2000s it was apparent that a third generation of healthcare data standards would be needed.

The Rise of the Web and What It Revealed

Between 2000 and 2010, the web underwent a transformation that reshaped the entire software industry and made the situation in healthcare IT look stranger with each passing year. In 2000, the web was primarily a document-delivery system. By

2005, REST (Representational State Transfer) had emerged from Roy Fielding's doctoral dissertation as a description of the architectural principles on which HTTP was built, and developers were beginning to use those principles to build APIs that enabled applications to communicate over HTTP using a small set of standard operations. By 2010, the REST API had become the default architectural pattern for web services, and the tooling ecosystem around it had grown to the point where a developer with no prior domain experience could build a functional API client in an afternoon.

The implications for healthcare IT were uncomfortable. The infrastructure the industry had built to exchange clinical data- the HL7 v2 interfaces, the MLLP transport protocol, the integration engines- required specialized expertise that took years to develop. A developer who had built consumer or enterprise software in other industries arrived at a healthcare IT job and found a technology stack architecturally alien to everything they already knew. REST APIs exchanged JSON. Healthcare exchanged pipe-delimited text or verbose XML over a proprietary network protocol. The gap between healthcare and the rest of the software world was not just a matter of tooling. It was a structural impediment to building healthcare applications, and it was growing wider every year.

Diagram 2.6: REST vs. Healthcare Integration Architecture

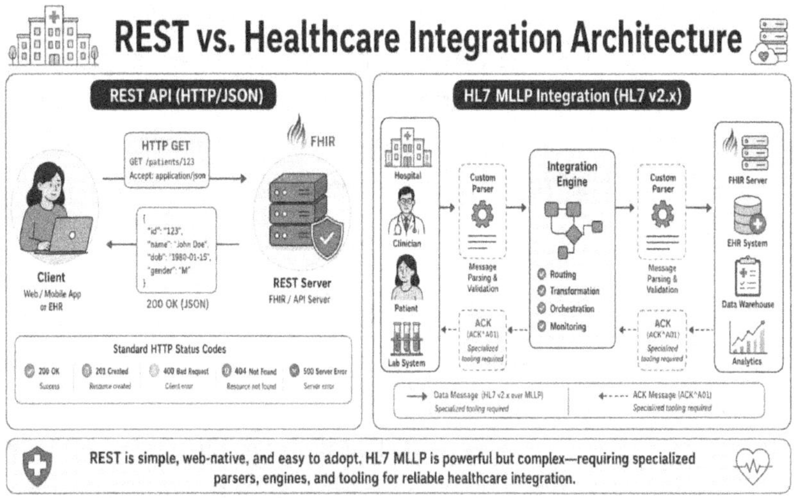

Dev had felt this gap personally in a way that shaped his view of FHIR before he ever read a FHIR specification. In 2018, Riverside's then-IT director proposed building a patient portal, and Dev was assigned to assess the integration requirements. The assessment took three weeks and produced a document embarrassingly long for what should have been a straightforward project. The problem was not the application layer; the web developer engaged for the project could have built the patient-facing interface in a week. The problem was data access: getting structured clinical data out of the hospital's clinical systems in a format the web application could consume required either expensive middleware, significant custom transformation logic, or both. The project was scoped down, and the patient portal that went live three months later was a PDF viewer.

The web developer on that project asked Dev a question on the last day that he still thought about. "Why doesn't the hospital just have an API? Every other service I work with has an API. Why does getting a patient's medication list take three months and a middleware product?" Dev gave her the historical explanation, the one about HL7 v2 and the 1980s. She listened politely and then said, "I understand why it is that way. I am asking why it is still that way." Dev had not had a good answer. That exchange was the reason he read Grahame Grieve's FHIR proposal with more than routine professional interest when it circulated the following year.

FHIR Is Born: The REST-ful Revolution

In 2011, Grahame Grieve, an Australian health informaticist who had spent years working on HL7 v3 implementations and understood both the standard's technical strengths and its adoption failures from the inside, published a proposal he called "Resources for Healthcare." The proposal argued that the approach underlying both v2 and v3, defining messages that carry clinical data from one system to another, was the wrong abstraction for the web era. Instead, it proposed treating each clinical concept, patient, medication, diagnostic observation, and encounter as a self-contained resource with its own permanent URL, identity, and defined set of operations. Resources would be exchangeable over standard HTTP, represented in JSON or XML without requiring specialized parsers, and designed to be accessible to developers who had never encountered an HL7 specification before.

The proposal was later renamed FHIR: Fast Healthcare Interoperability Resources. The word "fast" carried meaning beyond development speed. It meant accessible, low-barrier-to-entry, meaning that a web developer who knew nothing about healthcare data could build a working FHIR client in the same afternoon it would take to build a client for any other REST API. That accessibility was the central design objective distinguishing FHIR from every prior healthcare data standard. HL7 v2 was designed in 1987 for the people building hospital integration infrastructure. HL7 v3 was designed in 1999 for people building formal healthcare information models. FHIR was designed for the people building web applications in 2011, people who used HTTP every day and found the concept of a GET request to a URL completely natural.

Diagram 2.7: What Changes When a Message Becomes a Resource: FHIR's Shift From Event Stream to Addressable Object

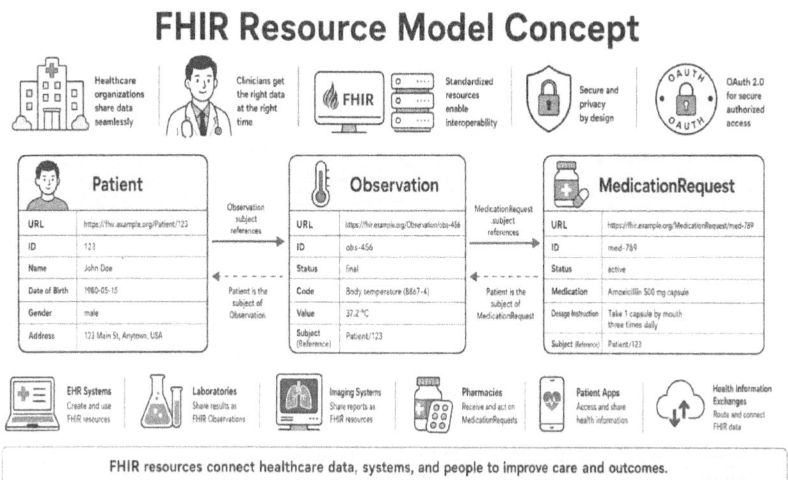

The HL7 community formally adopted the FHIR project in 2012, and the standard developed through a series of draft releases. DSTU 1 was published in 2014, DSTU 2 in 2015, and STU 3 in 2017. These trial-use releases were exactly what their names indicated: specifications good enough to build on and learn from, but acknowledged as provisional and subject to change based on implementation experience. Healthcare organizations that built FHIR implementations during the DSTU and STU phases were running an industry-wide proof of concept, and the feedback from those implementations drove the changes in each successive version.

FHIR Release 4, published in January 2019, was the moment the standard came of age. R4 was the first FHIR version to achieve normative status for a significant portion of its content, meaning the HL7 community had reviewed and balloted that content as stable, not subject to breaking changes without formal deprecation. R4 defined the foundation on which the US federal government's interoperability mandates would be built: the US Core Implementation Guide, which specified what an FHIR API had to expose, was developed to target R4, and the ONC's certification criteria specified R4 as the operative FHIR version. For US healthcare practitioners, FHIR R4 is not simply the current version of an evolving standard. It is the regulatory baseline against which certification requirements are measured, and vendor compliance claims are evaluated.

Dev had followed the FHIR releases from DSTU 1 onward with the particular attention of someone who could see both sides of the transition. His daily work was HL7 v2. His production environment was HL7 v2 all the way down. But he had been reading FHIR specifications and quietly building test implementations in a sandbox environment that nobody at Riverside had formally authorized. Nobody had formally objected to it, because he had learned years earlier that the most valuable integration work he did was the work he did before anyone asked him to.

The Regulatory Moment: The ONC Mandate

Technical merit had not driven HL7 v2 adoption: economic necessity had. Technical merit had not driven CDA adoption:

regulatory mandate had. And technical merit alone would not drive FHIR adoption at the scale the technology's designers hoped for. What would drive it was the regulatory environment that crystallized between 2016 and 2020.

The 21st Century Cures Act, signed into law in December 2016, included health IT provisions that went significantly further than any prior federal health IT legislation. The Act prohibited information blocking, defined as practices by health IT developers, health information networks, and healthcare providers that interfere with the access, exchange, or use of electronic health information unless the interference falls within a defined set of exceptions. The penalties were substantial: up to a million dollars per violation for health IT developers. The Act also directed the ONC to develop interoperability standards and certification criteria to give operational meaning to the information-blocking prohibition.

The ONC's 21st Century Cures Act Final Rule, published in May 2020, translated those provisions into specific requirements. Certified health IT products were required to support standardized FHIR-based APIs for patient access, provider directory, and subsequent payer-to-patient data exchange. The version of FHIR specified was R4. The required profile was the US Core Implementation Guide, which defined the specific FHIR resources, data elements, and search parameters that the API had to support. Health IT developers seeking or maintaining ONC certification were required to meet these requirements on a defined timeline.

Diagram 2.8: Regulatory Pathway to FHIR Mandate

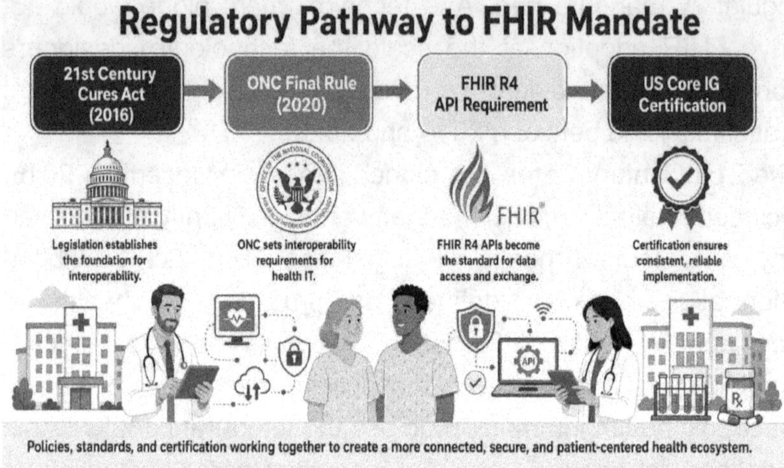

The CMS Interoperability and Patient Access Final Rule, published concurrently in May 2020, extended similar requirements to payers: Medicare Advantage organizations, Medicaid managed care plans, CHIP programs, and plans offered on federal and state exchanges were required to implement FHIR-based APIs allowing enrollees to access their claims and clinical data through third-party applications. The payer rule covered a distinct set of data holders operating in parallel with the ONC's provider-facing requirements.

Sandra Torres, Riverside's compliance and privacy officer, had been tracking both rules from their notice-and-comment periods, and she briefed Patrick and Dev together when the final rules were published. Her assessment was characteristically direct: "This is not a best practice anymore. This is a legal requirement with financial penalties attached. Dev's question is not whether to build FHIR APIs. His question is when and to what specification." Dev had already read the US Core Implementation Guide index. He had just not told anyone yet.

The information blocking prohibition changed the competitive calculus for health IT vendors in a way technical standards arguments alone could not have. Before the Act, there was a coherent business logic, however ethically questionable, to making data exchange difficult. The information blocking prohibition made that logic legally precarious. Practices that had been competitive strategies became potential federal violations, and vendors that had built their market position on data lock-in faced the choice of either genuinely supporting interoperability or facing enforcement scrutiny.

FHIR R4, R5, and the Version Map

Practitioners who encounter FHIR for the first time frequently find the version history confusing, because the terminology for release stages (DSTU, STU, Normative, Trial Use) does not map cleanly onto conventions familiar from other software versioning systems. Understanding what each release stage means in practice, which version is authoritative for which use cases, and what the differences between versions mean for implementation planning is a concrete skill every FHIR practitioner needs to interpret vendor compliance claims accurately.

FHIR has been published in the following releases: DSTU 1 in 2014, DSTU 2 in 2015, STU 3 in 2017, R4 in January 2019, R4B in May 2022, and R5 in March 2023. The progression from DSTU through STU and into R4's normative sections reflects increasing community confidence and stability. DSTU and STU content was explicitly provisional; implementers were encouraged to build on it but warned that content could change in breaking ways in future versions. Normative content, designated as such in R4 and expanded in R5, represents the community's commitment to backward compatibility: normative elements will not be changed in ways that break existing implementations without formal deprecation and a transition period.

For US healthcare organizations, FHIR R4 is the operative standard in nearly every compliance context. The ONC's certification criteria specify R4. The US Core Implementation Guide is developed against R4. The Da Vinci Project's payer-provider exchange guides target R4. Any vendor claiming FHIR compliance for ONC certification, Medicare Advantage payer API requirements, or Medicaid interoperability requirements is claiming compliance with R4, and assessing that claim must begin by asking which specific R4 profiles and implementation guides the vendor has implemented, not just which version number appears on their documentation.

FHIR R5, published in March 2023, introduces changes significant enough to matter for organizations planning new implementations, even if the majority of current integration work will remain in an R4 environment for the foreseeable future. The subscription framework in R5 is substantially redesigned: the R4 approach used a filter-based mechanism that was functional but limited in expressiveness; R5 replaces it with a topic-based subscription model that specifies what events the server publishes, allows clients to subscribe with well-defined filter criteria, and uses a notification bundle format that carries sufficient context for the subscriber to act without additional queries. For organizations building real-time event-driven integrations, the R5 subscription framework is a meaningful improvement.

Diagram 2.9: FHIR Version Timeline: DSTU to R5

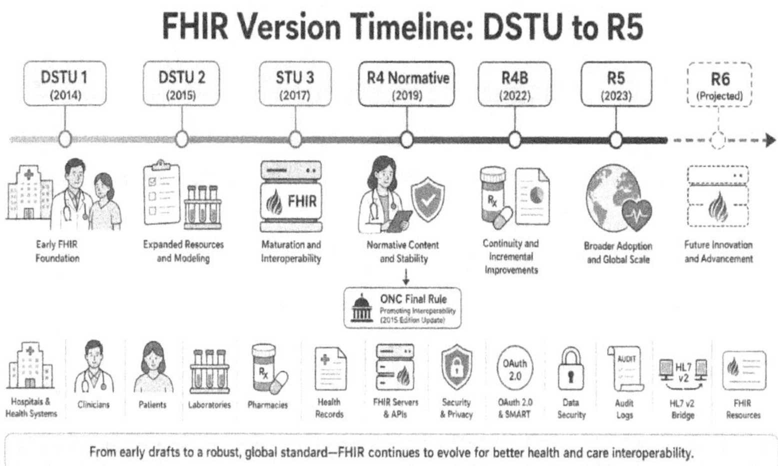

R5 also introduces improvements in bulk data operations, expanding the Bulk Data Access Implementation Guide to support more granular group exports and more flexible scheduling, which matters for payer-to-provider data exchange and population health analytics use cases. The clinical reasoning resources, including Measure, MeasureReport, Library, and PlanDefinition, are mature in R5 and support more sophisticated quality measurement and clinical decision support applications.

Dev's working assumption, documented in Riverside's integration architecture notes, was that the hospital would spend approximately two to three years completing its FHIR R4 compliance obligations before having the organizational bandwidth to evaluate R5 adoption seriously. That assumption was based on his assessment of the current implementation guide landscape: the US Core IG, the Da Vinci guides for prior authorization, and the CMS-mandated payer API requirements were all R4-based, and none had a published R5 migration path that would force Riverside's hand in the near term. When the time came, Dev intended to evaluate R5 adoption on its technical merits for specific use cases rather than treating it as a wholesale version upgrade. The history of healthcare standards had taught him that wholesale upgrades were almost always more expensive than they appeared in the planning document.

What This Chapter Covers

The arc from the 1987 founding of HL7 through the current FHIR version landscape is not background reading; each section contributes a specific piece to the foundational understanding that the rest of the book builds on. The founding of HL7 established both the first broadly adopted healthcare data standard and the fundamental tension between adoption accessibility and semantic precision that would shape every subsequent generation of standards. Seven vendors agreeing on a pipe-delimited text format was the precondition for the entire infrastructure that healthcare IT would build over the following four decades.

HL7 v2's compactness, composability, and legibility were responses to real constraints in operational environments, and they are why the standard has proven so difficult to displace. Understanding the design logic of v2 is the prerequisite for understanding why the integration infrastructure it created is as deeply embedded as it is, and why migration strategies that assume it can be replaced quickly tend to produce unpleasant surprises.

The HL7 v3 and CDA story is the cautionary tale that every standards organization needs to internalize. Architectural rigor is a virtue, but it is not sufficient. A standard that requires too much implementation effort to adopt widely will fail to achieve the network effects that make standards valuable. The CDA succeeded in contexts where regulatory mandates and constrained profiles reduced the implementation burden to a manageable scope. The broader v3 messaging framework did not achieve those conditions.

FHIR's REST architecture was a direct response to the gap the web had revealed between healthcare and other industries, and it was the right one. The regulatory moment of 2016-2020 was the adoption accelerant that turned FHIR from a promising technical standard into a compliance requirement, and understanding that regulatory environment is essential for anyone who needs to explain to a board why interoperability investment is a compliance obligation rather than an optional feature enhancement.

Takeaways

HL7 v2 was not a failure of imagination. It was a solution precisely calibrated to the problem of 1987, and its longevity is evidence of how well it was calibrated. The organizations that have struggled most with v2 legacy are not the ones that made bad decisions; they are the ones that built on v2 for thirty years without planning for what would come after it, because there was never a compelling reason to plan until the regulatory environment changed. Dev's 217 Mirth routes are a testament to the standard's resilience and a map of accumulated decisions that will eventually need to be unwound, one interface at a time, as Riverside migrates toward FHIR.

Understanding why HL7 v3 failed is as consequential as understanding why FHIR succeeded, because the v3 failure contains the design principles that FHIR explicitly reversed. The v3 framework's architectural completeness was its practical liability: it required implementation effort commensurate with its ambition, and the vendor community lacked sufficient incentive to sustain it without regulatory compulsion. FHIR succeeded in part because its designers had learned this lesson firsthand and made developer accessibility a first-order design objective. Every implementation team that gets a FHIR client working in an afternoon owes that experience to the decision Grahame Grieve made in 2011 to build on HTTP and JSON rather than the formal information model.

FHIR's REST architecture is a deliberate alignment with the broader technology industry's direction, and that alignment is the standard's single most important structural advantage. Every developer who already knows how to make an HTTP request can, in principle, build a FHIR application without years of specialized healthcare IT training. That accessibility compounds: more developers can build healthcare applications, more applications get built, more FHIR implementations generate implementation experience, which improves the standard and makes the next generation of applications easier to build.

Regulatory mandate, not technical merit, drove FHIR adoption to the scale it now operates at. The 21st Century Cures Act Final Rule and the CMS Interoperability Rules converted FHIR from an architectural choice into a compliance requirement, changing the conversations that IT departments have with hospital leadership, vendors, and payers. Sandra's observation at Riverside was the correct framing: the question is not whether to build FHIR APIs but when and to what specification. The regulatory environment does not leave room for a wait-and-see posture.

Version literacy is a practical skill that should be treated as a prerequisite for any FHIR implementation project or vendor evaluation. When a vendor says their product is FHIR-compliant, the immediately relevant questions are: which version, which profiles, which implementation guides, and what their conformance testing methodology looks like. A vendor fully conformant with US Core 4.0 is more useful for most US healthcare organizations than one that claims R5 support without a defined profile implementation.

The history of healthcare standards is a history of the recurring gap between what the industry needed and what it could actually build. FHIR is the current best answer to that gap. Still, it is not the final answer, and organizations that build their FHIR implementations as if they will never need to migrate again are repeating the mistake that made HL7 v2 so difficult to retire. Dev knew this, which is why his integration architecture notes included a section, still marked as draft, titled "Migration Preparation for What Comes After R4." He updated it every six months. Patrick had not yet asked about it. But Dev expected the question eventually, and when it came, he would have the folder ready.

3 FHIR Fundamentals

Scenario

Patrick Osei had been a hospital administrator for eleven years. He had reviewed hundreds of vendor contracts, negotiated service-level agreements with EHR companies, and sat through more technology demonstrations than he cared to count. But when FHIR came up in planning meetings, he noticed that everyone in the room assumed someone else had already explained it to him. So on a Tuesday afternoon, after the rest of the team had filtered out, he caught Dev Okafor at the door and said: "Five minutes. Explain FHIR without using the words 'standard,' 'specification,' or 'implementation.' I want to understand what we are actually building."

Dev set his laptop bag down and thought for a moment. Then he said: "Imagine that every piece of clinical information in a hospital has its own business card. A patient has a card. A lab result has a card. A prescription has a card. Each card has a permanent address on a server, a type that indicates what kind of thing it is, a set of structured fields that carry the actual data, and links to other cards. FHIR is the system that decides what goes on each card, how you ask for it, and how you send it to someone else."

Patrick nodded. "And the reason we are switching to it is that the old system did not have consistent cards."

"The old system had paper slips, cassette tapes, telegrams, and a few cards nobody agreed on. FHIR gives everyone the same card format, the same address system, and the same way to request cards. That is why it matters."

That exchange became the foundation for everything in this chapter. Dev would spend the next several weeks walking Patrick through the architecture behind the business card metaphor, and, in doing so, he discovered that explaining FHIR clearly to a non-technical colleague is one of the most effective ways to pressure-test his own understanding of a subject that rewards precision.

Why It Matters

FHIR's resource model, RESTful architecture, and reliance on familiar web conventions make it more accessible to modern developers than any prior healthcare data standard. A developer who already knows how to call a REST API and parse JSON can, with a reasonable time investment, build a functional FHIR client. The same developer approaching HL7 v3 or the full CDA specification would face a significantly steeper learning curve, requiring specialized parsing libraries, a working knowledge of the HL7 Reference Information Model, and expertise in XML schema validation that most web developers do not carry as standard equipment.

That accessibility is both FHIR's greatest strength and the source of its most persistent implementation challenges. Development teams sometimes underestimate the depth behind the approachable surface. It is entirely possible to build a client that fetches a Patient resource without understanding profiles, must-support obligations, terminology bindings, or the Capability Statement required for the server to claim conformance, resulting in software that passes basic connectivity tests but fails during real integration work.

This chapter gives readers the foundational vocabulary to avoid that trap. Riverside General's migration work touched on every concept introduced here, and the technical detail is the exact terrain that any integration team navigates for the first time when building against an FHIR server in a production environment.

The Business Card: What a FHIR Resource Is

Maya's question for this section is: What exactly is a FHIR resource, and how does it differ from a database record or a file? Every FHIR resource is a self-contained, addressable unit of clinical information. The business card metaphor is a useful starting point, but it understates the structure's precision. A FHIR resource has four distinct structural layers, each serving a specific function in the information model.

The first layer is resource identity and metadata. Every resource has a globally unique logical ID within its resource type on a given server, along with a meta element that contains version information, a last-updated timestamp, and optional security labels. The logical ID, combined with the server's base URL, produces the resource's canonical address: a URL that any authorized client can use to retrieve the resource at any time. This permanent addressability distinguishes a FHIR resource from a database record: a database record is retrieved by a query against a schema; an HTTP request to a URL retrieves a FHIR resource. That distinction runs through every design decision in the FHIR ecosystem.

The second layer is the human-readable narrative: an optional text element containing an HTML summary of the resource's content. This narrative is not the authoritative data; the structured domain elements are. But any system capable of rendering HTML can display a meaningful summary even without understanding the structured elements, making it a safety net in a world of deeply heterogeneous clinical systems.

The third layer is the extension mechanism. The base FHIR specification defines a fixed set of elements for each resource type, but clinical reality is richer than any fixed specification can fully capture. Extensions allow any resource to carry additional structured data elements using a defined URL as the extension's identifier. The US Core Implementation Guide uses extensions to carry race, ethnicity, and tribal affiliation data for Patient resources, elements absent from base FHIR but required for US regulatory reporting.

The fourth layer is the domain content: the clinical or administrative data fields specific to the resource type. For a Patient resource, this includes name, birthDate, gender, address, contact information, and external identifiers such as the medical record number. An Observation resource includes the observation code, the result value and unit, the reference range, the status, and references to the subject patient, performer, and encounter. Every resource type has its own precisely defined domain content layer, and understanding that content is the first prerequisite for correct implementation.

Diagram 3.1: Why “FHIR-Compliant” Does Not Mean “Interoperable”

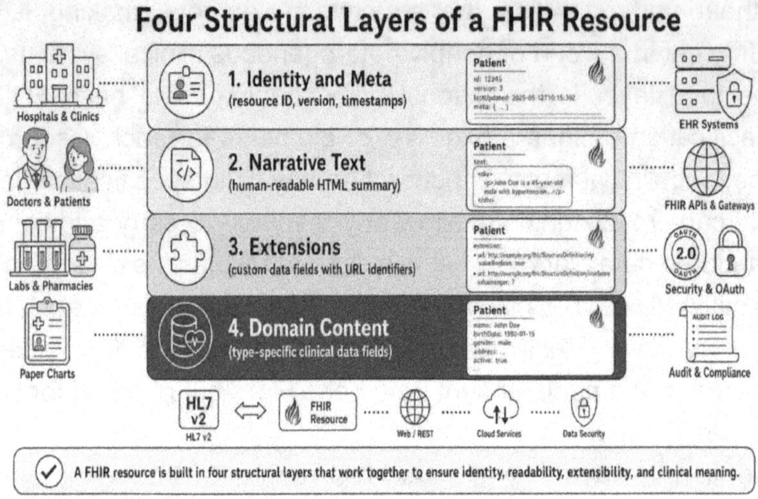

Dev drew this structure on the whiteboard during his first team walkthrough at Riverside General. Developers who skip this overview and jump directly to the API tend to build clients that handle the happy path correctly but fail in edge cases involving versioning, extensions, or narrative rendering. Patrick asked whether "permanent address" meant the URL never changed. Dev explained that responsible implementations maintain redirects across server migrations, and that the architectural point is universal: every design decision in FHIR assumes you can retrieve a resource by URL, a fundamentally different model from systems where data lives in a table and is retrieved by query.

Resource Types: The FHIR Vocabulary

Maya's question for this section is: How many types of resources are there in FHIR, and how does a team decide which to use?

FHIR R4 defines approximately 150 distinct resource types, organized into a taxonomy of domains that roughly map to the functional areas of a healthcare organization. The sheer number is initially intimidating, but most implementations work with a small subset, and the Capability Statement mechanism exists precisely to allow a FHIR server to declare which subset it supports.

The foundation domain contains infrastructure resources that every FHIR implementation uses regardless of clinical domain. The Bundle resource groups multiple resources for exchange. The OperationOutcome resource carries error and warning information in structured form. The CapabilityStatement describes what a server can do. The StructureDefinition defines the structure of other resources, data types, and profiles. These are the machinery that makes the clinical resources work.

The base domain contains foundational clinical and administrative resources that virtually every application needs. Patient, Practitioner, PractitionerRole, Organization, Location, and Encounter form the administrative backbone of any FHIR implementation. Almost every clinical resource links back to at least one of them through a Reference element.

The clinical domain contains resources for the actual clinical data that healthcare systems exist to manage. Condition carries diagnoses. Observation carries measurements and lab results. MedicationRequest carries prescriptions. MedicationAdministration records that medication was given. Procedure records a clinical procedure. AllergyIntolerance captures allergy information. DiagnosticReport groups observations into a clinical report. These are the resources that developers spend the most time on within clinical integration work.

The financial domain covers the business operations of healthcare: Coverage captures insurance information, Claim represents a billing claim, and ExplanationOfBenefit carries the payer's response. For revenue cycle systems and payer integrations, these resources are as central as Patient and Encounter are for clinical systems. The CMS interoperability rules that drove widespread FHIR adoption in the US market focused heavily on the Explanation of Benefits, because claims data often constitute the most complete longitudinal record of a patient's healthcare utilization.

Specialized domains extend FHIR into genomics, public health reporting, clinical reasoning, and medication knowledge management. Most implementers will use a fraction of the available resource types, and the Capability Statement mechanism ensures that a server can declare exactly which fraction it supports.

Diagram 3.2: Which FHIR Resource Carries Which Clinical Fact — and Where the Boundaries Between Domains Create Mapping Decisions

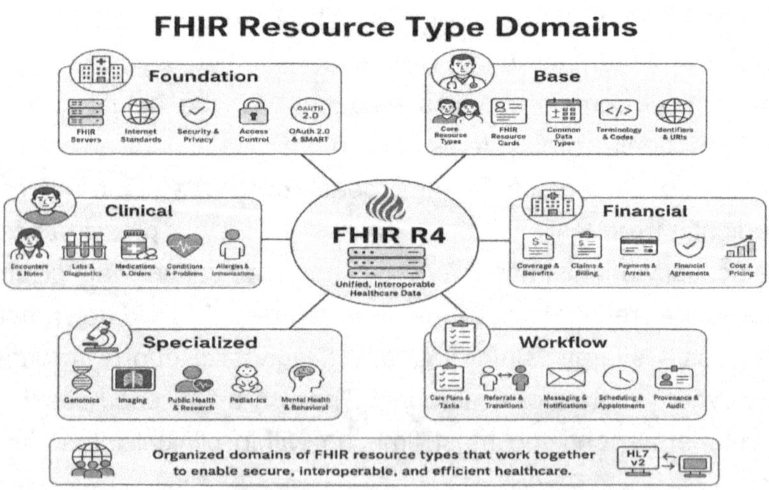

When Dev's team at Riverside General began their capability inventory, they identified seventeen resource types as essential to their initial scope: Patient and Practitioner anchored every interaction; Encounter linked clinical events to administrative context; Observation and DiagnosticReport carried lab data; MedicationRequest and MedicationAdministration covered the medication workflow; Condition and Procedure carried the problem list and surgical history; Coverage and ExplanationOfBenefit connected to the revenue cycle system; AllergyIntolerance fed the clinical decision support tool; and CapabilityStatement was the first document Dev requested from every vendor. Scoping to a defined subset of resource types is one of the most important project management decisions in FHIR implementation. Define the clinical use cases first, derive the minimum set of resource types required for those use cases, validate that subset, and expand scope only after the foundation is solid.

REST in Healthcare: Read, Write, Search

Maya's question for this section is: what does it mean to say FHIR is a REST API, and why does that matter for how we build systems?

REST is an architectural style for distributed systems that uses the HTTP protocol and its standard verbs (GET, POST, PUT, PATCH, DELETE) to perform operations on resources identified by URLs. Every developer who has built or consumed a web service in the last fifteen years has used REST, whether they recognized it as such or not. FHIR takes this universal pattern and applies it to clinical data.

The FHIR REST API defines five fundamental operations. Read (GET) retrieves a resource by URL. Create (POST) submits a new resource to a type endpoint, and the server assigns it a logical ID. Update (PUT) replaces the stored resource with the client's version and increments the version number. Patch sends only the changed elements, useful for high-frequency updates to large resources. Delete logically removes the resource; most FHIR servers maintain a history of deletions to comply with audit requirements.

Search is the sixth fundamental operation and arguably the most complex. A FHIR search query is a GET request to a resource type endpoint with query parameters that filter results. The patient search parameter filters by subject patient; the date parameter, with comparison operators, filters by date range; and the code parameter, with a system and value pair, filters by observation type. Chaining search parameters allows queries that span resource types. Implementing search correctly, particularly for complex chained queries involving terminology code systems, is one of the most technically demanding aspects of building a FHIR server.

Diagram 3.3: What Dev's API Call Actually Does — How GET, POST, PUT, and DELETE Map to Real Clinical Workflow Steps

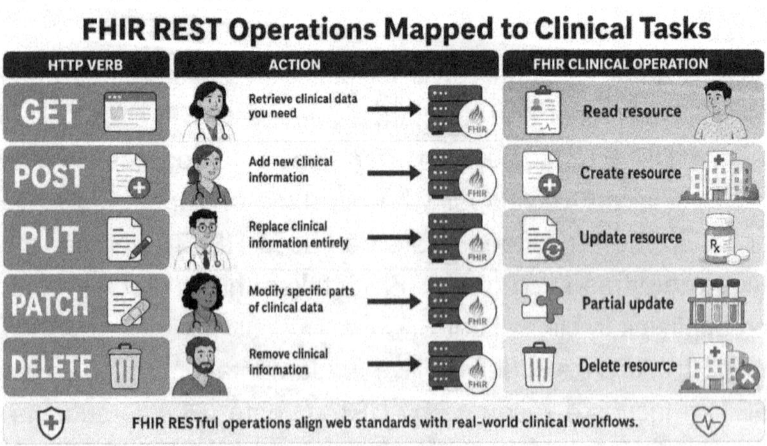

The pseudocode below illustrates the create operation. A developer submitting a new lab observation constructs a resource object, sets the required fields, and sends it via POST. The server's response carries the assigned ID and version information in the Location header.

POST to base_url + "/Observation" request body = { resourceType: "Observation", status: "final", code: { coding: [{ system: "http://loinc.org", code: "2339-0" }], text: "Glucose [Mass/volume] in Blood" }, subject: { reference: "Patient/12345" }, valueQuantity: { value: 94, unit: "mg/dL" } } response.status == 201 (Created) responseheaders["Location"] == base_url + "/Observation/67890/_history/1"

Dev used this pattern during the first integration sprint at Riverside General to connect the laboratory information system to the new FHIR server. The lab system's interface engine, previously sending HL7 v2 messages over TCP/IP, was reconfigured to transform each result message into an Observation resource and POST it to the endpoint. The transformation required mapping the v2 OBX segment's observation identifier to a LOINC code, the result value and unit to a FHIR Quantity, and constructing the reference to the correct Patient resource. Dev's team spent two weeks validating those mappings before enabling the live data feed.

The idempotency properties of the REST operations matter in healthcare integration. PUT is idempotent; POST is not: sending the same POST twice creates two separate resources. The conditional create operation (a POST with an If-None-Exist header containing a search query) provides idempotency by creating only if no matching resource already exists. HTTP response codes carry structured meaning: 200 OK for a successful read, 201 Created for a successful create, 400 Bad Request for validation failures, 404 Not Found for missing resources, and 422 Unprocessable Entity for business rule violations.

Bundles: Packaging Multiple Resources

Maya's question for this section is: if each resource is its own thing with its own address, when and why would you need to package multiple resources together?

A FHIR Bundle is a container resource that groups multiple resources into a single package for exchange, storage, or processing. It has a type element identifying what kind of Bundle it is, an optional total element carrying the count of matching resources, link elements carrying pagination URLs, and an entry array in which each element carries one resource along with processing metadata.

The transaction Bundle type is the most important for integration workflows that require atomicity: all operations in the Bundle succeed, or all fail, with no partial processing. This guarantee is essential for workflows where multiple resources must be created or updated together to maintain data consistency. Creating a new patient encounter may require creating an Encounter resource and simultaneously linking associated Condition and Procedure resources to the correct Patient, all of which must be treated as a single unit.

The searchset Bundle type is the package a FHIR server returns in response to a search query. When a client searches for all Observations for a patient, the server returns a searchset Bundle containing the matching resources, a total element giving the count of all matches, and link elements carrying pagination URLs. Clients that do not correctly handle pagination will silently miss resources when results span multiple pages, a failure mode easy to miss in testing with small datasets and dangerous in production.

Diagram 3.4: FHIR Bundle Types and Their Use Cases

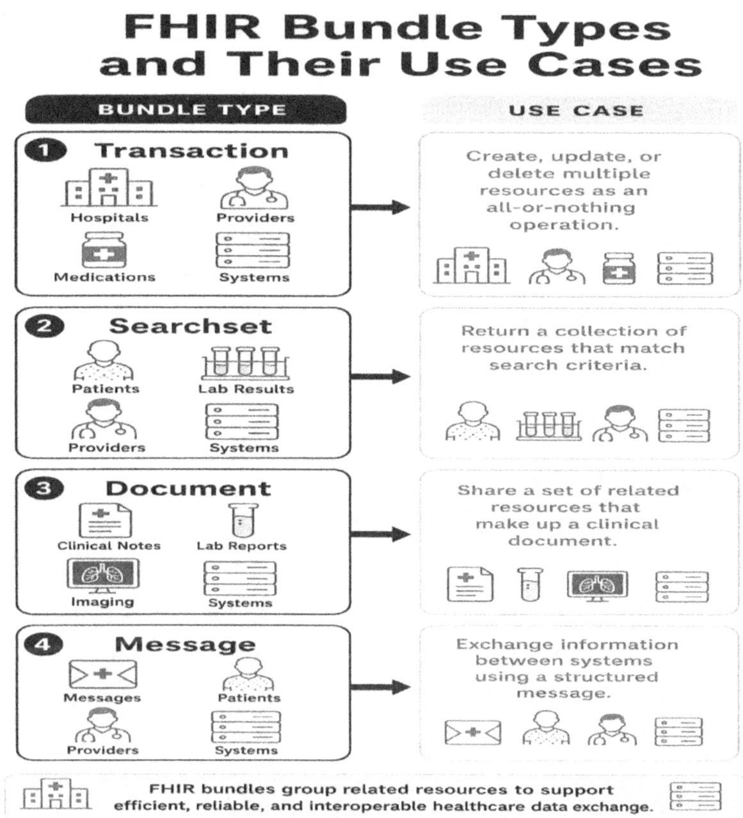

The document Bundle type packages a complete clinical document as a self-contained set of resources. A discharge summary, care plan, consultation note, or diagnostic report can be represented as a document Bundle in which a Composition resource acts as the index. The document Bundle includes all referenced resources within itself rather than pointing to external URLs, making it portable and archivable as a unit.

The message Bundle type supports asynchronous, event-driven messaging, in which a sender dispatches a message without waiting for an immediate response. A MessageHeader resource at the start of the Bundle identifies the event type, the sender, and the destination. The message Bundle type is FHIR's analog to HL7 v2 messaging, designed to support the same event-driven integration patterns in a FHIR-native format. Organizations migrating from HL7 v2 often find that their message-oriented workflows map more naturally to the message Bundle type than to direct REST operations, at least during a transitional period.

Dev's first encounter with Bundle type confusion came early in the Riverside General project. The pharmacy system's vendor had delivered a document Bundle when the EHR expected a transaction Bundle. Neither system was technically wrong; both were valid FHIR. The error was an architectural alignment failure that would have been caught before development began if the integration team had specified Bundle types explicitly in the interface contract. This experience became the basis for a standing rule in Dev's team: every interface contract must specify, by name, which Bundle type each endpoint accepts and which Bundle types each sender will produce.

Profiles and Implementation Guides

Maya's question for this section is: if FHIR is already a standard, why do we need profiles and implementation guides on top of it? The base FHIR specification is intentionally permissive. It defines the structure of each resource type, the data types for each element, and the cardinality constraints. But it leaves a large proportion of elements optional and allows any code system or value set to be used for coded elements unless a specific binding is declared. This permissiveness is a design choice: the FHIR specification is intended to be a foundation adaptable to an enormous range of clinical domains, national healthcare systems, regulatory environments, and technical use cases. A single prescriptive specification attempting to define exactly what every healthcare system in the world should do would be impossible to maintain.

A FHIR Profile is a set of constraints applied to a base resource type to make it suitable for a specific use case. A profile can declare that elements the base spec treats as optional are required. It can restrict the allowed values for a coded element to a specific value set. It can declare that certain extensions are expected and provide their definitions. It can add constraints expressed in FHIRPath, the expression language used in the FHIR specification for rules that cannot be expressed as simple cardinality and terminology constraints. A profile does not change the base resource type; a Patient resource conforming to a US Core Patient profile remains a valid FHIR R4 Patient resource that also satisfies the US Core constraints.

Diagram 3.5: How a Base Patient Resource Becomes a US Core Patient — and Why Each Profile Layer Narrows What the Receiving System Must Accept

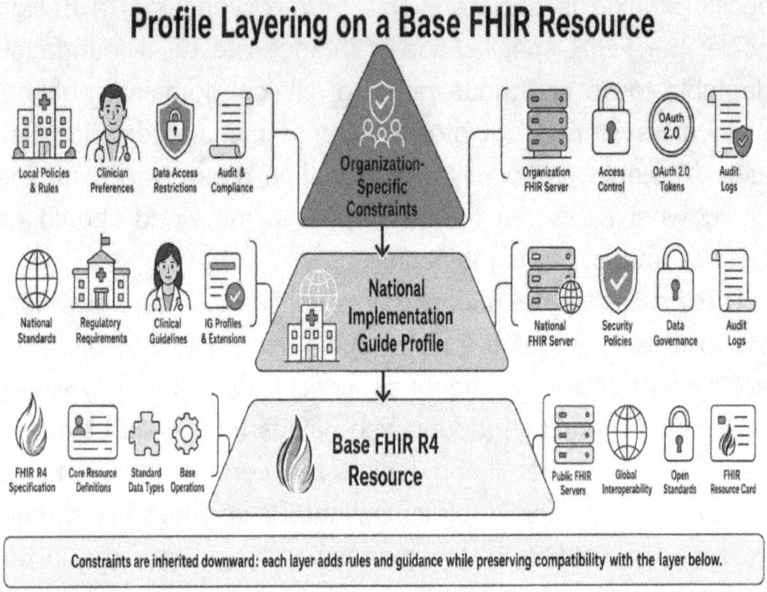

An Implementation Guide (IG) bundles profiles, extensions, value sets, code systems, capability statements, and usage guidance into a deployable package that defines how FHIR should be used in a specific context. The US Core Implementation Guide is the most important US-specific IG for clinical data exchange, defining profiles for core clinical and administrative resource types required to satisfy the ONC's interoperability requirements, including specific requirements for race, ethnicity, and sex at birth data elements not present in base FHIR. The Da Vinci Project IGs define patterns for payer-to-provider and payer-to-payer data exchange, and the Argonaut Project IGs defined the first widely adopted profiles for FHIR-based patient data access that directly influenced the CMS interoperability rules.

Saying a system is "FHIR compliant" without specifying which profiles and IGs it supports carries very little operational meaning. Two systems can both be fully conformant with base

FHIR R4 and fail to exchange a single useful data element, because one uses SNOMED CT codes for Condition code and the other uses ICD-10, or because one requires the CodeableConcept.coding element and the other only populates CodeableConcept.text. Profile conformance is where the work of real interoperability actually happens.

At Riverside General, Dev maintained a profile registry for each in-scope resource type, listing the base profile, any locally defined extensions, value set bindings, and must-support elements. The EHR vendor claimed US Core 4.0.0 conformance, the laboratory information system claimed an earlier version, and the pharmacy system had implemented base FHIR without any named profile. Dev's team spent the first three weeks mapping claimed conformance levels against actual Capability Statements, documenting gaps, and negotiating which gaps would be closed before go-live.

FHIR Data Types: Building Blocks of Resources

Maya's question for this section is: FHIR resources seem to contain many sub-structures. What are the most important ones to understand?

FHIR resources are built from a library of data types ranging from simple primitives to complex structured types designed specifically for clinical data. The primitive types (string, boolean, integer, decimal, dateTime, date, time, uri, url, and others) correspond closely to primitive types in JSON and XML. A Patient resource's birthDate element is a FHIR date primitive. An Observation resource's issued element is an FHIR instant primitive carrying a precise timestamp with a time zone offset. These primitives work as developers expect.

The complex data types are where FHIR's clinical specificity becomes apparent. The most important complex type for interoperability is CodeableConcept. A CodeableConcept has two main elements: a coding array carrying multiple coded representations of the same concept from different code systems, and a text element carrying a human-readable description. A single Condition resource's code element might carry a SNOMED CT code for a clinical finding, an ICD-10-CM code for billing, and a local code from the organization's internal system, all in the same CodeableConcept, along with the text description "Type 2 diabetes mellitus without complications." This structure allows a system that recognizes SNOMED CT to process the SNOMED code, a billing system that needs ICD-10-CM to use that code, and a system that recognizes neither to display the text.

Diagram 3.6: CodeableConcept Structure and Clinical Terminology Bridging

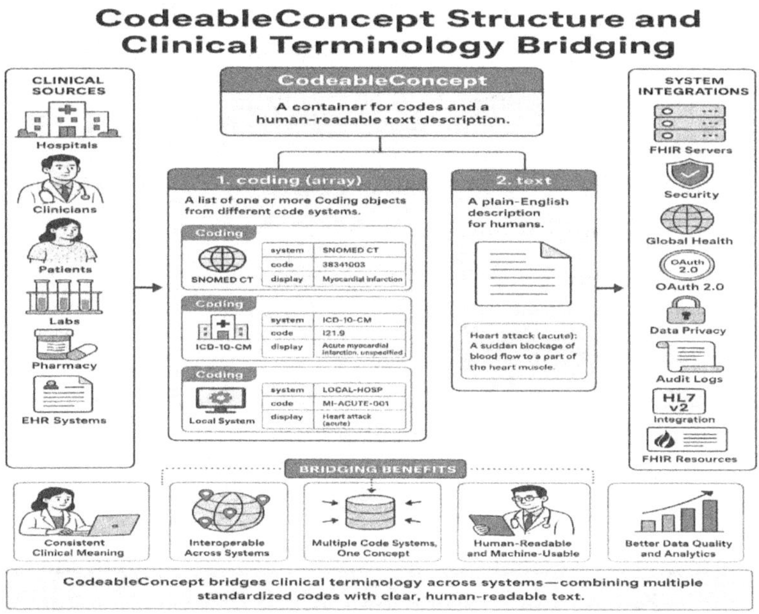

The Coding type is the component within a CodeableConcept that carries a single coded value. It has a system element (a URI identifying the code system, such as "http://snomed.info/sct" for SNOMED CT or "http://loinc.org" for LOINC), a code element (the actual code value), and a display element (the human-readable name for the code). Using the wrong URI for a code system causes terminology validation to fail and prevents automated systems from recognizing the code. This is one of the most common sources of interoperability failures in FHIR implementations.

The Quantity type carries a measured value: a numeric value element, a unit string for human display, and system and code elements that identify the unit using a standard unit code system (typically UCUM, the Unified Code for Units of Measure). An Observation carrying a blood glucose result uses a Quantity for its value[x] element. Consistent use of UCUM codes is a prerequisite for automated unit conversion and for population health analytics that aggregate observations across systems.

The Reference type is the mechanism used to connect FHIR resources. A Reference includes the URL of the referenced resource (either absolute or relative to the server's base URL), an optional type element, and an optional identifier element, which is useful when the full URL is not known. An Observation resource references its subject Patient, performer Practitioner, and encounter through Reference elements. Following these references allows a client to navigate from a single Observation to the full context of that measurement.

HumanName and Address are structured types accommodating naming and address conventions across cultures. HumanName carries a family name, given names as a repeating element, prefixes, suffixes, and a text element for the full name as a single string. Address carries structured elements (line, city, district, state, postalCode, country) and a type element distinguishing postal, physical, and combined addresses. US Core profiles apply specific constraints to both, requiring elements that base FHIR treats as optional.

Pseudocode: Reading and Searching FHIR

Maya's question for this section is: what does actual FHIR interaction look like in code, even at a conceptual level?

The following pseudocode illustrates the foundational FHIR client operations: reading a single resource by ID and executing a parameterized search. These two patterns underlie the majority of FHIR client interactions, from patient portals

displaying a single patient's record to population health tools aggregating data across thousands of patients.
Reading a single Patient resource by ID is the simplest FHIR interaction. The client constructs the URL from the server's base URL, the resource type, and the resource's logical ID; sets the Accept and Authorization headers; sends the GET request; and processes the response.

.

READ a Patient resource: url = base_url + "/Patient/" + patient_id headers = { Accept: "application/fhir+json", Authorization: "Bearer " + access_token } response = http_GET(url, headers) if response.status == 200: patient = parse_json(response.body) display patient.name, patient.birthDate, patientidentifier elif responsestatus == 404: display "Patient not found" else: log_error(response.status, parse_json(response.body))

Searching for all Observations filtered to glucose results in the last ninety days requires a parameterized search query. The client constructs the URL with query parameters identifying the patient, the observation code, and the date range.

SEARCH for recent glucose results: url = base_url + "/Observation" params = { patient: patient_id, code: "http://loinc.org|2339-0", date: "ge" + ninety_days_ago, _sort: "-date", _count: 50 } response = http_GET(url + "?" + encode_params(params), headers) if response.status == 200: bundle = parse_json(response.body) for each entry in bundle.entry: observation = entry.resource display observation.valueQuantity, observation.effectiveDateTime if bundlelink contains "next": fetch_next_page(bundle.link["next"].url)

The pagination loop at the end of the search pseudocode is critical and frequently omitted in prototype code. When the search returns more results than fit in a single Bundle page, the Bundle's link element carries a "next" URL. Clients that do not follow pagination links will silently receive an incomplete result set. In a clinical context, where a missing observation could affect a care decision, this silent incompleteness is a patient safety concern.

Diagram 3.7: Why Dev's First Query Returns Only Twenty Records — How Pagination Tokens Control What the Server Releases and When

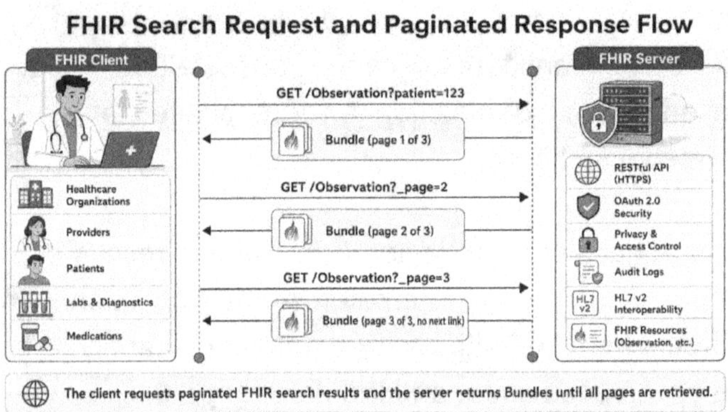

The conditional create operation extends the basic REST pattern to handle idempotency requirements. A conditional create sends a POST with an If-None-Exist header carrying a search query; the server creates the resource only if the search returns no existing matches, preventing duplicate creation on retry.

```
CONDITIONAL CREATE (upsert-like): headers["If-None-Exist"]
= "identifier=MRN|" + mrn response = http_POST(base_url +
"/Patient", headers, patient_resource) if response.status ==
201: log "Patient created: " + responseheaders["Location"] elif
```

response.status == 200: log "Patient already exists, no duplicate created"

Dev's team used the conditional create pattern extensively when loading historical patient data from the legacy system. Each patient record was submitted as a conditional create using the medical record number as the deduplication key. The first submission created the resource; subsequent submissions for the same MRN returned a 200 without creating a duplicate, avoiding the need for a mapping table between legacy IDs and FHIR logical IDs during the initial load.

The Capability Statement: A Server's Self-Description

Maya's question for this section is: how do we find out what a FHIR server can actually do, before we start building against it? The CapabilityStatement resource is the formal self-description of a FHIR server: a machine-readable document that declares which resource types the server supports, which HTTP operations it supports for each resource type, which search parameters are available, which profiles it claims conformance with, and which FHIR version it implements. Every FHIR server is required to serve its CapabilityStatement at the metadata endpoint, the URL formed by appending "/metadata" to the server's base URL. The rest element in the document contains the most operationally important information: the security settings (OAuth2, SMART on FHIR, or other mechanisms), the resource types supported, including their interactions, and the search parameters available for each resource type.

```
FETCH the server's CapabilityStatement: url = base_url +
"/metadata" response = http_GET(url, { Accept:
"application/fhir+json" }) capability =
parse_json(response.body) supported_resources = [] for each
resource in capability.rest[0].resource:
supported_resources.append({ type: resource.type,
interactions: [i.code for i in resource.interaction],
searchParams: [sp.name for sp in resource.searchParam] })
display supported_resources
```

Dev's standing rule for every new vendor engagement was to retrieve the CapabilityStatement before any other technical conversation. He kept a template evaluation checklist comparing the claimed CapabilityStatement against the project's requirements: which resource types were listed, which interactions were supported, which profiles were referenced, and whether SMART on FHIR authorization was declared in the security element. The checklist had saved the Riverside General project from two significant integration failures, both involving vendors whose sales teams had claimed US Core conformance but whose CapabilityStatements listed only base FHIR R4 profiles.

Patrick reviewed one of these checklist outputs during a vendor evaluation meeting. Dev pointed to the profile element of the Patient resource entry in the CapabilityStatement JSON. "The vendor told us they support US Core," he said. "US Core requires that entry to reference the US Core Patient profile StructureDefinition URL. It does not. What they actually support is base FHIR R4 Patient, which is missing the race and ethnicity extensions, the sex at birth element, and several must-support fields required for ONC certification. We cannot satisfy our information blocking obligations with what they have."

Patrick considered that. "So the CapabilityStatement is the server's transcript. It shows what it actually studied, not just what it claims to know."

"That is a better analogy than the business card," Dev said.

Diagram 3.8: Reading a Server's CapabilityStatement Before Writing a Line of Code — What Each Section Tells Dev About What the Endpoint Will and Won't Do

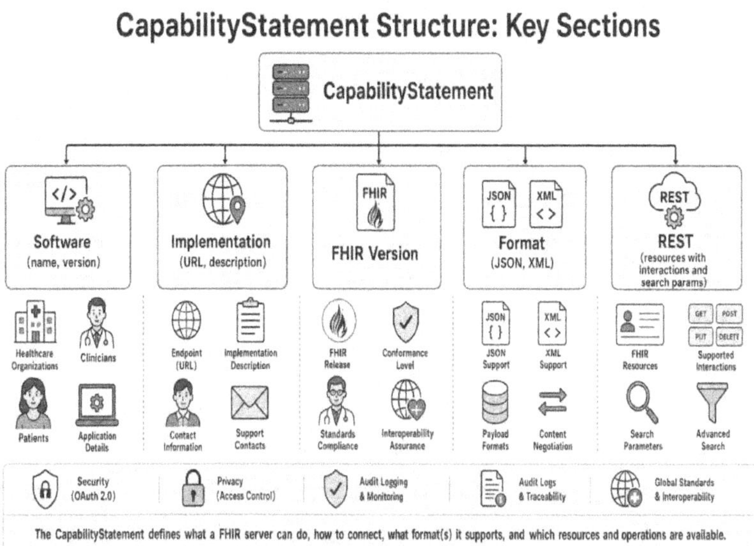

The must-support concept is defined within Implementation Guides. A must-support element is one that, if a system has the data, it must populate, and that a receiving system must process and store without discarding. The CapabilityStatement is the mechanism by which a server declares which profiles, and therefore which must-support obligations, it has implemented. The ONC Inferno testing tool evaluates a FHIR server's actual behavior against its claimed CapabilityStatement and the requirements of the profiles it supports. ONC certification requires passing Inferno tests, turning CapabilityStatement claims from vendor assertions into externally validated facts.

Versioning and Resource History

Maya's question for this section is: what happens to a resource when it gets updated, and how can we look back at what it used to say?

FHIR's versioning model is built into the resource identity system. Every time a resource is updated, the server increments the resource's version number in its meta.versionId element and updates the meta—lastUpdated timestamp. The versioned URL of a resource combines the resource URL with a "/_history/" segment and the version number: retrieving this URL returns the resource as it existed at that specific version, even if it has been updated many times since. This immutable history is not optional for FHIR servers in clinical settings; it is an auditability requirement rooted in the same regulatory framework governing medical record retention and amendment.

The vread operation (versioned read) retrieves a specific historical version of a resource. The history operation retrieves all historical versions as a Bundle, ordered from most recent to oldest. The _since parameter on a history query requests only changes after a specified timestamp, enabling incremental synchronization without re-fetching unchanged resources. Version-aware updates use optimistic locking through the ETag and If-Match headers: the server returns the current version ID in the ETag response header, and a client that sends an update with a matching If-Match header is guaranteed the update applies only to the version it read. If another client has updated the resource in the meantime, the server returns a 409 Conflict status code, preventing the lost-update problem in concurrent systems.

Diagram 3.9: How the Server Tracks Every Change to a Resource — and What the Version History Reveals When a Patient Record Is Quietly Overwritten

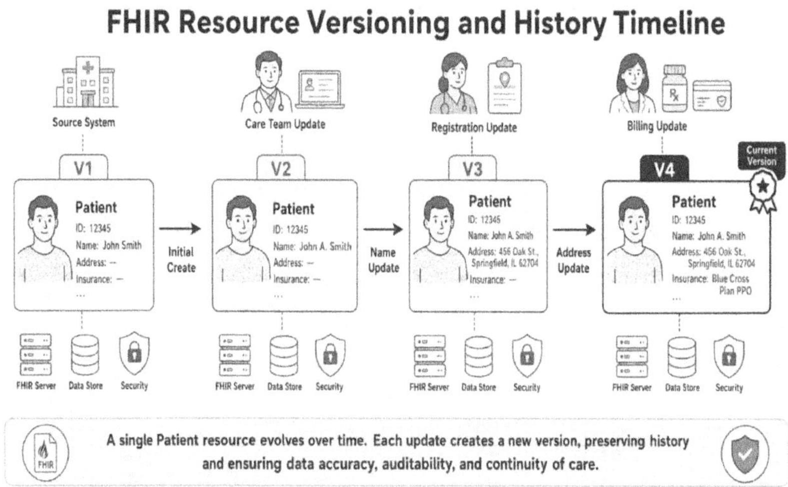

At Riverside General, the versioning system became significant during a medication reconciliation audit. A patient's MedicationRequest had been updated three times in a forty-eight-hour period. Dev retrieved the full history Bundle, which contained all four versions, each with its versionId and lastUpdated value. The audit was completed in minutes; the same work in the legacy system would have required the database administrator to review transaction logs manually.

"So every change is preserved," Patrick said. "You cannot just overwrite the record."

"Correctly implemented, no," Dev said. "The history is always there for audit purposes. The same regulation that says you cannot delete a medical record applies here."

Organizations designing FHIR server infrastructure must account for the storage costs of full resource history, particularly for high-frequency resources like Observation. History retention is the audit foundation that enables retrospective clinical and quality analyses.

Takeaways

The weeks Dev spent walking Patrick through FHIR's architecture revealed a layered structure: each concept was a prerequisite for the next. Resources before Bundles. Base resources before profiles. REST operations before the Capability Statement. The architecture as a whole is coherent and carefully considered, but coherence only becomes visible once you have worked through each layer in sequence.

A FHIR resource is a self-contained, addressable unit of clinical information, not a database record or a file. Its four structural layers (identity and metadata, narrative, extensions, domain content) and its permanent URL together define the foundational architectural choice that distinguishes FHIR from every prior healthcare data standard. Thinking of FHIR resources as database records leads to incorrect design decisions; understanding them as addressable objects on a server leads to correct ones.

Bundle type selection is an architectural decision that requires explicit agreement among integration partners. Transaction Bundles for atomic operations, searchset Bundles for query responses, document Bundles for portable clinical documents, and message Bundles for event-driven messaging are four distinct mechanisms with four distinct processing contracts. Interface specifications must name Bundle types before development begins, not after the first 422 error appears in the logs.

"FHIR compliant" is a meaningful claim only when followed by which profiles, which IGs, and which version. Base FHIR R4 conformance is the floor, not the ceiling. US Core, Da Vinci, and other IGs define the specific constraints, extensions, and value set bindings required for clinically meaningful exchange. Every vendor evaluation should start with the Capability Statement and a checklist that maps its contents to the project's profile requirements.

CodeableConcept is the architectural bridge between FHIR's resource model and the clinical terminology systems that give healthcare data its semantic meaning. A FHIR implementation that populates CodeableConcept.text but not CodeableConcept.coding cannot participate in automated clinical decision support or population health analytics. The system URI within each Coding element must match the canonical URI defined by the terminology authority; a mismatch that appears to be a formatting issue in development becomes a semantic failure in production.

Versioning and resource history are built into the FHIR identity model to satisfy the same regulatory requirements governing medical record retention and amendment. Each update increments the version ID and preserves the previous version in the history. This design anticipates compliance requirements rather than leaving them to implementers. Organizations planning FHIR server infrastructure must account for the storage and query costs of full resource history, particularly for high-frequency resource types.

The business card metaphor held up; Patrick told Dev at the end of the last walkthrough session, but it needed a card catalog to back it up. The cards were the resources. The REST API was the catalog query system. The profiles were the rules for what had to be on each card in a specific context. The Capability Statement was the catalog's own description of its contents. And the version history was the archive ensuring no card ever truly disappeared.

Chapter 4 returns to HL7 v2, examining its message structure and segment anatomy through Dev's eyes as he maps legacy interface logic onto the FHIR resources that will eventually replace those interfaces. Understanding the contrast between v2's flat, pipe-delimited messages and FHIR's structured, addressable resource model is one of the most effective ways to understand why both standards coexist in real healthcare IT environments, and why migration is measured in years rather than months.

4 HL7 v2 in Depth: Reading the Pipes

Scenario

Dev Okafor pulled a second chair up to his workstation and gestured for Patrick Osei to sit. On the monitor in front of them, the Mirth Connect console was alive with motion: a scrolling feed of pipe-separated text, each line representing something that had just happened to a patient somewhere in Riverside General. A new admission from the emergency department. A completed chemistry panel from the lab—a transfer from the surgical floor to the ICU. The screen showed 217 active routes, and the messages kept scrolling.

"This is what runs the hospital," Dev said. "Every patient admission, every lab result, every transfer, every medication order that goes from the EHR to the pharmacy. It all comes through here."

Patrick leaned forward. The text was dense and unfamiliar: each line began with a three-letter code, followed by a long string of fields separated by vertical bars, with carets and ampersands nested inside. He could see timestamps, names that might be patient, numeric values that might be lab results. But the syntax was opaque. "I have no idea what any of that means," he said.

Dev nodded. "Most people don't. Even some people in health IT for years never really learn to read it. It stays a black box." He pulled up a single message from the feed: an ADT^A01, a patient admission notification. "By the end of this chapter," Dev said, "you'll be able to read this."

Dr. Maya Chen had joined them at the doorway. She had ordered hundreds of labs and written hundreds of admission notes, and every one of those clinical actions had generated exactly this kind of message. She stepped into the room. This chapter belongs to Dev, but it answers questions that both of them have been carrying for a long time.

Why It Matters

HL7 v2 literacy matters even in a FHIR-forward world for a reason that is easy to overlook. The majority of hospitals across the United States still exchange the majority of their clinical transactions in HL7 v2 format, and that will remain true for at least the next decade. A FHIR migration does not begin with a blank slate. It begins with a current state of v2 interfaces that must be inventoried, understood, and either migrated or bridged. A developer who cannot read an HL7 v2 message cannot assess the scope of the migration they have been asked to plan, cannot validate that the translation logic is correct, and cannot diagnose the integration failures that will inevitably occur during the coexistence period when both v2 and FHIR messages are flowing simultaneously.

There is also a practical career argument that Dev understands, having lived it. HL7 v2 expertise is a differentiating skill in health IT precisely because the industry is trying to move away from it even as it depends on it. The developer who can fluently read and write both v2 and FHIR, who understands the mapping between an ADT^A01 and a FHIR Encounter, between an OBX result and a FHIR Observation, is the developer who can maintain the systems running today while building the systems that will run tomorrow. This chapter is the beginning of developing that capability.

At Riverside General, the stakes are concrete. The hospital runs 217 active integration routes. Every one of them depends on HL7 v2 messages flowing correctly between the EHR, the laboratory information system, the pharmacy, the radiology information system, the dietary system, the ADT system, and the billing system. None of those connections can be safely touched, migrated, or replaced by a team that does not understand what is flowing through them. The v2 literacy this chapter builds is not academic preparation for a future migration. It is the minimum competency required to work safely with the integration infrastructure currently running.

The Message in the Pipe: Anatomy of a v2 Transaction

Maya's question for this section is: what exactly is an HL7 v2 message, and how does it travel from one system to another?

An HL7 v2 message is a sequence of named segments, each beginning with a three-character segment identifier, followed by a pipe character, and a series of fields separated by additional pipe characters. Each complete message represents a clinical event: a patient admission, a lab result, a medication order, a discharge. The message flows from a sending application to one or more receiving applications via a transport mechanism, and the receiving application parses the message's raw text to reconstruct the clinical event it describes.

The transport mechanism HL7 v2 was designed to use is MLLP (Minimal Lower Layer Protocol), a simple framing protocol that runs over a persistent TCP connection. Each HL7 v2 message is wrapped in framing characters: a vertical tab (ASCII 0x0B) marks the start, and a file separator (ASCII 0x1C) followed by a carriage return marks the end. These framing characters allow the receiving system to identify the exact boundaries of each message in the byte stream, even if multiple messages arrive back-to-back.

The MLLP connection is persistent, not request-response. The sending system holds a TCP connection open across multiple messages. This differs from HTTP, where each request opens and closes its own connection. The persistent connection allows the receiving system to send acknowledgments over the same connection without connection setup overhead, which matters in high-volume environments where messages can flow at several hundred per minute during peak periods.

Diagram 4.1: The Message That Moved Data for Thirty Years — and the Failure That Finally Made the Cost Visible

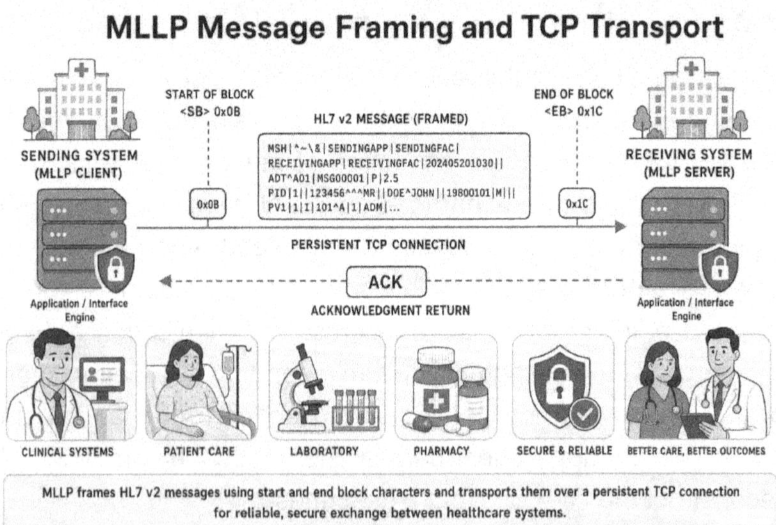

The anatomy of the message can be grasped from a single example. The following is a simplified ADT^A01 message representing a patient admission at Riverside General. Every

field reflects actual v2 structure, though the values are illustrative:

```
MSH|^~\&|RIVERSIDE_EHR|RIVERSIDE|LAB_SYS|RIVERSID
E|20240315093045||ADT^A01^ADT_A01|MSG000412|P|2.5.1
EVN|A01|20240315093045 PID|1||RGH-
10042^^^RIVERSIDE^MR||Morales^Elena^R||19851204|F|||12
20 Elm St^^Baltimore^MD^21201||(443)555-
0198^^PRN|||||123-45-6789
PV1|1|I|MEDSURG^312^A^RIVERSIDE|||1234^Singh^Priya^A^
MD^^^RIVERSIDE^L|||||||||1234^Singh^Priya^A^MD^^^RIVERS
IDE^L|||||||||||||||||||||||||20240315093045
```

Each line is a segment. The first three characters name the segment type. Everything after the first pipe is data. The structure is hierarchical, the delimiters are fixed, and the v2 specification defines the field positions. A developer who knows the specification can read this message as fluently as a prose sentence.

The MSH segment tells the integration engine who sent the message (RIVERSIDE_EHR), who is supposed to receive it (LAB_SYS), what event it represents (ADT^A01, an admit), when it was created, and what version of the v2 standard it uses (2.5.1). The PID segment carries the patient's identity: her medical record number, name, date of birth, gender, address, and phone number. The PV1 segment carries the visit context: the room and bed she has been assigned, her attending physician, and when the visit began.

Patrick was following along. "So the pipe is like a comma in a CSV file," he said. Dev nodded. "Exactly. Except the columns aren't named in the file. The specification tells you what each position means. The standard fixes that, and it's also why v2 can be brittle: if a sending system puts data in the wrong field, the message parses fine, and the data ends up in the wrong place silently."

Encoding Characters and the Delimiter Hierarchy

Maya's question for this section is: why do HL7 v2 messages contain so many different special characters, and what does each one actually do?

HL7 v2 encoding is hierarchical, resembling nested containers. At the outermost level, the message is divided into segments separated by carriage returns. Within each segment, the content is divided into fields separated by the pipe character. Within each field, the content may be divided into components separated by the caret character. Within each component, the content may be divided into subcomponents separated by the ampersand character. The tilde character serves a different purpose: it separates repeated values in a field that can hold multiple values.

These five characters form the complete delimiter vocabulary of a v2 message. Every level has its own separator, and the levels nest cleanly: split by carriage return to get segments, by pipe to get fields, by tilde to get repetitions, by caret to get components, and by ampersand to get subcomponents.

The encoding characters are declared in MSH-2, always the literal string "^~\&" in a standard-compliant message. The backslash is the escape character, used to represent any delimiter character as literal data within a field value. If a patient's name contains a caret, the sending system must encode it as an escape sequence; a literal caret inside a PID name field would be misread as a component separator. Without the escape mechanism, any delimiter character in actual clinical data would corrupt the message structure.

Diagram 4.2: The Four Characters That Hold an HL7 v2 Message Together — and What Riverside's Lab System Does When One Appears Inside a Data Value

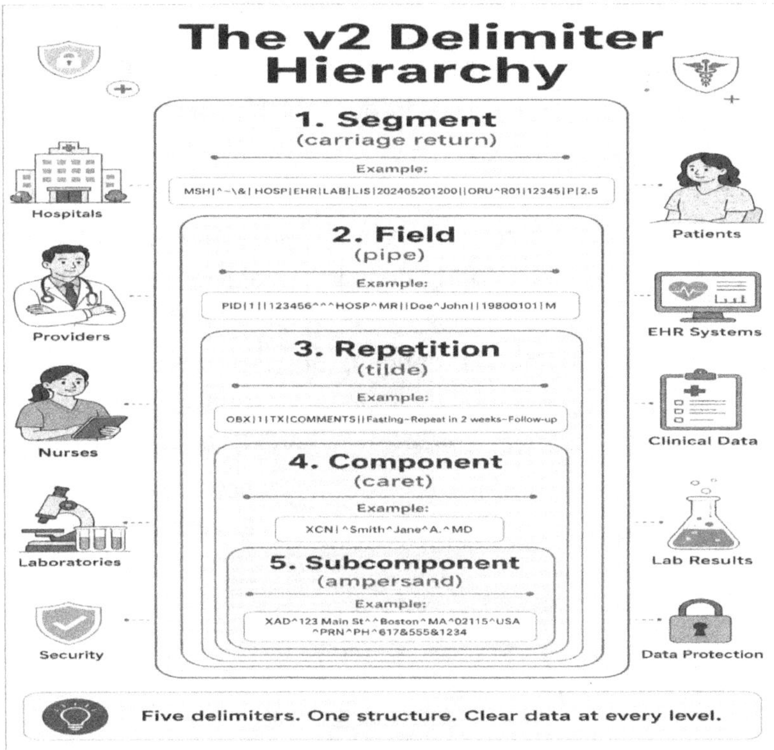

The patient name field in the PID segment illustrates the component structure. A PID-5 patient name might appear as:
Morales^Elena^R^^Ms
The family name is the first component (Morales). The given name is the second (Elena). The middle initial is the third (R). The fourth component, empty here, would hold a suffix. The fifth holds a prefix (Ms). All of this travels in a single field, and the

receiving system splits the field by caret to retrieve each name part individually.

Dev showed Patrick a more complex example: a PID field with multiple patient identifiers, each one a separate repetition, each broken into components for the identifier value, the assigning authority, and the identifier type code. "This is why you need a parser," Dev said. "You could split this by hand, but the moment you get a name with a hyphen or an unusual character, you'd be chasing bugs for hours." The proper tool is an HL7 v2 parsing library that understands the full delimiter hierarchy and handles escape sequences correctly.

The most common parsing errors in custom v2 implementations come from exactly this kind of delimiter confusion. A developer who manually splits a message on pipe characters without accounting for escape sequences, or who assumes that a component is always a simple value, will produce a parser that works for ninety percent of messages and silently misreads the remaining ten percent. In a clinical integration environment, a ten-percent misread rate is not a performance metric. It is a patient safety risk.

MSH, PID, PV1: The Registration Segments

Maya's question for this section is: what does the administrative information in a hospital admission message look like at the data level, and why does it matter for clinical care?

While HL7 v2 defines dozens of segment types, three segments carry the registration and visit information that anchor every clinical transaction to a specific patient and encounter. The MSH segment is the message envelope, the PID segment is the patient identity record, and the PV1 segment is the visit record.

Together they form the demographic and administrative foundation on which every other segment in the message depends.

The MSH segment is mandatory in every HL7 v2 message and is always the first segment. Its fields carry the sending application, the sending facility, the receiving application, the receiving facility, the message creation timestamp, the message type, a unique message control identifier, the processing identifier (P for production, T for test), and the v2 version number. The integration engine uses the sending and receiving application identifiers, the message type, and the version number to route and process the message. A misconfigured MSH is the most common cause of messages arriving at the wrong destination or being rejected by a receiving system that expects a different message type.

Diagram 4.3: What the First Segment of Every HL7 v2 Message Declares — and Which Fields the Receiving System Uses to Decide Whether to Accept or Reject

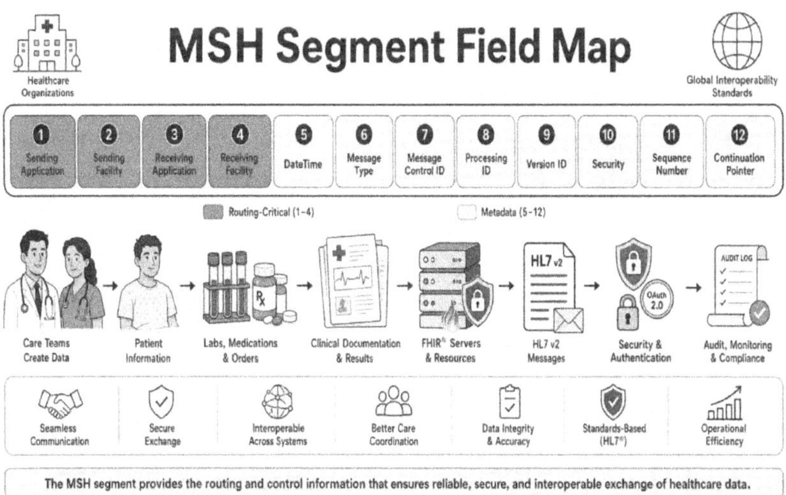

The PID segment carries the patient's identity. Its most critical fields are the patient identifier list (PID-3), the patient name (PID-5), the date of birth (PID-7), the administrative sex (PID-8), and the patient's address (PID-11). The patient identifier list warrants particular attention because it may contain multiple identifiers for the same patient, each assigned by its authority and identifier type code. A typical Riverside General patient might have a medical record number, a Medicare beneficiary identifier, a state Medicaid identifier, and an insurance subscriber identifier, each carried in a different PID-3 repetition. Downstream systems use the assigning authority and type code to determine which identifier to use for their own internal linkage.

Patient matching based on PID data is one of the most consequential operations in a hospital integration environment. When a PID arrives with an identifier the receiving system does not recognize, it must decide whether to create a new patient record, match the patient demographically to an existing record, or reject the message for manual resolution. A false match links two patients' records together, potentially causing one patient to receive another's allergies or problem list. A failed match creates a duplicate record, fragmenting the patient's history.

The PV1 segment carries the visit context: the patient class (inpatient, outpatient, emergency, recurring), the assigned patient location (composed of point of care, room, bed, and facility components), the admission type, the attending physician, the referring physician, the hospital service, the admission date and time, and the discharge date and time. Dev walked Patrick through the PV1 segment of the on-screen message. He pointed to the location field: MEDSURG^312^A^RIVERSIDE. "That's the floor, the room, the bed, and the facility," he said. "When the bed management system gets this, it marks bed 312-A as occupied. When dietary gets it, it starts sending meal trays to that room. When pharmacy gets it, it links medication orders to that location. This one field drives all of that."

OBR and OBX: When Results Flow Back

Maya's question for this section is: How does a completed lab result get back to my screen after I order a test?

The OBR and OBX segments carry laboratory orders and results, and they are the two segments that matter most for understanding what happens between when a physician places a lab order and when the result appears in the EHR. The flow is bidirectional and involves at least two separate message transactions: an ORM message carrying the order from the EHR to the laboratory information system, and an ORU message carrying the result from the laboratory back to the EHR.

The OBR (Observation Request) segment describes the order itself. Its key fields are the placer order number (OBR-2), assigned by the ordering system; the filler order number (OBR-3), assigned by the laboratory when it accepts the order; the universal service identifier (OBR-4), the coded identifier for the test (typically a LOINC code or a local code); the observation date and time (OBR-7), representing the specimen collection time; and the result status (OBR-25), indicating whether the results are preliminary, final, or corrected.

Diagram 4.4: ORM to ORU Order-Result Roundtrip

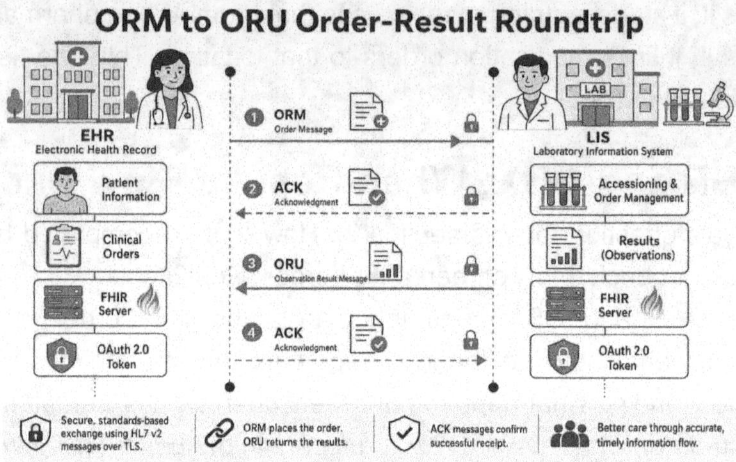

The OBX (Observation Result) segment carries the actual result: the observation identifier (OBX-3, always a coded value), the value type (OBX-2), the observation value (OBX-5), the units (OBX-6, for numeric results), the reference range (OBX-7), the abnormality flags (OBX-8), and the observation result status (OBX-11). A single OBR segment can be followed by multiple OBX segments, one for each result in a panel. A basic metabolic panel might generate eight OBX segments, one for each component result.

The OBX value type field controls how the observation value in OBX-5 should be interpreted. The most common types are NM (numeric), ST (string), and TX (text). Still, the specification also defines types for coded values (CE and CWE) and encapsulated data (ED), which are used to carry binary objects, such as PDF documents encoded in base-64, within the OBX segment. The range of value types is part of what makes v2 flexible enough to handle diverse clinical use cases, and part of what makes v2 parsing non-trivial.

Maya had been thinking about something Dev said earlier. "So when I order a sodium, and it comes back flagged as low," she said, "there is literally an OBX segment somewhere with 'L' in the abnormality flag field?" Dev nodded. "OBX-8. The sending system puts L for low, H for high, LL for critically low, HH for critically high, A for abnormal. Your EHR reads that field and decides whether to display the result in red." She absorbed that for a moment. Thirty years of lab results, thousands of sodium values, all of them ultimately a capital letter in field eight of a segment she had never seen.

ADT, ORU, ORM, MDM: The Four Message Types

Maya's question for this section is: if every clinical event generates a message, what are the different kinds of messages and what does each one actually do?

HL7 v2 defines dozens of message types, each corresponding to a clinical event or a workflow trigger. Still, four message types account for the great majority of traffic at a typical community hospital. Understanding these four types means grasping the event-driven architecture underlying a hospital's information systems. Every major clinical action produces at least one of these messages, and every downstream system that needs to know about that action receives a copy.

ADT messages (Admit, Discharge, Transfer) are the administrative heartbeat of a hospital information system. An ADT^A01 is a patient admission, ADT^A02 is a transfer, ADT^A03 is a discharge, and ADT^A08 is a patient information update. Each trigger event causes the EHR or ADT system to broadcast a message to every downstream system that maintains a patient census or record. When Elena Morales is admitted to the med-surg unit, an ADT^A01 flows to the laboratory, pharmacy, dietary, radiology, billing, and nursing documentation systems. When she transfers to the ICU, an ADT^A02 flows, and each system updates her location. When she is discharged, an ADT^A03 closes her encounter record across all systems.

Diagram 4.5: ADT Event Flow Across Hospital Systems

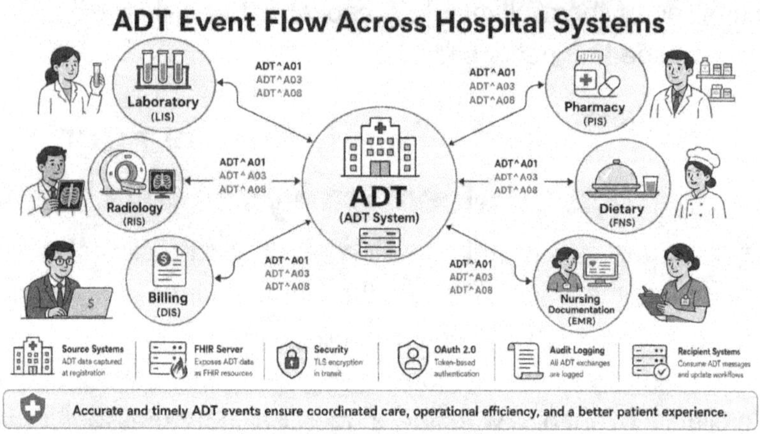

ORU messages (Observation Result Unsolicited) carry completed results from producing systems to consuming systems. The "unsolicited" in the name reflects that the receiving system does not explicitly request the result; it receives it because the producing system determined that the result was ready and pushed it. When the laboratory completes a metabolic panel, the LIS generates an ORU^R01 containing the OBR segment and one OBX per result component. The ORU flows to the EHR, which matches it to the original order using the placer and filler order numbers. The same ORU may also be distributed to a clinical data repository or a population health platform, depending on the routing configuration.

ORM messages (Order Messages) flow in the opposite direction, from the ordering system to the fulfilling system. An ORM^O01 carries a new order. The same message type also carries modifications, cancellations, and status requests, with the order control code (ORC-1) distinguishing these actions: NW for new order, CA for cancel, XO for change. The ORC (Common Order) segment that precedes each order group carries the order control code, the placer order number, the filler order number, and the order status. A developer who can read the ORC segment can determine at a glance whether a message is placing a new order, canceling an existing one, or requesting a status update.

MDM messages (Medical Document Management) carry clinical documents between systems, including discharge summaries, operative reports, and consultation notes. Document content is carried in an OBX segment with a value type of TX for plain text or ED for formatted documents such as PDFs. MDM messages are important for understanding the FHIR migration scope because documents that currently flow as MDM messages need to be represented as FHIR DocumentReference resources that point to FHIR Binary resources. This transformation is non-trivial when document content contains its own internal structure.

ACK and NACK: Handshake That Guarantees Delivery

Maya's question for this section is: how does the sending system know that the receiving system actually got the message and didn't just lose it?

One of HL7 v2's most important features for clinical environments is its built-in acknowledgment mechanism. Every message that requires guaranteed delivery is followed by an

ACK message from the receiving system. A NACK (negative acknowledgment) indicates the message was received but could not be processed, and the MSA segment carries a code identifying the type of failure. The distinction between these codes matters for how the sending system handles the response.

An ACK with MSA-1 value AA (Application Accept) means the receiving system accepted the message without errors. An ACK with MSA-1 value AE (Application Error) indicates that the receiving system encountered an error during processing, typically a business-logic error, such as an unrecognized patient identifier or an invalid order code. The message was not lost but was not successfully processed, and the sending system must decide whether to retry, queue for manual review, or alert an administrator. An ACK with MSA-1 value AR (Application Reject) indicates that the receiving system refused to process the message, typically due to a structural problem such as an unrecognized message type or a missing required field.

Diagram 4.6: ACK/NACK Acknowledgment Decision Tree

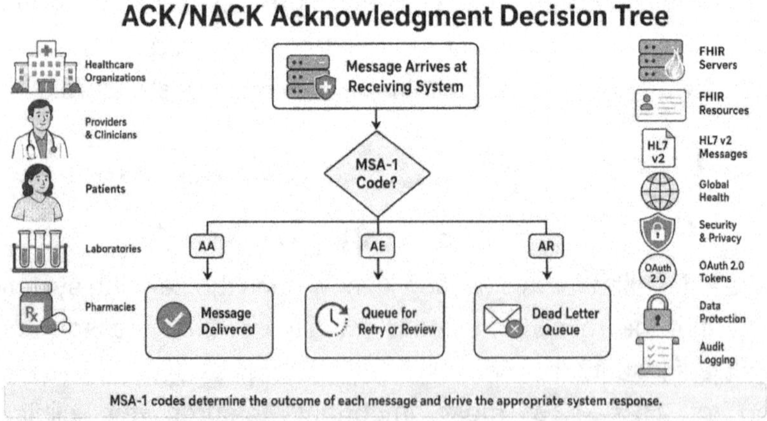

The MLLP protocol was specifically designed to support the ACK/NACK handshake. Because the TCP connection is persistent, the receiving system can send the ACK back immediately after processing, without the overhead of establishing a new connection. The sending system waits for the ACK before sending the next message in synchronous mode, which is the standard configuration for clinically critical interfaces. A sending system that does not receive an ACK within the configured timeout period treats the message as undelivered and either retries the send or routes the message to a dead-letter queue for manual intervention.

Dev pulled up the dead-letter queue for the pharmacy interface. It had four messages sitting in it from the previous evening. "These all got AR responses," he said. "The pharmacy system was updated overnight, and somebody changed a field mapping. These four orders were sent after the update went live but before we caught the error." Four medication orders, each representing a prescribed treatment that had not reached the pharmacy system. The message had sat in a queue for which nobody had configured an alert. "I'm setting up an alert for any dead-letter queue depth over zero," Dev said. "This should not require a manual check."

The NACKs that integration engines generate and receive are one of the richest diagnostic signals in a health IT environment. An integration engineer who monitors NACK rates has an early warning system for interface failures, misconfigurations, and data quality issues. A sudden spike in AE responses on a lab results interface almost always indicates a patient identifier or test code mismatch. A sudden spike in AR responses almost always indicates a structural change on the sending or receiving side, such as a system upgrade that changed the message type or version.

The ACK/NACK mechanism is one of the most important features of v2 that FHIR does not replicate at the protocol level. REST-based FHIR interactions return HTTP status codes, but the HTTP response is connection-scoped: once returned, the connection closes, and neither party retains an obligation to the other. There is no built-in retry mechanism, no dead-letter queue, and no equivalent of the MLLP persistent connection. Organizations building FHIR-based integrations that require guaranteed delivery must implement these mechanisms explicitly, through a message broker, a FHIR subscription service, or a middleware layer that adds delivery guarantees on top of the FHIR API.

Z-Segments: Custom Extensions and Hidden Data

Maya's question for this section is: if the standard covers all the common things hospitals need to share, why do hospitals keep adding their own custom fields?

HL7 v2's flexibility was both its greatest strength and one of its most persistent liabilities. The standard allowed vendors and healthcare organizations to extend messages with custom segments called Z-segments, because their names always begin with the letter Z. A segment named ZPI might carry custom patient insurance information; ZAL might carry a custom allergy representation; ZPAT might carry additional patient attributes defined by the vendor outside the standard. Z-segments were widespread in the 1990s and 2000s as vendors competed to offer features the base standard did not support, and they remain prevalent in many production integration environments today.

The problem with Z-segments is both simple and serious. They are invisible to any receiving system that was not specifically configured to process them. A receiving system that encounters an unrecognized Z-segment will skip it without generating any error. The message parses successfully, the ACK goes out, and the data in the Z-segment is silently and permanently lost—no NACK. No alert. The clinical user who depends on that data, perhaps an insurance coordinator who needs the coverage information in ZPI to authorize a procedure, will not discover the problem until they look for the data and find it absent.

Dev opened the Z-segment documentation folder on his workstation. Riverside General had seventeen Z-segments in active use. He had cataloged them during a data governance project the previous year: each had been added by a different vendor at different times, and documentation for four was incomplete. "This is what I think of when somebody says 'simple FHIR migration,'" he said. "These seventeen Z-segments contain data that clinical workflows depend on right now. We have to figure out what to do with each one before we can migrate the interfaces that carry them."

Diagram 4.7: Z-Segment Visibility Problem

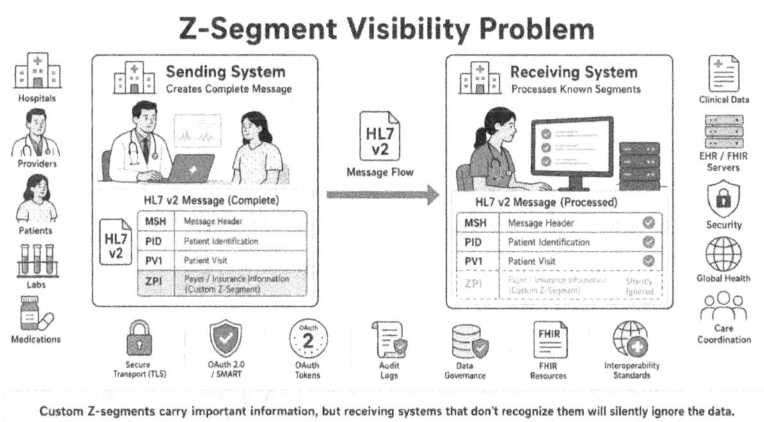

Patrick had a follow-up question. "If a Z-segment gets silently ignored, how do you even know what data you're losing?" Dev's answer was methodical: start with the interface specification documents; examine Z-segment definitions in the sending system's technical documentation; run sample messages through a parser that captures all segments including unknowns. And then, critically, talk to the clinical users who work with the data and trace each piece back to its source segment. That last step is the one most migration projects skip, and it is the one that produces the surprises.

Z-segments also create a mapping problem for FHIR migrations. The v2-to-FHIR mapping guide covers standard segments, but Z-segments by definition are not in the standard. Each Z-segment requires a custom mapping decision: either a FHIR resource element can hold the data, a custom FHIR extension must be defined and agreed upon with all receiving systems, or the data has no FHIR equivalent and a business decision must be made about whether to preserve it. All three outcomes require effort, documentation, and stakeholder agreement on migration planning timelines.

v2 to FHIR: The Translation Conceptually

Maya's question for this section is: when a system converts an old HL7 v2 message into a FHIR format, what actually changes, and what can go wrong?

The conceptual mapping between HL7 v2 and FHIR is well-documented. The HL7 organization maintains a v2-to-FHIR mapping guide that specifies how each v2 segment and field corresponds to FHIR resource elements. The guide covers the major segment types in detail: PID maps primarily to Patient;

PV1 maps primarily to Encounter; OBR and OBX map primarily to ServiceRequest and Observation; AL1 (allergy) maps to AllergyIntolerance; DG1 (diagnosis) maps to Condition. At the conceptual level, the mapping is clear and well understood. At the practical level, the translation is never as clean as the guide suggests.

The first complexity is cardinality. A PID segment may contain multiple patient identifiers in PID-3, each with its own assigning authority and identifier type code. In FHIR, each of these identifiers becomes a separate Identifier element within the Patient resource, and the assigning authority and type code need to be translated into FHIR identifier system URIs and type codings. A PID-3 carrying four identifiers (an MRN, a Medicare identifier, a Medicaid identifier, and an insurance identifier) produces a Patient resource with four Identifier elements, each mapped from a different repetition of the same PID field. The translation requires a mapping table that maps each assigning authority code to the corresponding FHIR identifier system URI. That table must be created, tested, and maintained for each deployment environment.

Diagram 4.8: What Gets Lost and What Gets Transformed When Riverside's PID Segment Becomes a FHIR Patient Resource

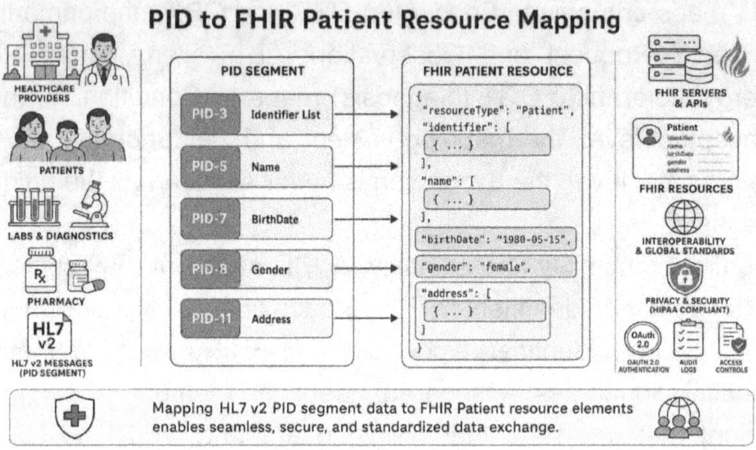

The second complexity is structure. A PID-5 patient name field carries name components as positional elements within a single field. A FHIR HumanName element carries the same information in explicitly labeled child elements: family, given (an array for first and middle names), prefix, and suffix. The translation is straightforward when the v2 data is well-formed, but real-world v2 data often is not. Names with non-standard component placements, Unicode characters in escape-sequence encoding, or multiple given names encoded inconsistently across messages all require transformation logic that goes beyond simple field-to-element mapping.

The third complexity is coded values. HL7 v2 messages carry coded values using a variety of coding systems: some fields use HL7 user-defined tables, some use SNOMED CT codes, some use LOINC codes, some use local codes specific to the sending institution. FHIR resources expect coded values to be represented as FHIR Coding elements with an explicit system URI, a code value, and a display string. Translating from a v2 coded value to a FHIR Coding element requires knowing which coding system the v2 code belongs to, what the corresponding FHIR system URI is, and whether any code translation is needed when the sending system used local codes rather than standard ones. This coded data translation problem is particularly acute in the v2-to-FHIR translation layer and directly intersects with the clinical terminology work covered in Chapter 6.

Dev walked Patrick through a simplified translation on the whiteboard: PID-3 becomes the identifier array; PID-5 becomes the name array; PID-7 becomes birthDate; PID-8 becomes gender; PID-11 becomes the address array. "This is the easy part," he said. "The hard part is when you get to PID fields that FHIR represents differently, like PID-10 for race and PID-22 for ethnic group. FHIR has a US Core extension for race and ethnicity, but the v2 codes and the FHIR value sets aren't the same, so you need a code translation step in the middle." Patrick was beginning to appreciate why migration projects ran over budget.

What Runs Riverside's 217 Routes

Maya's question for this section is: with so many different systems needing to communicate, how does a hospital keep track of what should flow where?

Riverside General's 217 active integration routes are managed through a centralized integration engine, Mirth Connect, which Dev has configured and maintained over several years of incremental additions. Each route defines a source channel (where the message comes from), a transformation (what needs to be done to the message before it reaches its destination), and a destination channel (where the message is going). The transformation step is where the business logic lives: field mappings, data translations, routing conditions, and error handling policies.

The 217 routes break down into categories: ADT distribution routes that fan out admission and discharge events; result routing routes that carry lab, radiology, and pathology results; order routing routes that carry orders from the EHR to fulfilling systems; document routing routes for clinical documents; and a small number of specialty routes for the cardiac monitoring system, the ventilator data interface, and the bed management system. Each category has its own patterns and failure modes, and a developer who understands the categories can approach a new routing problem with a framework rather than starting from scratch.

The ADT fan-out is the simplest category to understand and one of the most consequential to maintain. When the EHR sends an ADT^A01, Mirth Connect receives it, parses it, and sends copies to ten separate destination systems. Each destination may require a slightly different version of the message: some systems expect v2.3, some expect v2.5.1, some need fields normalized to local code tables, some need Z-segments stripped, and some need additional data added from a lookup table. The transformation logic that produces ten different outbound messages from a single inbound message is the kind of configuration that accumulates complexity over time as each destination system's requirements change independently.

Diagram 4.9: Riverside's Integration Engine Routing Architecture

Riverside's Integration Engine Routing Architecture

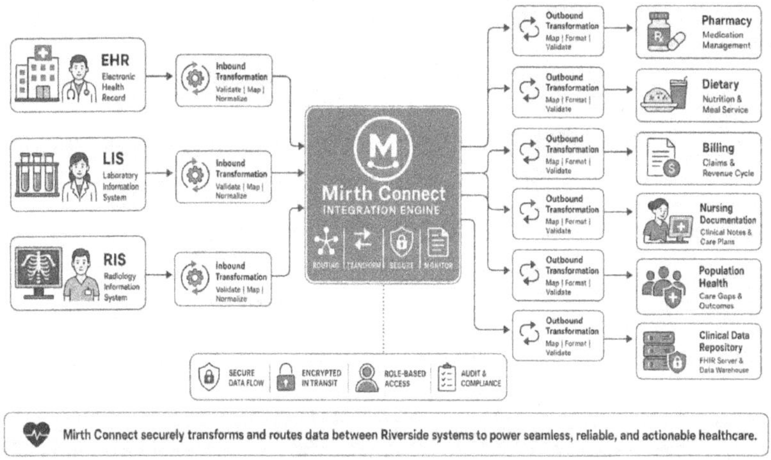

Dev had a practice he had developed over years of managing integration environments: he documented every route in a spreadsheet capturing the source system, destination system, message types handled, transformation logic applied, current status, and date of last validation. He called it the interface inventory and updated it whenever a route was added, modified, or decommissioned. Patrick had seen the spreadsheet and had been struck by how much of the hospital's clinical operations depended on configurations managed by a single engineer. "What happens to all of this if you get hit by a bus?" he had asked. Dev took the question seriously. "That's why I'm training Keisha," he said. "And why I document everything."

The interface inventory also serves as the starting point for the migration scope assessment. Every row represents an interface that needs a migration decision: migrate to FHIR now, temporarily bridge with a translation layer, or keep it native v2 indefinitely. Without the inventory, the migration team has no baseline. With it, they have a structured starting point for the analysis that Chapter 7 and Chapter 13 will return to in detail.

Takeaways

Dev closed the Mirth Connect console. The messages were still scrolling. They would be scrolling when he went home, when the hospital opened its new outpatient building and added another forty routes, and when the FHIR migration eventually began, and the first interfaces were cut over. The 217 routes did not care about the architecture discussions in conference rooms. They just kept flowing, one pipe-delimited segment at a time.

HL7 v2 literacy is a prerequisite for any realistic FHIR migration planning. An organization that cannot read and interpret its own v2 messages cannot accurately scope the migration, validate the translation logic, or diagnose the integration failures that will occur during the coexistence period. Investing in v2 literacy before beginning FHIR adoption pays dividends throughout the migration and beyond. The fluency this chapter introduced is the foundation that prevents expensive late-stage discovery during a migration project, when the stakes are highest.

The five core segments, MSH, PID, PV1, OBR, and OBX, account for the majority of clinical data in most integration environments. A developer who can fluently interpret these segments can diagnose the majority of production integration issues, validate mapping logic, and communicate meaningfully with clinical staff about what data is and is not flowing correctly.

Maya's observation about the OBX abnormality flag was not incidental. When a clinician understands what the integration layer is doing to their clinical data, conversations between clinical and technical staff become substantially more productive.

The ACK/NACK mechanism is the feature that makes v2 trustworthy in clinical environments. Its absence from the FHIR protocol layer is one of the most important architectural gaps in REST-based FHIR interactions. Implementing equivalent delivery confirmation in a FHIR environment requires explicit architectural design, using a message broker, a subscription framework, or a middleware delivery layer, rather than relying on a protocol-level guarantee. Every FHIR integration architect should understand this gap and plan for it deliberately.

Z-segments are migration scope risks that must be inventoried before a migration begins. Every Z-segment represents a data element that may not have a direct FHIR equivalent, may require a custom extension, and may carry data that clinical workflows depend on in ways not documented anywhere. Riverside General's seventeen Z-segments represent seventeen separate discovery and decision tasks that must be resolved before any of the interfaces carrying those segments can be migrated. That work is not glamorous, but it is load-bearing.

The v2-to-FHIR mapping guide is a starting point, not a complete specification. The practical translation requires handling cardinality mismatches, structural differences, coded-value translation, and Z-segment content that the guide does not address. Every mapping decision must be documented, tested, and validated against clinical requirements. The mapping work is where the conceptual migration plan meets the actual data, and the gap between them is always larger than initially estimated.

HL7 v2 will remain relevant for as long as it takes to replace every system that currently produces or consumes it, a timeline that will likely outlast any current FHIR roadmap. Organizations that plan to decommission v2 infrastructure on a two-year timeline should build in significant contingency for the discovery that the legacy v2 environment is more deeply embedded in clinical operations than initial assessment revealed. The pipes are older than most of the people who maintain them, and they carry more than the architectural diagrams show. Chapter 5 turns to the integration engine itself: the systematic tools and frameworks for managing that complexity rather than being managed by it.

5 The Integration Engine

Scenario

Patrick Osei spent eleven years approving IT budgets without ever asking a fundamental question about the infrastructure that held his hospital together. He knew the names of the vendors whose logos appeared on the walls of the server room. He could recite the annual EHR licensing fees from memory. What he could not have told you on the morning he walked into Dev Okafor's office was the name of the software that connected all those systems.

Dev was at his workstation reviewing the overnight error log. The screen showed a green dashboard with one amber warning about a laboratory acknowledgment delay that had self-resolved at 3:47 a.m. Patrick sat down across from him with the careful posture of an administrator who has a question he has been embarrassed not to ask for too long.

"Hypothetically," Patrick said, "what would happen if someone turned off Mirth Connect for an hour?"

Dev did not answer immediately. There was a pause, the kind that means the answer is complicated and slightly terrifying.

"Labs stop flowing. Admissions stop posting. Pharmacy stops getting orders." He gestured at the dashboard. "The hospital keeps running, but it goes blind."

Patrick was not a technical person, but he was a precise administrator, and administrators are trained to feel the weight of words spoken quietly by people who know what they are talking about. He had always understood the EHR as the central system, the one everything else orbited. What Dev was telling him was that the integration engine was not a supporting player in that architecture. It was the architecture.

The conversation that followed lasted three hours and covered the ground that this chapter will cover more efficiently. By the end of it, Patrick had a legal pad of notes and an understanding of why the medication synchronization failure that had nearly harmed a patient six months earlier was not caused by the EHR, or by the pharmacy system, but by a gap in the invisible layer between them.

Why It Matters

The integration engine, the middleware layer that sits between clinical systems and routes, transforms, and delivers messages, is the most underappreciated infrastructure in healthcare IT. It is invisible when it works and catastrophic when it fails. Most hospital administrators can name their EHR vendor. They could not name the integration engine that connects it to the laboratory, the pharmacy, the radiology system, and the patient portal, nor tell you what would happen to clinical operations if it went offline.

That invisibility is, in a specific and ironic way, earned. A well-configured integration engine runs without requiring attention. It processes tens of thousands of messages per day and delivers them correctly without triggering alerts or requiring human intervention. The people who built it are usually the same people who would need to explain its importance to an executive who has never had reason to think about it.

The near-miss medication event at Riverside General was not caused by the EHR or by the pharmacy system. It was caused by a synchronization gap in the integration layer between those two systems: a routing delay that allowed a medication reconciliation update to reach the pharmacy 47 minutes after a dosing order had already been initiated. Both systems were functioning correctly by their own logic. The problem was in the channel connecting them, in a route configuration that had not been reviewed since the pharmacy system was upgraded eighteen months earlier.

Any FHIR strategy that ignores the existing middleware layer is destined to fail. Most hospitals will not replace their integration engines overnight. They will extend them to support FHIR alongside legacy formats, running v2 and FHIR messages simultaneously through the same routing infrastructure for years. Organizations that understand this approach their migrations with realistic timelines. Organizations that ignore it tend to discover their mistake at go-live, when the laboratory information system that has been sending HL7 v2 messages for fifteen years has no idea what a FHIR Observation resource looks like, and the integration engineer who knew how to build that translation channel left the organization eight months ago.

Patrick's three-hour conversation with Dev is the conversation every administrator needs before approving a migration plan. This chapter is the condensed version.

Diagram 5.1: Why the Hospital Goes Blind When the Integration Engine Fails

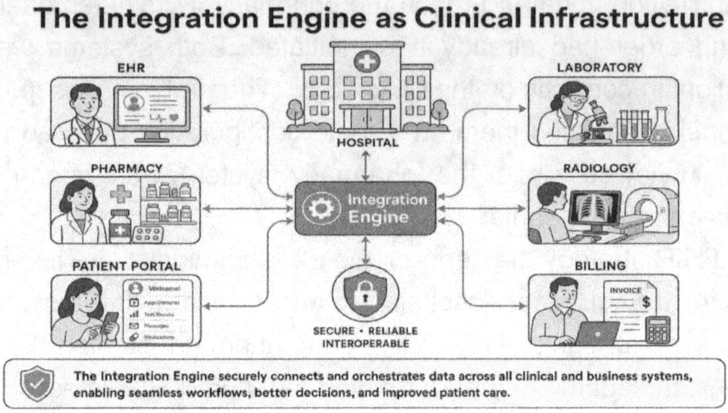

The Hospital Postal Service

Maya's question for this section is: what exactly does an integration engine do, and why can't clinical systems connect directly?

The question Maya Chen asked when Dev walked her through the architecture is the same question most technically curious non-engineers eventually ask. Clinical systems are sophisticated software applications. They have databases. They have APIs. Why does a hospital need a separate piece of infrastructure to move data between them?

The answer begins with a number. At Riverside General, the clinical IT environment includes thirty-seven distinct applications. Each was built by a different vendor, in a different decade, using a different data model, messaging format, and assumptions about how patient identifiers are structured and how clinical events are represented. The EHR uses one patient identifier namespace. The laboratory uses another. The radiology system uses a third. The pharmacy system was integrated against the previous EHR's identifier format and has never been fully updated.

If each system needed to communicate directly with every other system it depended on, the organization would need hundreds of point-to-point interfaces, each custom-built, each independently maintained, with no shared logging, no shared monitoring, and no central visibility when something goes wrong. This "spaghetti architecture" approach characterized hospital IT in the 1990s, and its legacy still lives in forgotten configurations in older integration environments.

Think of an integration engine as a postal sorting facility for clinical messages. It receives incoming messages in multiple formats from multiple sending systems, reads the addressing information on each message, transforms the message into the format the receiving system expects, and routes it to the correct destination, all within milliseconds, without human intervention, twenty-four hours a day. A well-configured integration engine handles HL7 v2 ADT messages, ORU messages, FHIR resources, custom XML documents, and CSV files through the same routing framework, applying different transformation rules to each type while maintaining a consistent logging posture across all of them.

Dev illustrated this with a concrete example. Every time a patient is admitted to Riverside General, the registration system generates an HL7 v2 ADT^A01 message. That single admission event needs to reach twelve downstream systems: the laboratory, the pharmacy, the dietary system, the radiology system, the billing system, the ICU monitoring system, the bed management system, the housekeeping system, the security system, the patient portal, the case management system, and the HIE interface. Each of those systems has different requirements: different v2 versions, different patient identifier formats, one requires a custom Z-segment, two now accept FHIR resources. The integration engine receives the single message, creates twelve copies, applies the appropriate transformation to each, and delivers them all before the patient has finished handing over an insurance card.

Diagram 5.2: Message Fan-Out from a Single Admission Event

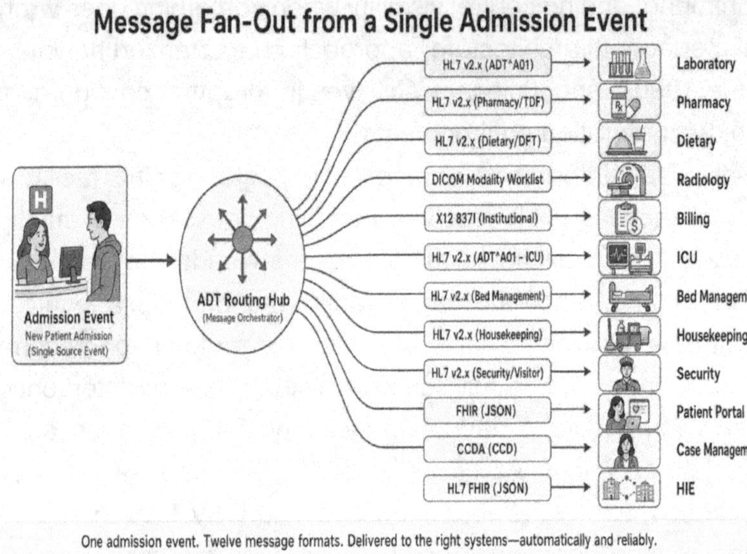

The postal analogy extends to accountability. When a package is lost, the postal service has a tracking record showing where it was last seen. When a clinical message fails to reach its destination, the integration engine's audit log is that tracking record: the message arriving, being parsed, having a transformation applied, being routed, and either being acknowledged or entering an error state. Without that audit log, a failed delivery is invisible. With it, the failure is diagnosable and correctable.

This is what Dev means when he says the hospital goes blind. The clinical systems do not stop functioning. The EHR still has the data it already held. The pharmacy still processes orders it already received. But the flow of new information stops, new admissions do not propagate, new lab results do not reach clinicians, and in a clinical environment, stopped flow is a patient safety event waiting to happen.

The Integration Engine Landscape

Maya's question for this section is: there must be different tools for this, so how does a hospital choose, and does the choice matter?

It matters considerably, and the choice is not purely technical. The integration engine market presents options that differ in cost structure, deployment model, feature depth, scalability, and the level of internal expertise required to operate them. Dev's blunt assessment is that the best integration engine is the one your most experienced integration engineer knows how to use, which is accurate as far as it goes, but the full picture requires understanding the major options and their trade-offs.

Mirth Connect is the platform running Dev's 217 routes at Riverside General. It is the most widely deployed open-source integration engine in healthcare, originally developed by Mirth Corporation and now maintained by NextGen Healthcare. It is free to use under an open-source license, highly configurable, and has a large community of experienced healthcare integration developers. Its configuration model is based on channels: each channel defines a source connector, a destination connector, and the transformation logic applied in between. A Mirth Connect installation with hundreds of channels represents years of accumulated configuration work, but each channel is relatively straightforward to understand and debug once you know the tool.

The commercial side of the market is dominated by platforms that offer enterprise support contracts, prebuilt interface libraries for major EHR systems, and advanced monitoring capabilities. InterSystems HealthShare is deployed at large academic medical centers needing deep integration with the InterSystems IRIS data platform. Rhapsody offers a graphically oriented configuration environment and robust prebuilt EHR connectivity. Infor Cloverleaf provides extensive interface accelerators for common clinical messaging scenarios. Licensing costs are substantially higher than open-source alternatives, but for organizations with very large message volumes, complex multi-site environments, or stringent uptime requirements, enterprise support and built-in capabilities may justify the investment.

Diagram 5.3: How Mirth, Rhapsody, and Azure Integration Services Differ on the Question That Matters Most to Dev: Build Complexity vs. Operational Control

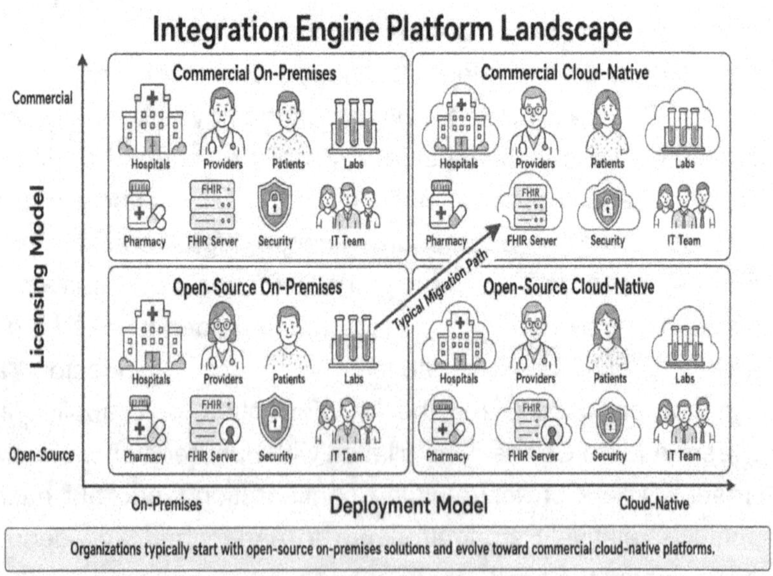

The cloud-native generation of integration infrastructure represents a fundamentally different deployment model. Azure Health Data Services, AWS HealthLake, and Google Cloud Healthcare API provide managed FHIR servers, along with integration and analytics services, that run entirely on cloud infrastructure. They abstract away infrastructure management at the cost of dependency on a specific cloud provider's service model, pricing structure, and data residency policies. For organizations already invested in a particular cloud ecosystem, this trade-off may be favorable. For organizations cautious about health data residency in cloud environments, on-premises or hybrid models remain the default.

Dev's actual advice to organizations evaluating platforms is to resist the temptation to select before auditing the existing integration estate. Until an organization knows how many active channels it runs, what message types they carry, which clinical workflows depend on them, and what the transformation logic looks like, it cannot meaningfully assess platform fit. He has consulted for organizations that spent months evaluating cloud-native platforms, only to discover during the audit phase that a significant portion of their integration logic depended on Z-segments not supported by any of the platforms under evaluation. Inventory first. Platform selection second.

Transformation Logic

Maya's question for this section is: when a message goes through the integration engine, what actually changes, and how does the engine know what to do?

The transformation layer is where the actual translation work happens, and it accounts for the majority of the time, cost, and risk in any v2-to-FHIR migration project. The postal analogy covers routing, the concept that messages are addressed and sorted. Transformation is different. It is the process of opening the envelope, reading its contents, rewriting the letter in another language, and resealing it in a different envelope, all while preserving the original's accuracy.

Consider a concrete scenario. The Riverside laboratory information system sends an HL7 v2.5.1 ORU^R01 message containing a complete blood count result. The new patient portal expects to receive this data as a FHIR DiagnosticReport resource containing embedded FHIR Observation resources, one per result component, in JSON format. The transformation script that converts one to the other must perform all of the following operations:

RECEIVE v2 ORU^R01 message PARSE MSH segment: extract sendingApp, messageControl, timestamp PARSE PID segment: extract patientId from PID-3 look up FHIR Patient.id in cross-reference table if not found: route to dead-letter queue with error "patient not found" PARSE OBR segment: extract orderId from OBR-3 extract testCode from OBR-4 (map to FHIR code system) look up FHIR Practitioner.id from OBR-16 ordering provider NPI FOR EACH OBX segment: extract observationCode from OBX-3 (verify LOINC code is valid) extract value and units from OBX-5 and OBX-6 extract referenceRange from OBX-7 map abnormalFlag from OBX-8 to FHIR interpretation code BUILD Observation resource with assigned Observation.id. BUILD DiagnosticReport resource referencing Patient, Practitioner, Observations. VALIDATE DiagnosticReport against FHIR R4 schema. IF validation fails: route to dead-letter queue with validation errors POST to FHIR server endpoint

Each step can fail, and each failure mode requires a specific handling decision. What happens if the patient identifier in PID-

3 does not match any patient in the cross-reference table? What if OBX-3 contains a local laboratory code instead of a LOINC code? What if OBX-5 contains a text narrative rather than a numeric value? A transformation script that does not handle these edge cases will produce either invalid FHIR output that the server's validation layer rejects, or technically valid but clinically incorrect output, which is far more dangerous because it generates no error.

Diagram 5.4: HL7 v2 ORU to FHIR DiagnosticReport Transformation

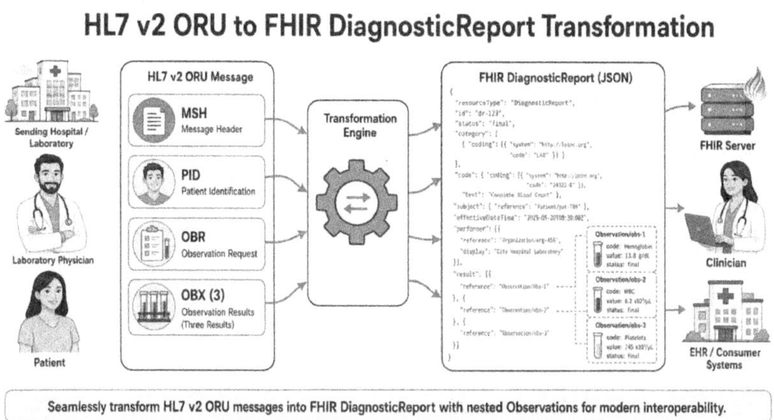

The patient cross-reference problem deserves special attention because it is the most common source of transformation failures at Riverside and at most other hospitals. Clinical systems do not share a single patient identifier. The EHR assigns one. The laboratory uses another. The integration engine must maintain, or have access to, a master patient index mapping identifiers across systems. If that cross-reference is incomplete or out of date, which happens during EHR upgrades, patient merges, and demographic corrections, the transformation script will fail to match the patient and route the message to the dead-letter queue.

Dev showed Maya the cross-reference failure queue. On a typical day, Riverside sees between eight and fifteen such failures, resolved by end of business. But in the months following the EHR upgrade two years earlier, that number spiked to over three hundred per day. The upgrade had changed the patient identifier format in a way not documented in the vendor's release notes. Transformation scripts were looking for the old format. Every lab result for a newly registered patient was going to the dead-letter queue. Clinical teams, unaware of the dead-letter queue, assumed the laboratory system was slow.

That story captures why transformation logic must be validated against real production messages, not test messages constructed to match the specification.

Message Routing

Maya's question for this section is: if the integration engine manages hundreds of routes, how does it know which message goes where, and what happens if the routing rules are wrong?

Dev's 217 routes at Riverside accumulated over fourteen years, one interface at a time, as new systems were added, new workflows created, new message types introduced, and organizational decisions changed the routing topology. Each route was built to solve a specific problem at a specific point in time. The aggregate is, in effect, a map of every clinical data dependency in the hospital: every system that needs to receive a specific type of information from another system to perform its function.

A route in an integration engine is a directed path from a source to one or more destinations, governed by filtering rules and transformation logic. The source definition specifies which sending application the route listens to, which message type it expects, and which field values optionally trigger it. The destination definition specifies the endpoint, the delivery protocol (MLLP over TCP/IP for most v2 interfaces, HTTP/HTTPS for FHIR), and the acknowledgment mode. A conceptual representation of routing evaluation

:

RECEIVE message EXTRACT messageType, sendingApp, fieldValues FOR EACH route in routeTable: IF route.sourceApp MATCHES sendingApp: IF route.messageType MATCHES messageType: IF route.filterConditions ALL PASS: APPLY route.transformationLogic DELIVER to route.destination ON success: log delivery, return ACK to sender ON failure: increment retryCount IF retryCount < maxRetries: re-queue for retry ELSE: move to deadLetterQueue with errorDetails IF no route matched: log unrouted message, alert operator

On Dev's hardware, the median time from message receipt to delivery acknowledgment for a standard v2 ADT message is under 80 milliseconds. The speed obscures the complexity.

Diagram 5.5: Route Dependency Graph for a Single ADT Event

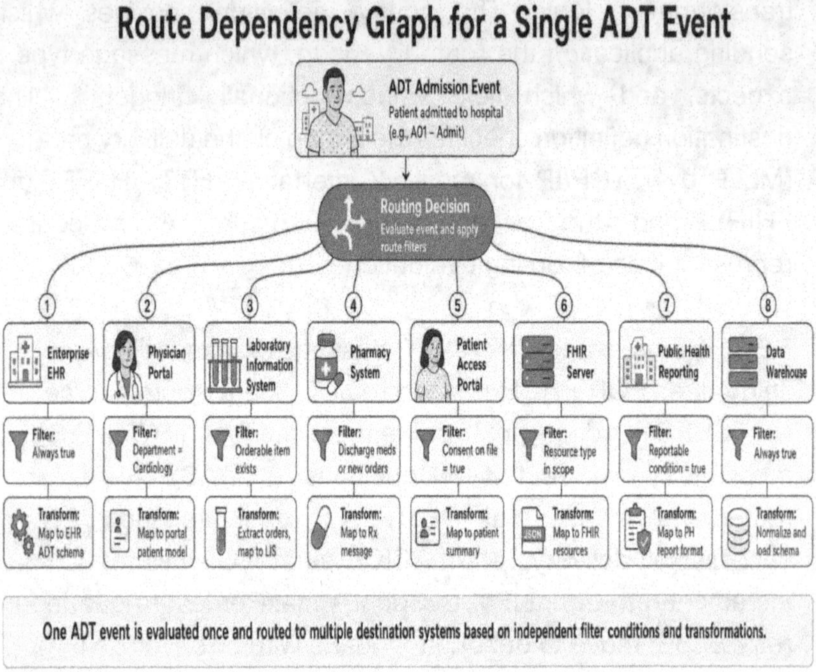

When Patrick asked Dev why he would not let anyone else modify the routing configuration, the answer was not about trust. It was about the invisibility of downstream dependencies. Changing a routing configuration without understanding its downstream clinical effects is among the most operationally dangerous actions in a hospital IT environment.

Eighteen months before the FHIR migration planning began, a member of Dev's team modified the filtering logic on the dietary notification route to reduce noise: the dietary system was receiving admission notifications for observation-status patients who were often discharged within hours. The engineer changed the filter to exclude messages where PV1-2 (patient class) was "O" for outpatient/observation. The change took effect at 9 a.m. on a Thursday.

By 11 a.m., three patients on the fifth floor had not received their noon meal trays. The investigation took six hours and four people to complete. What nobody had documented was that Riverside classified post-surgical observation patients as observation status during their first twenty-four hours before converting them to inpatient status. The filter change had inadvertently removed meal tray notifications for every post-surgical patient during their most nutritionally sensitive recovery window.

Dev now requires any proposed route modification to be accompanied by a written impact assessment describing every clinical workflow that depends on it, every system receiving messages through it, every filter condition and what it excludes, and a sign-off from clinical informatics before the change reaches production. That process adds two days to what would otherwise be a ten-minute configuration change. Every one of those days is worth it.

A complete route inventory must capture, for each route: the unique identifier, source application and message type, all filter conditions, all transformation logic with field-level mapping documentation, all destination endpoints, expected message volume per hour, maximum acceptable delivery latency, retry configuration, dead-letter queue assignment, date of last review, and name of person who last modified it. At Riverside, roughly forty percent of Dev's routes had never been formally documented before the FHIR migration planning process began.

Error Handling and Dead-Letter Queues

Maya's question for this section is: what happens to a clinical message when something goes wrong during delivery, and how does the hospital ensure nothing gets permanently lost?

The dead-letter queue is one of those concepts that sounds like an operational technicality but, in the context of clinical messaging, is a patient safety feature. The name comes from postal terminology: the practice of holding undeliverable mail in a separate facility where it can be examined and redirected rather than destroyed. In a clinical integration environment, the dead-letter queue is the holding area for messages that could not be delivered, preserved with their full content and error context, waiting for a human operator to review them and decide what to do.

A production integration engine encounters errors continuously. Dev's categories at Riverside are: malformed messages that fail initial parsing because they contain syntax errors, missing required segments, or encoding issues introduced by the sending system; transformation failures where the message is syntactically valid but contains data values the script cannot process (a missing patient identifier, an unrecognized code value, a null field where a value is required); delivery failures where the message transforms correctly but the destination endpoint is unavailable or returns an error; and validation failures where the transformed message is submitted to the FHIR server's validation layer and rejected.

Each error category requires a different handling strategy. A malformed message cannot be corrected automatically. It must be reviewed by a human who can determine whether it represents a sending system bug, a data entry error, or a format change requiring a script update. A delivery failure may resolve itself if the destination comes back online, which is why it is retried before being moved to the dead-letter queue. A validation failure usually indicates a deeper incompatibility between what the sending system produces and what the receiving system expects.

Diagram 5.6: Dead-Letter Queue Processing Flow

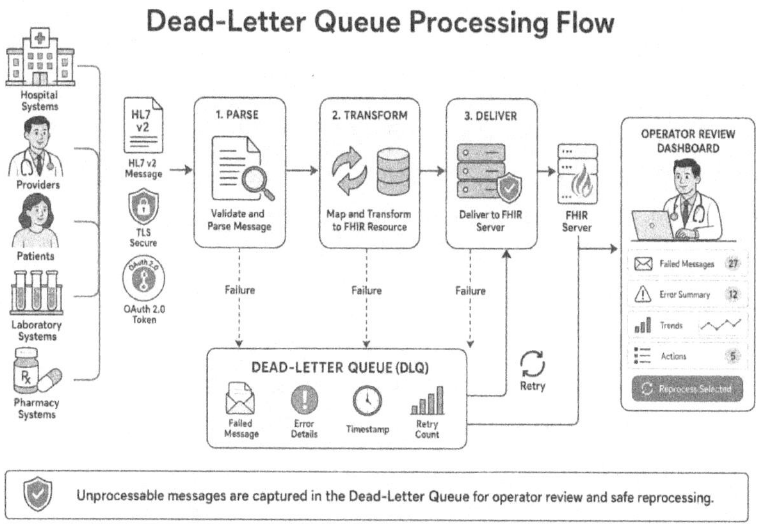

Dev's retry logic for delivery failures uses exponential backoff: the first retry occurs 30 seconds after the initial failure, the second 90 seconds after that, and the third 4 minutes later. If the third retry fails, the message is moved to the dead-letter queue, and an alert is triggered. This gives a destination system roughly six minutes to recover from a transient issue, sufficient for brief outages while short enough to detect sustained ones quickly.

The alert policy governing dead-letter queue notifications is as important as the retry policy. At Riverside, dead-letter queue entries for laboratory result messages generate an immediate SMS alert to the on-call integration engineer, because an undelivered lab result is a direct clinical safety concern. Dead-letter queue entries for billing messages generate an alert to the operations queue. Dietary notification failures go to the morning review queue. The tiered policy reflects a deliberate risk assessment: not all message failures carry the same clinical weight, and the operational response should be proportionate to the clinical risk.

The absence of a dead-letter queue, which is the configuration state of a surprising number of legacy integration environments, means that undeliverable messages disappear. The sending system receives an acknowledgment (because in many legacy v2 configurations the ACK is sent before delivery is confirmed). The destination system never receives the message. No error is logged. No alert fires. The message is gone. The first indication that something went wrong is a clinical event: a patient who does not receive a medication. This discharge does not trigger a follow-up appointment or an abnormal lab value that nobody follows up on.

Diagram 5.7: Tiered Alert Policy for Integration Errors

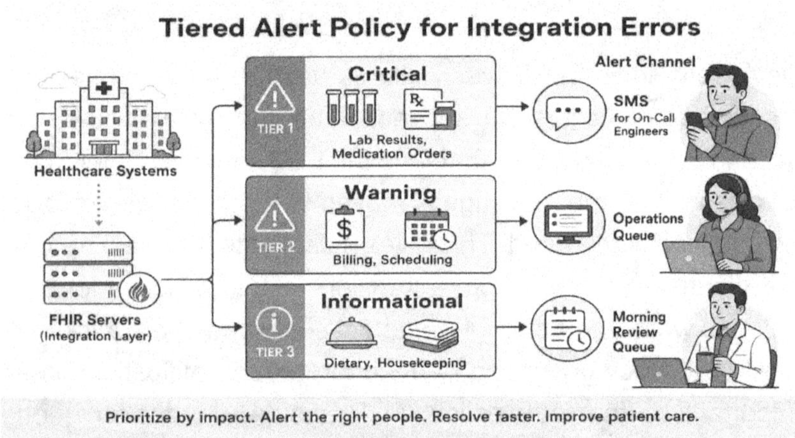

Dev refers to the dead-letter queue as the hospital's confession booth for integration failures: the place where everything that went wrong comes to be examined and corrected. Organizations without one are not running a healthcare integration environment. They are running a system that believes it is delivering messages and has no mechanism to discover otherwise.

Monitoring and Alerting

Maya's question for this section is: how does the integration team know in real time that messages are flowing correctly, before a problem becomes a clinical incident?

Dev monitors his integration engine the way an air traffic controller monitors radar: continuously, with visual dashboards, layered alert logic, and the discipline to distinguish signal from noise. When he pulled up his monitoring dashboard during the conversation with Maya, she expected something like an EHR dashboard, with patient names and clinical data. What she saw was a grid of green, yellow, and amber indicators, each representing a channel, with numeric throughput counts updating in real time, latency histograms, and a sidebar showing the last five alerts and their resolution status.

"This is how I know the hospital is talking to itself," Dev said.

A mature monitoring posture tracks four categories of metrics across every active channel. Throughput metrics measure how many messages per hour are flowing through each channel. Latency metrics measure how long messages take from receipt to delivery acknowledgment. Error rate metrics measure the percentage of messages that fail at each processing stage. Queue depth metrics measure how many messages are waiting to be processed or delivered at each point in the routing path.

Each metric tells a different story. Throughput drops are often the first indicator of a sending system failure. If the laboratory information system suddenly stops generating ORU messages at 2 a.m., the throughput metric for the laboratory result channel falls to zero. Distinguishing "channel is quiet because it is 2 a.m." from "channel is quiet because the laboratory has stopped sending" requires threshold-based alerting that accounts for expected traffic patterns by time of day and day of week. Latency spikes are often the first indicator that a destination system is becoming slow, frequently preceding delivery failures by 30 to 60 minutes and giving the team time to investigate before messages begin to accumulate in the retry queue.

```
FOR EACH channel in activeChannels: throughputRatio =
messagesPerHour(channel) / historicalBaseline(channel) IF
throughputRatio < 0.10: RAISE alert(channel, level=CRITICAL,
reason="throughput near zero") ELIF throughputRatio < 0.50:
RAISE alert(channel, level=WARNING, reason="throughput
below baseline")
p95Latency = latencyPercentile(channel, p=95,
window=last15min) IF p95Latency >
channel.latencyThresholdCritical: RAISE alert(channel,
level=CRITICAL, reason="latency critical") ELIF p95Latency >
channel.latencyThresholdWarning: RAISE alert(channel,
level=WARNING, reason="latency elevated")
errorRate = errorCount(channel) /
max(messageCount(channel), 1) IF errorRate >
channel.errorRateThreshold: RAISE alert(channel,
level=CRITICAL, reason="error rate exceeded")
ROUTE each alert based on alert. level and
channel.clinicalPriority
```

Diagram 5.8: The Six Numbers Dev Checks First When an Alert Fires — and How the Dashboard Reveals Whether a Spike Is a Retry Storm or a Dead Channel

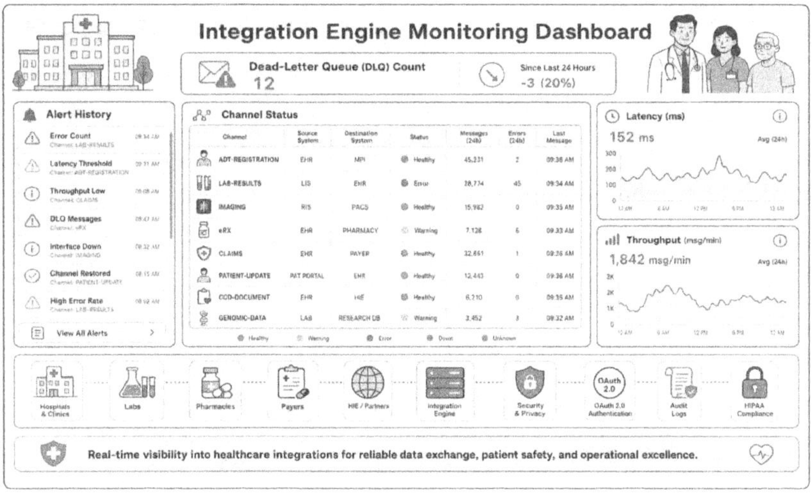

The alert routing policy determines who gets called at 2 a.m. and who waits until 8 a.m. This is a clinical governance decision, not merely an operational preference, because the integration team's fatigue is itself a patient safety factor. An integration team that receives too many low-priority alerts at night will habituate to the alert noise and begin treating warnings as background events, including those that matter. A clean critical alert channel ensures that when a genuinely critical alert fires, the on-call engineer takes it seriously.

Dev calibrates alert thresholds iteratively: establish initial thresholds from thirty days of baseline data, run the alert policy for two weeks in logging mode without sending notifications, review which alerts would have fired, classify each as correctly calibrated or noise, adjust thresholds, then activate in production. He revisits calibration every six months and after any major system change.

Channel documentation is the complement to monitoring. It is the reference that tells the on-call engineer at 2 a.m., with a critical alert in their inbox, what the channel does, which systems it connects to, what message types it carries, what the expected throughput is, and how to resolve the most common error conditions. Before FHIR migration planning began, 43% of channel documentation pages at Riverside were blank or contained only the channel name and creation date. Completing them required tracking down three engineers who had left the organization over five years ago to recover institutional knowledge that existed only in their memories.

Integration Engine and FHIR

Maya's question for this section is: if the hospital is moving to FHIR, does it need a whole new integration engine, or can the existing one handle FHIR too?

The realistic end state for most hospitals over the next five to ten years is not "FHIR replaces HL7 v2." It is "FHIR coexists with HL7 v2 in a hybrid integration architecture that the existing integration engine manages." This is not a pessimistic assessment of FHIR adoption. It is a technically honest description of the migration path that healthcare organizations follow when they approach the problem from the integration layer outward.

Dev laid this out for Maya on a whiteboard that has since become the official architecture reference for Riverside General's FHIR migration project. On the left: the existing legacy systems. Laboratory, pharmacy, radiology, dietary, bed management, billing. All connected to Mirth Connect via MLLP, all sending and receiving HL7 v2 messages. These systems will not be replaced in the near term. Their EHR interfaces are certified for specific v2 message types. Their internal databases are structured around v2 data models. Migrating them to FHIR would require upgrades that, in several cases, vendors are not prepared to deliver within timelines that align with the patient portal project.

On the right: the new FHIR-native components. The upgraded patient portal consumes FHIR resources from an FHIR R4 server. The new mobile application for patients with chronic conditions uses SMART on FHIR for authentication and FHIR APIs for data retrieval. The regional HIE has adopted FHIR as its exchange standard. These components were built FHIR-native and do not speak HL7 v2.

In the center: Mirth Connect, extended with new channels that translate between the two worlds. When the laboratory sends an HL7 v2 ORU^R01 message with a lab result, a new channel transforms it into a FHIR DiagnosticReport resource and posts it to the FHIR server, where the patient portal and mobile application can retrieve it. When the patient portal triggers a medication request, the integration engine translates the FHIR MedicationRequest into an HL7 v2 ORM^O01 message and delivers it to the pharmacy system's MLLP endpoint. The legacy systems never know FHIR exists. The FHIR-native systems never know HL7 v2 exists.

Diagram 5.9: Hybrid v2 and FHIR Integration Architecture

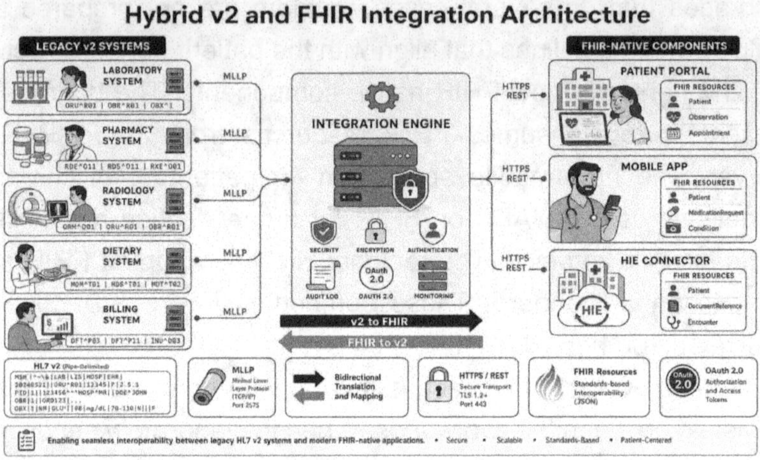

The hybrid architecture resolves Maya's question but introduces new operational requirements. Running FHIR channels through the integration engine requires the team to learn FHIR transformation patterns alongside v2 patterns. The conceptual model is the same: extract data from a source format, apply mappings, produce a target format. But the target is now a JSON document that must conform to an FHIR profile, with specific element names, data types, code system bindings, and required fields. A FHIR server's validation layer is considerably stricter than most v2 receiving systems' validation logic, which means FHIR transformation scripts require more careful edge-case handling and more thorough testing against production message samples.

The FHIR server is a new infrastructure component the integration team must treat as a managed endpoint with its own availability requirements, capacity planning, and operational monitoring. When it is slow or unavailable, channels that post to it experience delivery failures that trigger the same retry and dead-letter queue logic as failures to any other destination.

Operational procedures for FHIR server failures must be defined before go-live.

Terminology management is the third new requirement. HL7 v2 messages often carry local codes: laboratory result codes specific to the sending LIS, medication codes specific to the pharmacy formulary, diagnosis codes entered using the hospital's legacy code system. When those codes need to be represented in FHIR resources, they must be mapped to standard terminologies: LOINC for laboratory observations, RxNorm for medications, SNOMED CT for clinical findings. The integration engine performs these mappings as part of the transformation logic, requiring an accurate, up-to-date, and audited terminology mapping table. That subject occupies Chapter 6.

Organizations that successfully navigate the hybrid architecture phase treat it as a permanent operating model rather than a transition state to be tolerated until legacy systems are replaced. The hybrid model requires route governance covering both v2 and FHIR channels, monitoring treating FHIR server endpoints with the same operational priority as legacy MLLP endpoints, and documentation capturing clinical dependencies of FHIR channels with the same rigor as legacy channel documentation. Dev's observation is direct: "You do not get to have a well-governed FHIR environment and a poorly-governed v2 environment at the same time. They share infrastructure. The discipline is either organization-wide, or it is not discipline."

Takeaways

The conversation between Patrick and Dev in October was the beginning of eighteen months of work: a full integration engine audit, a route governance framework, a FHIR hybrid architecture design, and a migration plan built on the actual state of Riverside General's integration estate. None of that work could have started without the moment of recognition that Patrick described afterward as "the conversation where I realized I had been approving budgets for years without understanding what I was paying for."

The integration engine is the circulatory system of a hospital's clinical data environment, and its failure has consequences comparable to circulatory failure in any organism. Connected systems may still be running. Data in their databases may still exist. But the flow has stopped, and stopped flow in a clinical environment is a patient safety event.

Any FHIR migration strategy that does not begin with a complete inventory of the integration engine is incomplete. The inventory must capture every active channel, the source and destination systems it connects to, the message types it carries, the transformation logic it applies, the clinical workflows that depend on it, the error-handling configuration, and the date of last review. At Riverside, that inventory had to be constructed from server configuration exports, institutional memory interviews, and four weeks of log analysis. Organizations that complete it before planning begins save the time, cost, and risk of discovering gaps after go-live.

Transformation logic is not automatable at the quality level that clinical data requires. Automated mapping tools can generate useful first drafts of a transformation specification. But every mapping decision between a v2 field and a FHIR element must be reviewed by a person who understands both the clinical meaning of the data and the technical constraints of the target resource. The edge cases that automated tools miss- locally modified code values, Z-segment extensions, fields used in non-standard ways by specific vendors- are precisely where transformation errors produce clinically incorrect output. The human review step is not optional.

Dead-letter queues are a patient safety feature, not an operational convenience. Every message that fails delivery and is not captured for human review is a potential clinical error: a lab result that did not reach the EHR, a medication order that did not reach the pharmacy, an admission notification that did not reach the dietary system. The dead-letter queue strategy must be defined before the first message passes through the integration engine. It must cover every channel, define the alert policy for every error category, and include documented operational procedures for reviewing entries within timeframes proportionate to the clinical sensitivity of each message type.

The hybrid v2+FHIR architecture is the technically responsible plan, not a compromise. Organizations that insist on fully replacing HL7 v2 before considering FHIR adoption will either delay their FHIR work by years because legacy systems will not be replaced within the timeline required for FHIR adoption, or they will take clinical risks not justified by the benefits of a cleaner architecture. The hybrid model, properly governed, delivers the benefits of FHIR for systems and workflows that can consume it while preserving the stability of the legacy environment for systems that cannot yet be migrated.

Integration engine expertise is among the scarcest and most consequential skills in healthcare IT. Dev's knowledge of his 217 routes, the clinical workflows they support, and the decisions made when each one was built is irreplaceable organizational knowledge. It cannot be fully reconstructed from configuration exports or log files. Succession planning for integration engine expertise is not a human resources exercise. It belongs on the same risk register as any other single point of clinical failure in the organization, because that is precisely what it is.

6 Clinical Terminologies

Opening: Two Printouts on a Desk

Dr. Maya Chen did not make a habit of delivering paper to Patrick Osei's office. She was a digital-first clinician who had spent a decade arguing for EHR documentation over paper, and she practiced what she preached. When she appeared in his doorway with two printouts and set them side by side on his desk without a word, Patrick understood that something had gone badly wrong.

Both printouts were laboratory result reports for the same patient on the same day, both showing a serum potassium of 2.9 milliequivalents per liter. One had come from Riverside General's in-house laboratory. The other had come from a regional reference laboratory used for overflow and specialty testing. The ordering physician had initiated the clinical decision support protocol for hypokalemia after seeing the first result. The second result, arriving four hours later as confirmation, had not triggered the same protocol. It sat in the chart unacknowledged for 11 hours before a nurse reviewing it caught it.

Maya pointed to the code string beneath each result. One read "2823-3." The other read "K-SER-ISE." Patrick recognized neither. She told him they meant the same thing: serum potassium, same analyte, same specimen type, same clinical interpretation. The EHR treated them as different tests because it had no shared reference telling it they were the same. The clinical decision support rule had been written against LOINC codes. The in-house laboratory result carried a LOINC code, having been translated by the integration engine's mapping table. The reference laboratory result carried a LOINC code natively. But a configuration update six weeks earlier had deleted the mapping table entry for K-SER-ISE, so the in-house result was entered into the EHR without any LOINC translation, and the clinical decision support rule stopped finding it.

The eleven-hour delay was not a near-miss in the dramatic sense. The patient recovered without incident. But the logic that produced the delay was one that, under different circumstances and with a more critical analyte, could produce a fatal outcome. The root cause was not a broken interface or a failed server. It was a vocabulary problem, and the solution was not a better network. It was better terminology use, governed by a named owner with an explicit process. That morning in Patrick's office was the beginning of Riverside's terminology governance initiative, and this chapter is the account of what Maya explained before that initiative could begin.

Diagram 6.1: Two Lab Results, Two Codes, One Clinical Failure

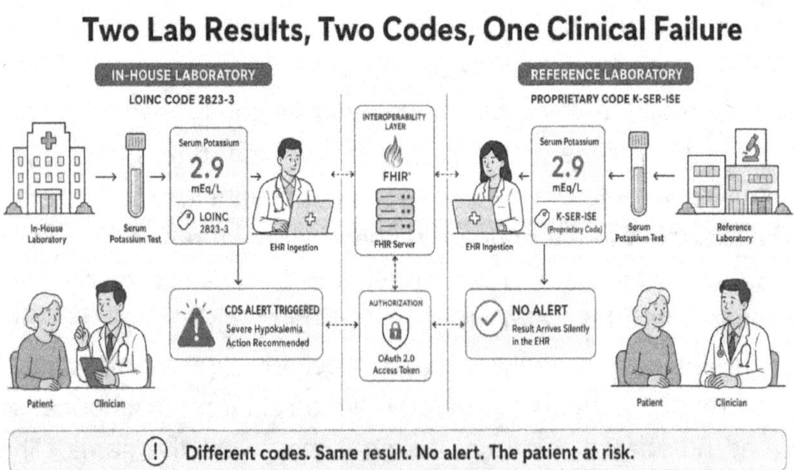

The Rosetta Stone Problem

In 1799, soldiers in Napoleon's Egyptian campaign discovered a granodiorite stele inscribed with the same royal decree in three scripts: Ancient Egyptian hieroglyphics, Demotic, and Ancient Greek. The Greek was readable; the hieroglyphics were not. Because the same content appeared in all three scripts, scholars could use the readable Greek to decode the symbols, and that decoding effort unlocked millennia of Egyptian history that had been inaccessible simply because the reference layer connecting symbols to meanings had been lost.

Clinical terminologies are healthcare's Rosetta Stone. Every EHR, laboratory information system, pharmacy system, and clinical decision support engine has its own internal representation of clinical concepts. The in-house laboratory at Riverside used a system implemented in 2011, when the vendor used a proprietary code vocabulary. The code "K-SER-ISE" was that vendor's identifier for serum potassium measured by ion-selective electrode. The reference laboratory used a different system, implemented by a different vendor, and that vendor had used LOINC from the start. The code "2823-3" is the LOINC identifier for "Potassium [Moles/volume] in Serum or Plasma," a code that any LOINC-aware system anywhere in the world interprets identically.

Without a shared reference layer, a serum potassium result in System A and a serum potassium result in System B are two different character strings representing the same clinical measurement. No technical integration will make System B's clinical decision support rules fire on System A's data without that shared reference. This is not a data transmission problem; it is a semantic problem, and its scale in real hospital systems is not a metaphor. A single hospital may have thousands of codes, across dozens of concept domains, stored in hundreds of systems, each using a different vocabulary or a different version of a shared vocabulary.

Patrick grasped the issue readily. "It's like one nurse charting blood pressure as 'BP' and another as 'blood pressure,' and the alert system only recognizing 'BP,'" he said. Maya told him that was right in principle and wrong in scale: the problem was not one ambiguous abbreviation but potentially thousands of mappings, maintained across dozens of systems, degrading silently every time a vendor released an update, a laboratory added a test, or a configuration change was deployed without a terminology review.

The Serum Potassium Journey

Maya and Dev used a single clinical scenario to illustrate how the terminology systems in this chapter work together in a real FHIR data flow. The scenario was the one that had brought Maya to Patrick's office: a serum potassium of 2.9 milliequivalents per liter traveling from a laboratory instrument through multiple systems to a clinical decision support rule that needed to fire an alert.

The journey begins at the laboratory analyzer. The instrument records the result in the laboratory information system using its internal code, "K-SER-ISE," the vendor's proprietary identifier for serum potassium by ion-selective electrode. The laboratory system sends the result to the integration engine in an HL7 v2 ORU message. The OBX-3 field carries the proprietary code. The engine applies a terminology mapping step: it looks up "K-SER-ISE" in a mapping table, finds its LOINC equivalent 2823-3, and substitutes the LOINC code into the observation identifier field before forwarding the message to the EHR.

The EHR stores the result and, when a FHIR client requests the patient's observations, generates a FHIR Observation resource. The code element is a CodeableConcept, which can carry multiple simultaneous codings. The EHR populates it with two codings: LOINC code 2823-3 with the standard LOINC system URI, and the original proprietary code K-SER-ISE with the local system URI. Both travel in the FHIR resource, preserving the local code for systems that need it while ensuring the LOINC code is present for systems that require it.

The clinical decision support rule queries the FHIR server for Observation resources with LOINC code 2823-3. It finds the resource, reads the valueQuantity element, and evaluates the result against the hypokalemia threshold. The threshold itself is defined using SNOMED CT concept 43339004 ("Hypokalemia") — the CDS engine uses SNOMED CT to represent the clinical finding that the rule is designed to detect. The rule fires, generating an alert notification through the FHIR Subscription mechanism. The physician is notified and acts.

Three points emerge from this worked example. First, terminology mapping is a transformation step requiring a maintained mapping table, a correctly configured integration engine, and ongoing oversight as laboratory methods change. Second, the CodeableConcept design explicitly supports multiple simultaneous codings, allowing local codes to be preserved alongside standard codes without forcing a choice. Third, the clinical decision support rule depends entirely on the LOINC code being present; if the mapping table entry is missing, the rule does not fire, and the patient safety function it serves is silently disabled without generating any error to alert the integration team. A well-designed terminology architecture would detect the absence of a LOINC code in an incoming observation, generate an alert, and route the uncoded result to a review queue rather than allowing it to enter the EHR silently.

Why Terminologies Exist

Before explaining the major clinical terms, Maya needed to explain to Patrick why they existed at all, since the answer is not obvious. Most industries that exchange structured data use relatively simple shared identifiers: a product SKU, a stock ticker, a postal code. Healthcare needs elaborate vocabulary systems with hundreds of thousands of codes, hierarchical structures, licensing arrangements, and versioning schedules because clinical information is inherently multidimensional.

A serum potassium value is not just a number. It is a number measured in a specific specimen type, using a specific analytical method, under specific conditions. "Potassium in serum measured by ion-selective electrode" and "potassium in whole blood measured by atomic absorption spectroscopy" are not the same test. They produce results that are not directly comparable because the reference ranges and clinical interpretations differ. A code system that treats them as the same test will introduce errors into clinical decision support and population health analytics that are difficult to trace because the data will appear superficially correct.

Clinical diagnoses present a further challenge. "Diabetes mellitus" is not a single disease; it is a family of related conditions with distinct pathophysiology and treatment implications. Type 1, Type 2, gestational, and monogenic diabetes must all be distinguishable in clinical data systems. At the same time, a query asking how many diabetic patients were admitted last quarter must aggregate across all of those conditions. Clinical terminology must support both precision and aggregation simultaneously, which requires a hierarchical structure in which specific concepts are organized as descendants of more general ones, allowing queries to be written at any level of the hierarchy.

Medications add a third dimension of complexity. A drug can be known by its generic name, brand name, chemical name, National Drug Code, or dozens of other identifiers, depending on which system records the information. Without a shared vocabulary mapping all those names to the same underlying concept, automated drug interaction checking and cross-system medication reconciliation fail silently. The major vocabulary systems in US healthcare represent decades of collective effort by clinicians, informaticists, government agencies, and professional organizations to build shared reference layers that enable independently developed systems to recognize the same clinical facts.

Diagram 6.2: Why Clinical Terminologies Are Necessary

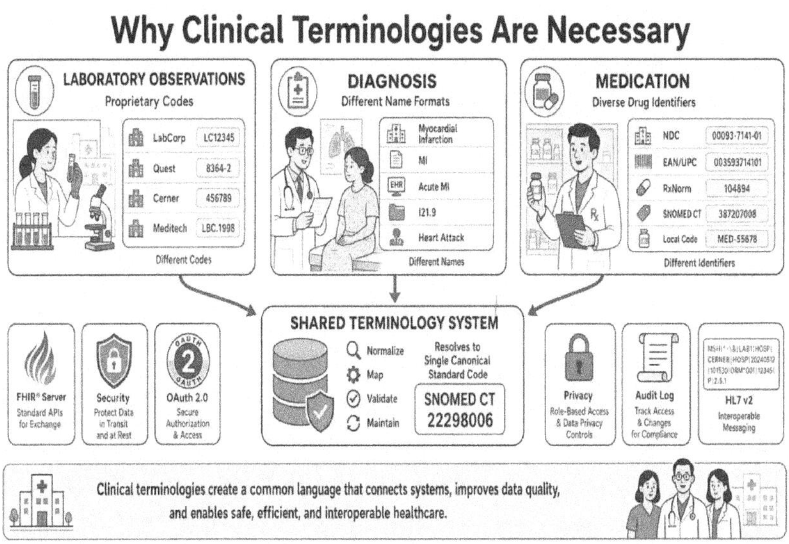

SNOMED CT: The Dictionary of Clinical Concepts

SNOMED CT, the Systematized Nomenclature of Medicine Clinical Terms, is the most comprehensive clinical terminology in existence. It contains more than 350,000 active concepts spanning clinical findings, diseases, procedures, body structures, organisms, substances, and pharmaceutical products. Each concept has a unique, permanent numeric identifier, a set of synonyms, including a fully specified name that distinguishes it from all other concepts, and a set of logical relationships that connect it to other concepts within a formal hierarchical structure.

Those relationships are what make SNOMED CT qualitatively different from a lookup table. A concept is a node in a directed acyclic graph, connected to parent concepts via "is a" relationships and to related concepts via named attribute relationships. "Type 2 diabetes mellitus" is a subtype of "diabetes mellitus," which is a subtype of "endocrine disorder," which is a subtype of "metabolic disease." That hierarchical chain allows a clinical system to answer not just "does this patient have this exact condition" but "does this patient have any condition in this clinical family," a capability essential for clinical decision support and population health analytics.

In the FHIR context, SNOMED CT is the preferred code system for clinical findings and diagnoses when precision matters more than billing alignment. US users access it at no charge through the NLM's national license. The Condition resource uses SNOMED CT in many US Core and international profiles, and the same hypokalemia example from the Serum Potassium Journey illustrates why: SNOMED CT gave the CDS engine a named clinical concept to act on, not just a threshold number.

Diagram 6.3: How SNOMED CT Connects Hypokalemia to Its Parent Concepts and Clinical Attributes — and Why the CDS Rule Fires at the Right Level of Specificity

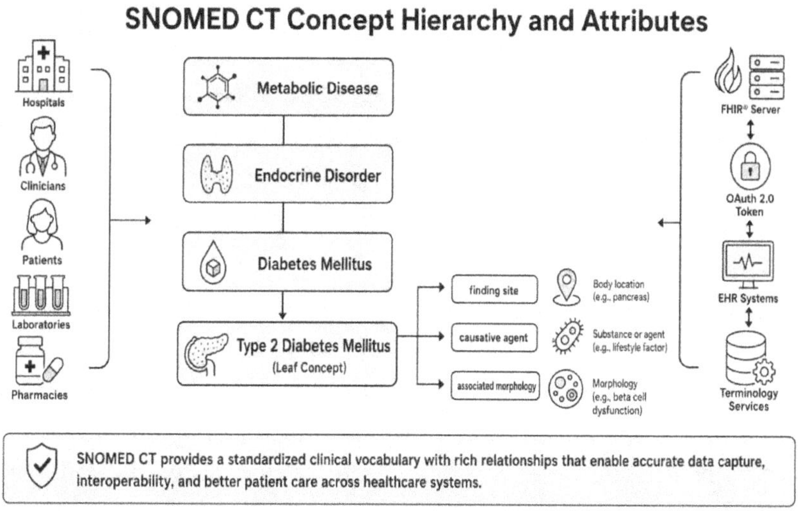

Dev Okafor's practical concern when Maya walked him through SNOMED CT was version management. SNOMED International publishes releases twice annually, in January and July. Each release may add concepts, retire obsolete ones, update descriptions, or modify relationships. A clinical decision support rule written against a SNOMED CT concept retired in a subsequent release will stop firing without generating any error. Terminology version management is an ongoing operational responsibility, not a one-time implementation task, and it carries resource implications that must be built into the maintenance budget from the beginning.

LOINC: The Language of Observations

LOINC, the Logical Observation Identifiers Names and Codes, is the standard vocabulary for laboratory test results, clinical measurements, vital signs, clinical documents, and standardized survey instruments. The Regenstrief Institute at Indiana University develops and maintains LOINC, which is freely available without a licensing fee, subject to a copyright notice requiring attribution. LOINC contains more than 90,000 active codes and grows with each release, published twice annually.

Each LOINC code encodes six attributes — component, property, time aspect, specimen system, scale, and method — making each code precise enough that no two LOINC codes describe the same test in the same specimen by the same method.

LOINC code 2823-3 decodes as follows: the component is potassium; the property is substance concentration in moles per volume; the time aspect is a point measurement; the system is serum or plasma; the scale is quantitative; and the method is unspecified because common serum potassium methods are clinically interchangeable. That specificity is what allows LOINC to serve as a universal reference: any system receiving an Observation resource with code 2823-3 knows exactly what was measured, in what specimen type, in what units, without additional context.

That same specificity makes LOINC mapping difficult. A hospital laboratory with 1,000 to 3,000 active test codes requires a trained clinical informaticist to evaluate each mapping. An incorrect match — coding a whole-blood potassium test with the serum potassium LOINC code — produces data that appears valid but is semantically wrong. CDS rules and population health queries built on those codes will silently produce incorrect results.

Diagram 6.4: The Six LOINC Parts Applied to Code 2823-3

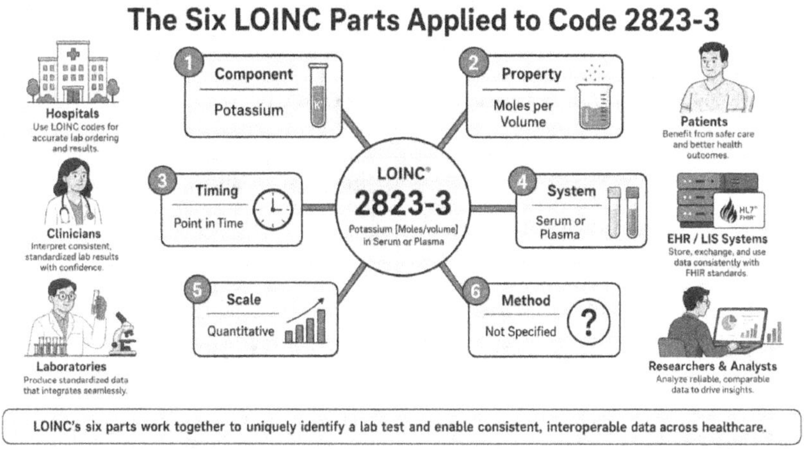

FHIR uses LOINC pervasively. The Observation resource uses LOINC codes for clinical measurements, the DocumentReference resource uses LOINC document type codes, and the DiagnosticReport resource uses LOINC to identify diagnostic panels. US Core profiles require LOINC codes for laboratory observations, vital signs, and clinical notes, making LOINC compliance a prerequisite for US Core conformance. Dev noted that Riverside's laboratory information system had 1,847 active test codes, in production since 2011, with no current team member involved in the original implementation. He did not know how many codes had LOINC mappings, who had created them, or whether those mappings had been reviewed as laboratory methods changed. Maya told him he had just described the terminology audit Riverside needed to conduct.

ICD-10-CM and the Path to ICD-11

ICD-10-CM is a statistical classification, not a clinical terminology. Its ~70,000 diagnosis codes were designed for epidemiological reporting and have been repurposed in the United States as the required code system for billing — a use case they serve adequately but imprecisely. A clinician selecting E11.65 (Type 2 diabetes mellitus with hyperglycemia) is choosing an administrative reimbursement category, not encoding a clinical concept.

This distinction has direct implications for FHIR implementations. The Condition resource can carry ICD-10-CM codes, and in US healthcare environments it frequently does. But an ICD-10-CM code on a Condition resource describes what billing category was selected, not what the clinician diagnosed with clinical precision. Two patients with E11.65 may have very different clinical presentations, treatment histories, and prognoses because the ICD-10-CM category is too coarse to capture these differences. Clinical decision support built against ICD-10-CM codes will be less precise than decision support built against SNOMED CT codes, a tradeoff that implementation teams must understand and document explicitly.

Diagram 6.5: How a Seven-Character ICD-10-CM Code Encodes Category, Etiology, and Encounter Type — and Where That Code Lands in a FHIR Condition Resource

ICD-10-CM Structure and FHIR Condition Resource

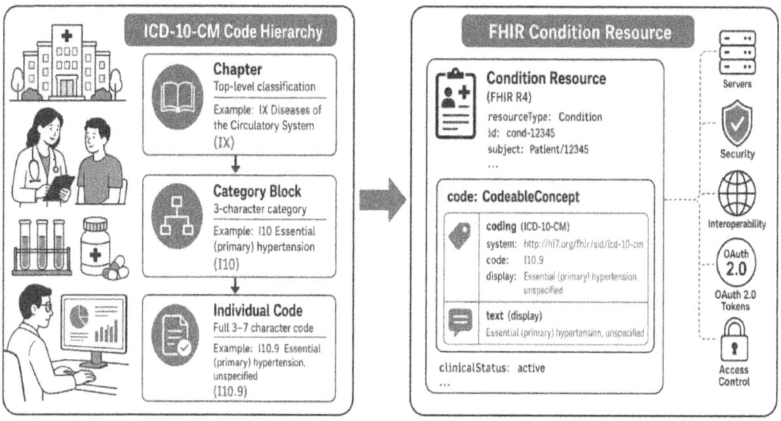

ICD-10-CM provides a standardized clinical vocabulary; FHIR Condition resources enable secure, interoperable exchange.

ICD-11, published by the WHO in 2019, offers better alignment with SNOMED CT and a more granular clinical structure, but the United States had not mandated adoption at the time of this writing. Sandra noted that any change to Riverside's coding practice required contracts-team sign-off before touching EHR configuration.

RxNorm: The Standard for Medication Identification

RxNorm, maintained by the National Library of Medicine, provides a unique numeric identifier for every clinical drug formulation in US pharmacy systems — from raw ingredient (metformin) to branded pack (metformin 500mg oral tablet [Glucophage]). RxNorm concept 861007, "metformin 500 MG oral tablet," means the same thing to any FHIR-capable pharmacy, EHR, or CDS engine without translation. The NLM maintains an RxNorm-to-NDC mapping table that insulates systems from NDC changes when manufacturers repackage products.

Diagram 6.6: Why the Pharmacy System and the EHR Describe the Same Drug at Different Levels — How RxNorm's Concept Tiers Resolve the Reconciliation Gap

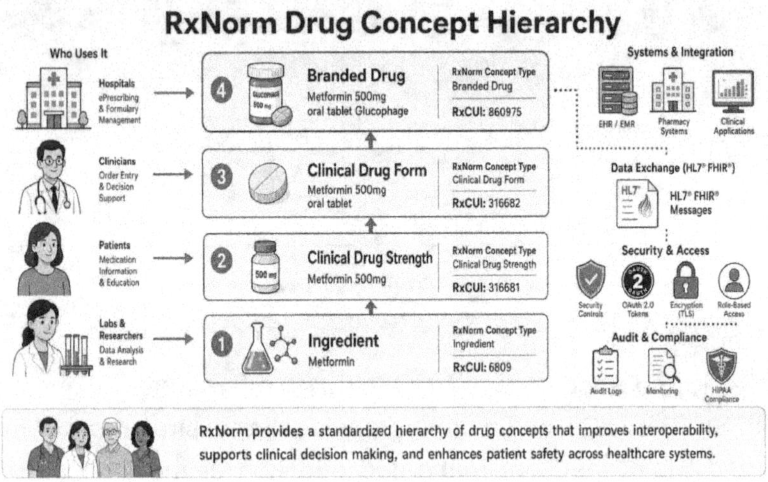

Dev confirmed that Riverside's EHR stored RxNorm codes internally, but the EHR-pharmacy interface used NDCs — unstable identifiers that change when a manufacturer repackages a product. Maya's concern was unambiguous: she considered automated medication reconciliation a patient safety function, and RxNorm the vocabulary infrastructure it depended on.

CPT and HCPCS: Procedures and Reimbursement

Current Procedural Terminology, published by the American Medical Association, is the primary vocabulary for physician procedures, services, and diagnostic tests submitted to payers for reimbursement in the United States. CPT codes are five-character alphanumeric identifiers organized into three categories: Category I for established procedures and services, Category II for performance measurement tracking, and Category III for emerging technologies not yet assigned Category I codes. CPT appears on virtually every physician claim submitted to any US payer, and any FHIR implementation touching claims processing or prior authorization will encounter CPT codes throughout the financial data flow.

Unlike LOINC and RxNorm, CPT codes are not freely available — the AMA holds copyright and actively enforces it. Any system that processes, displays, or transmits CPT codes requires a license, and that obligation extends to downstream applications consuming a FHIR API. Medicare claims also require HCPCS Level II codes, a CMS-maintained extension set for equipment, drugs, and services outside CPT's scope.

Diagram 6.7: CPT and HCPCS in FHIR Financial Workflows

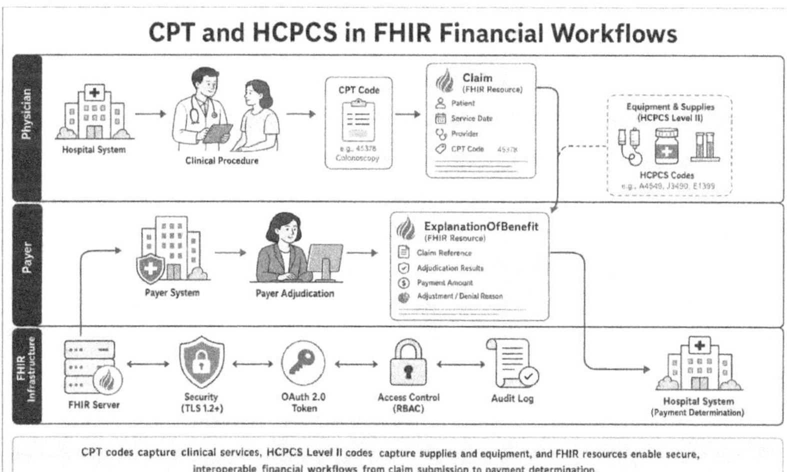

Sandra's concern was scope: Riverside's EHR vendor license covered in-EHR use, but the new Patient Access API ran on separate infrastructure and could be consumed by external apps. Whether those apps needed their own AMA licenses was unresolved. Sandra scheduled a call with AMA's licensing team; Dev confirmed the API could be architected to minimize CPT exposure if the constraint was established at the design stage.

ValueSets and Binding Strength

The connection between FHIR and clinical terminologies is mediated through ValueSets and binding strength. Understanding both is a prerequisite to understanding why a FHIR resource that appears structurally well-formed can fail validation on a server implementing a particular profile, and why the choice of terminology for a given resource element is a conformance requirement rather than a preference.

A ValueSet is a curated collection of codes drawn from one or more terminology systems. A ValueSet might contain all LOINC codes representing laboratory tests relevant to metabolic panels, all SNOMED CT concepts that are subtypes of "infectious disease," or all ICD-10-CM codes in a range relevant to a population health reporting use case. A ValueSet has a canonical URL that serves as its permanent identifier, a version, and rules that define which codes are included, either by explicit enumeration or by inclusion rules based on hierarchy membership or code system filters. In FHIR, ValueSets are resources in their own right, stored and served by a terminology server, and queryable by name or version.

Binding strength is the mechanism by which a FHIR profile attaches a ValueSet to a specific resource element and specifies how strictly that ValueSet must be honored. FHIR defines four binding strengths. A "required" binding means the element must contain a code from the specified ValueSet; any other code value is a validation error. An "extensible" binding means the element should use a code from the ValueSet, but a code from another system is permitted when no appropriate ValueSet code exists. A "preferred" binding encourages, but does not require, the use of the ValueSet. An "example" binding illustrates what codes might appear in the element, with no conformance expectation.

The practical consequence is most visible in "required" bindings. The US Core Laboratory Result Observation Profile binds the code element of the Observation resource to a LOINC-based ValueSet with "required" strength. Any FHIR Observation submitted to a US Core-compliant server with a code outside that ValueSet will fail validation and be rejected, regardless of how technically correct the resource is in every other respect. The validation failure is not a server bug. It is the terminology binding working exactly as designed.

Diagram 6.8: What Happens When the Code Sandra's System Sends Isn't in the ValueSet — How Binding Strength Determines Whether the Server Accepts, Warns, or Rejects

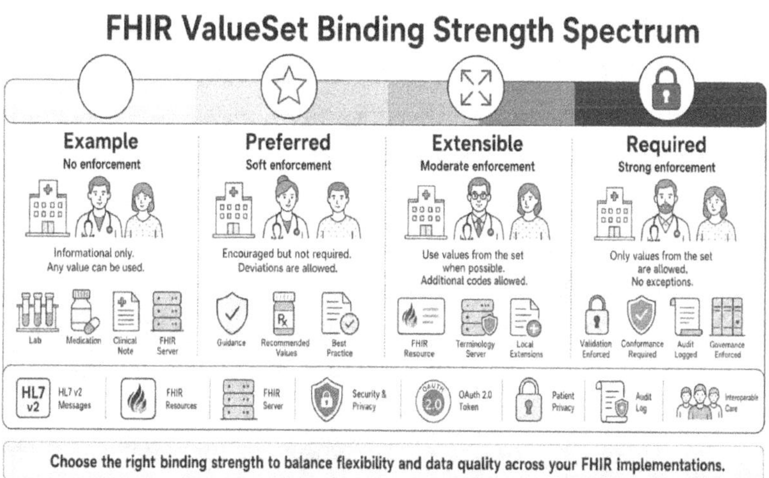

Dev described the ValueSet binding as one of the most common sources of FHIR implementation surprises. Teams building FHIR-producing systems focus on structural conformance: does the resource have the right elements, in the right data types, at the right cardinality? They sometimes overlook terminology conformance: are the codes drawn from the right ValueSets with the right binding strength? The result is a system that passes structural validation and fails terminology validation, indistinguishable from a wrong-code system from the server's perspective. Teams that review binding requirements in the design phase can plan necessary transformation steps as part of the initial architecture. Teams that discover binding mismatches during integration testing will spend additional weeks adding terminology mapping logic to a system not designed to accommodate it.

Terminology Licensing in Practice

Not all clinical terminologies are freely available, and the licensing landscape is complex enough that many implementation teams encounter its constraints later than they should. The major terminology systems used in United States healthcare fall into three licensing categories: freely available without restriction, freely available with attribution requirements, and commercially licensed with fee structures that vary by organization type and use case.

LOINC is freely available in any context, including commercial software development, under a copyright notice requiring attribution to the Regenstrief Institute. The LOINC license does not require a signed agreement, cap the number of codes that may be used, or restrict the types of applications. This permissive licensing has contributed significantly to LOINC's widespread adoption: any organization can implement it in any system without incurring licensing costs or entering into negotiated agreements.

US government agencies maintain ICD-10-CM and HCPCS and are in the public domain in the United States. They may be used freely without attribution or fee obligations. ICD-11 is maintained by the WHO, which makes it available for commercial use with attribution. RxNorm is maintained by the National Library of Medicine as a federal government product and is freely available to any user without licensing restrictions, accessible through the UMLS Metathesaurus and as a standalone NLM download.

SNOMED CT requires a license. In the United States, the National Library of Medicine holds a national license with SNOMED International covering use by US-based individuals, organizations, and their information systems. US users can access SNOMED CT through NLM's UMLS distribution at no charge under the UMLS license terms. Organizations outside the United States must obtain a license from SNOMED International or their country's National Release Center, and the fee scales with organization size and use case.

Diagram 6.9: Which Terminology Licenses Riverside Already Pays For — and Which Ones Create a Compliance Exposure When Embedded in a Patient-Facing App

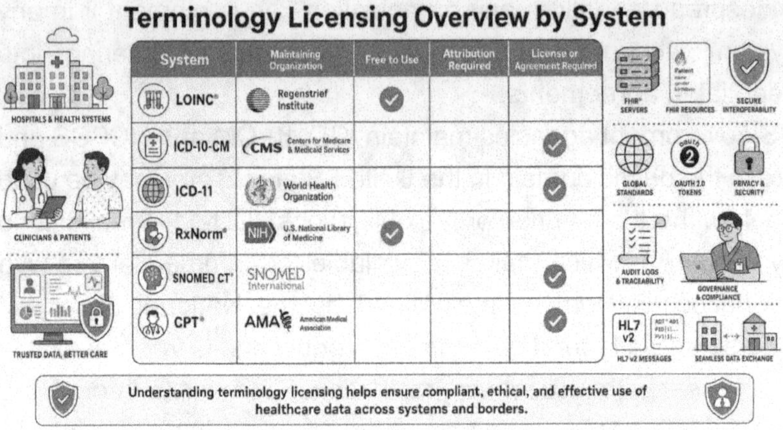

CPT requires a license from the American Medical Association. Any health IT vendor, health plan, or healthcare organization that generates, processes, displays, or transmits CPT codes in an electronic system must have a current AMA license. The cost varies by organization type: vendors pay a flat fee plus a royalty, and healthcare organizations pay based on physician count and scope of use. A hospital deploying an FHIR API that serves patient applications must determine whether its existing license covers CPT codes in the resources returned by that API or require a separate license. These are legal and contractual questions that must be answered before an API goes into production. Maya summarized the licensing picture for Patrick: LOINC, ICD, RxNorm, and HCPCS were effectively free; SNOMED CT was free in the US because NLM paid the national license; CPT was not free, and its cost was not trivial. Patrick added CPT licensing review to the legal checklist Sandra was maintaining for the FHIR project.

Terminology Governance and Ongoing Maintenance

Terminology management is not a project with a completion date. It is an operational function that must be staffed, budgeted, and governed as a permanent part of the health IT operational model. This lesson arrives for many healthcare organizations the hard way: after implementing a terminology-dependent system without a plan to maintain the terminology, and discovering months or years later that the vocabulary has drifted out of alignment with current releases and real-world usage, without anyone noticing.

Terminology drift has multiple sources. Publishers release updates on regular schedules: SNOMED CT and LOINC twice annually, ICD-10-CM annually, RxNorm monthly. Each release may add concepts, retire obsolete ones, or modify relationships. A clinical decision support rule written against a SNOMED CT concept retired in a subsequent release will stop firing without generating any error. A LOINC mapping table accurate at initial deployment accumulates inaccuracies as the laboratory adds new tests, changes analytical methods, or adopts new instruments, all of which may require new or updated LOINC codes.

When Riverside's laboratory added a new high-sensitivity troponin assay, the IT team configured the test in the laboratory information system and set up the HL7 interface. Nobody reviewed whether the new test had a correct LOINC mapping, assuming the existing table would handle it. It did not: the high-sensitivity assay required different LOINC codes than the conventional troponin assay it was replacing. The EHR received high-sensitivity troponin results coded with the conventional troponin LOINC code for several months before a clinical informaticist reviewing cardiac care documentation noticed the discrepancy. The affected data had to be reviewed and, in many cases, manually corrected.

A terminology governance program prevents these failures by establishing organizational ownership, review processes, and operational procedures to ensure accurate, up-to-date terminology use. The minimum viable structure includes a named terminology steward with clinical informatics training, a change management process requiring LOINC mapping review whenever laboratory tests are added or modified, a periodic audit cycle comparing active mappings against current terminology releases, and a documentation requirement recording every mapping decision with its rationale, date, author, and next review date.

Maya proposed a Terminology Stewardship Council with herself as chair, drawing members from the laboratory, pharmacy, clinical informatics, IT, and compliance. The council's mandate: review the impacts of terminology releases, approve changes to the mapping table, investigate data quality incidents, and report quarterly to the clinical quality committee. Patrick approved the council in the same meeting at which he approved the FHIR project budget. His rationale was simple: the council's operating cost was measurably smaller than the cost of the incident that had brought Maya to his office, and the potential cost of a more serious failure, one that reached a patient before being caught, was not a number he was prepared to accept without mitigation. The council held its first meeting three weeks later, and its opening agenda item was the laboratory LOINC mapping audit that Dev had already begun scoping.

Toward a Terminology-Aware Implementation

The practical lesson of this chapter is that terminology decisions are design decisions with long-term consequences that compound with time. A mapping table entry missing at go-live creates a silent gap in clinical decision support on day one. If unnoticed, the gap persists. Six months later, a cohort pulled for quality improvement is incomplete, and the analyst does not know it. A year later, a patient safety measure uses a denominator calculated from data that was silently miscoded for 12 months. The original cause, one missing table row, is three causal steps removed from the visible outcome, and the connection is invisible without a thorough terminology audit.

Maya's first conviction: use the terminology designed for the purpose, not the one that is already in the system. LOINC for laboratory observations, SNOMED CT for clinical findings, ICD-10-CM for billing, RxNorm for medications, CPT and HCPCS for procedures in financial workflows. Substituting a local code or a billing code in a clinical slot does not fail silently — it fails when a CDS rule that expected SNOMED CT receives an ICD-10-CM code and has nothing to act on.

Dev's first rule for Riverside: map early and treat the mapping table as a living system, not a project deliverable. Terminology releases publish twice a year. Lab instruments retire codes. Vendors push updates. A mapping table that was accurate at go-live accumulates errors at every change event that isn't reviewed. The correct default assumption is that an unreviewed mapping table is gradually becoming wrong.

Dev's second rule: never discard source codes. The FHIR CodeableConcept structure exists specifically to carry both the local code and the standard code. Dropping the source code during transformation destroys the provenance chain. When a mapping error surfaces six months later, the only way to trace it is backward through that chain.

Sandra’s requirement before any go-live: a named terminology steward with a documented review schedule. Terminology governance is an operational role, not a project task, and organizations that assign it to no one until an incident forces the question will spend far more building governance retroactively against a backlog of silent errors than they would have spent building it first. Chapter 7 addresses FHIR Profiles and Implementation Guides — the mechanisms that specify and enforce exactly these terminology bindings in operational systems.

Maya put the principle more plainly when she walked out of Patrick's office that morning: a code is a clinical instruction. If the instruction is wrong, the care is wrong, and no amount of network reliability changes that. Dev's version, written on the whiteboard before the Stewardship Council's first meeting, was shorter: "A missing row in a mapping table is a clinical defect." Sandra's version was the one Patrick remembered when he signed the council's charter: the cost of governance is always smaller than the cost of the first incident that forces it. None of the three was the official takeaway of the chapter. All three were what each of them meant by it.

7 FHIR Profiles, IGs, and US Core

A Printout from the ONC Website

Sandra Torres did not knock. She walked directly into Patrick Osei's office on a Tuesday morning in late October, set a single sheet of paper on his desk, and pointed to a sentence she had highlighted in yellow. Patrick leaned forward and read it aloud: "Certified Health IT must support the US Core Implementation Guide." He looked up. Sandra's expression made clear that the sentence was not a suggestion.

Patrick forwarded the printout to Dev Okafor by email twenty minutes later, after a brief conversation with Sandra covering the regulatory timeline and the phrase "information blocking." Dev's reply arrived in two minutes: "We support FHIR. That's not the same thing."

Patrick read the reply, then the highlighted sentence again, then sent Dev a meeting invitation for that afternoon. He had spent enough time in healthcare administration to know that when his IT lead said something was "not the same thing," the difference would cost him money, time, or political capital. He wanted to understand the gap before he had to explain it to the board.

Dev arrived at Patrick's office with a laptop and the patience he had cultivated over years of translating technical architecture into language administrators could act on. He did not begin with the specification. He began with a question: "Patrick, do you know what it means when two people both speak English but still can't understand each other?" Patrick said he did. Dev nodded. "FHIR is the language. US Core is the grammar rulebook. We have been speaking the language. We have not adopted the grammar book."

That conversation became the foundation for a six-month implementation effort that would touch every interface at Riverside General and result in successful ONC certification renewal. This chapter is the record of what Dev explained, what Patrick understood, and what both learned about the distance between a vendor's marketing claim and a regulatory requirement.

Diagram 7.1: The Gap Between FHIR Compliance and US Core Compliance

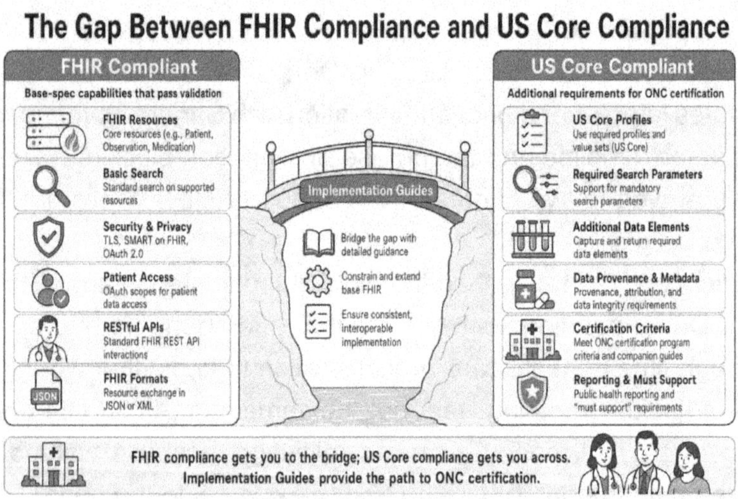

The gap Dev described was not a vendor failure or an implementation bug. It was a structural feature of the FHIR specification's design. FHIR was built to be internationally applicable, flexible enough for healthcare systems in a hundred different countries. Achieving that flexibility required keeping the core specification permissive: most clinical data elements are optional in the base resource definitions. Any system that populates the required elements and validates them against the base schema is technically FHIR-compliant. But "technically FHIR compliant" and "exchanging clinically meaningful data" are not the same achievement, and the gap between them is exactly where profiles and implementation guides live.

The Base Specification Problem

The FHIR base specification defines over 140 resource types covering the clinical, administrative, and financial dimensions of healthcare. The Patient resource contains elements for name, identifier, birthDate, gender, address, telecom, marital status, communication preferences, and more. The Observation resource supports laboratory results, vital signs, and social determinants of health through a flexible coding system that accommodates nearly any clinical concept.

That richness creates an immediate practical problem. When nearly every element is optional, every system makes independent choices about which optional elements to populate. The result is a collection of FHIR-compliant systems that speak the same technical language but carry different vocabularies: System A populates race and ethnicity on the Patient resource because its registration workflow captures those elements from a federally mandated form; System B omits them because its vendor made a different design decision in 2018 and has not revisited it. Both systems are FHIR compliant. Neither can exchange complete patient demographic data with the other without a custom mapping layer.

The problem compounds across resource types. A Condition resource from System A might include the clinicalStatus, verificationStatus, category, and code elements required by clinical decision support. System B's Condition resource might include only the code because the other elements were optional, and the vendor chose not to implement them. At each point of exchange, the gap between what was populated and what is needed creates a manual reconciliation burden—the administrative friction Dev had been describing to Patrick for three years.

Dev walked Patrick through a concrete example from Riverside General's own data. When the hospital's EHR sent a patient summary to the regional health information exchange, the exchange's analytics platform flagged a 31 percent error rate on patient matching. The root cause: Riverside's FHIR Patient resources were missing the race, ethnicity, and preferred language extensions that the exchange's patient-matching and equity analytics workflows depended on. The resources were valid FHIR. They were not US Core compliant. The difference cost the HIE analytics team six weeks of manual remediation work.

Diagram 7.2: Optional Elements and the Data Quality Gap

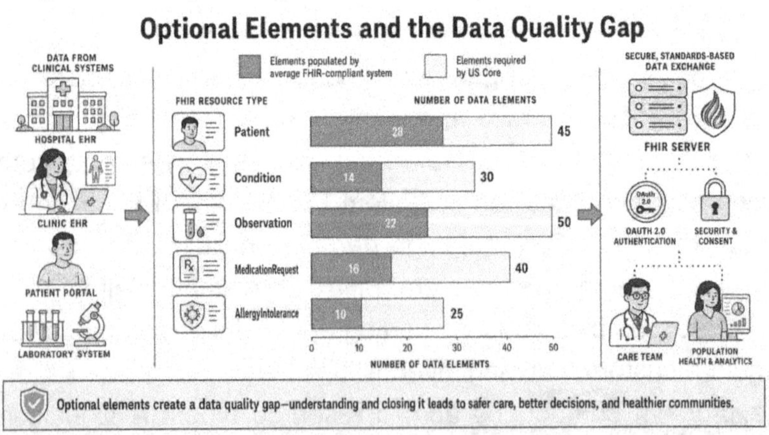

Patrick asked whether other hospitals in the exchange had the same problem. Dev said yes, almost universally, among hospitals that had implemented FHIR without profiles. The base specification problem was not specific to Riverside General; it was endemic across the industry and precisely what ONC had attempted to address through the mandate Sandra had highlighted. The mandate was not punitive. It was a regulatory acknowledgment that the industry would not voluntarily converge on consistent data quality, and that a federal floor was necessary to make interoperability real rather than theoretical.

Sandra added a dimension Patrick had not considered: the liability implications of inconsistent data. When a patient's race and ethnicity were missing from a clinical record that traveled across organizational boundaries, the receiving organization could not run the equity analytics required by its value-based care contracts. When preferred language was missing, the receiving organization could not fulfill its Title VI obligations. The missing elements were not only a data-quality problem but also compliance exposures for every organization in the exchange chain. From Sandra's perspective, US Core was not primarily an interoperability standard. It was a risk management framework.

Profiles Explained

A FHIR profile is a StructureDefinition resource that applies constraints to a base resource type. Profiles can make optional elements required, restrict element values to specific terminology systems or value sets, add extensions for data elements not in the base resource, and provide implementation guidance for populating elements correctly in context. A profile does not replace the base resource; it constrains it, and any instance conforming to a profile also conforms to the base resource from which the profile was derived.

Dev opened the US Core Patient Profile page on HL7.org and pointed out three key views: the narrative description of the profile's purpose, the differential view showing which constraints the profile adds to the base Patient resource, and the snapshot view showing the complete constrained resource definition. The differential was what the profile author wrote. The snapshot was what a validation engine used.

The US Core Patient Profile's differential showed several categories of constraints. Some elements had their cardinality changed from optional (0..1 or 0..*) to required (1..1 or 1..*): every US Core Patient resource must have at least one identifier and at least one name. Other elements were marked with the Must-Support flag, which imposed an obligation distinct from a cardinality constraint: a conformant system must be capable of populating those elements when the data becomes available and must process them without error when they appear in a received resource. Still other elements were bound to specific value sets: the race and ethnicity extensions were bound to CDC vocabularies, ensuring that any organization implementing race and ethnicity data used the same coding vocabulary as every other organization.

Patrick asked a predictable question: "So once we add the profile, are we done?" Dev said no. Implementing a profile was not a single technical action; it was a commitment to a data model with implications for every system that produced or consumed the profiled resource. If the US Core Patient Profile required a race element and Riverside General's registration system did not collect race data, then implementing the profile required either modifying the registration workflow or populating the profile with a code indicating the data was not captured -- both requiring workflow changes, staff training, and governance decisions far beyond the IT department.

Diagram 7.3: What Every StructureDefinition Must Declare Before a System Can Claim US Core Conformance — and the Fields Where Implementers Most Often Fall Short

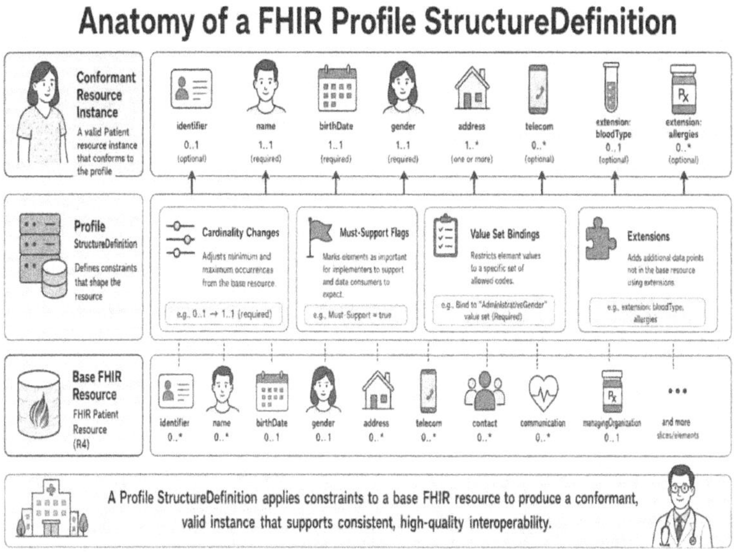

Dev had a term for the accumulated remediation work: "data completeness debt." Every optional element skipped in the original FHIR implementation represented a gap between what the system could produce and what the profile required. The technical mapping was typically the smallest part of the remediation project. The workflow analysis, gap assessment, stakeholder alignment, and data governance decisions were more extensive and more difficult to make.

The extension mechanism deserved particular attention, Dev explained, because it was the pathway by which data elements not in the base FHIR specification could be added in a standards-compliant way. Race and ethnicity were the canonical US example: the FHIR R4 Patient resource did not include native elements for those concepts, because our federal data collection requirements defined them and did not generalize globally. The US Core Patient Profile added them as extensions—named, typed, vocabulary-bound additions to the resource. Extensions defined in profiles could be validated, indexed, and searched just like native resource elements.

The alternative was proprietary extensions: vendor-specific additions that carried the same data but used names and structures no other system would recognize. A proprietary race extension in System A was invisible to System B unless System B had specifically implemented a mapping for it. US Core's standardized extensions solved this by defining a single extension structure every conformant system used, making the data universally readable rather than vendor-specific.

Must-Support vs. Required

The distinction between Required elements and Must-Support elements was the concept Dev returned to most often, because it generated the most confusion among developers implementing FHIR profiles for the first time. Both described conformance obligations. But they described obligations for different things: one for the resource instance, the other for the system.

A Required element was a cardinality constraint on the resource instance. If a profile specified cardinality 1..1, any resource instance that did not include that element was invalid, and a validation tool would flag it as an error. The obligation was absolute: no instance conforming to a profile with a 1..1 cardinality constraint could ever omit that element. This was the concept most developers understood immediately, because it behaved like a required field in any data model.

Must-Support was different. A Must-Support flag on an element did not change the element's cardinality; the element could retain its base cardinality of 0..1 or 0..*. What the flag said was not "this element must be present in every instance" but "a system claiming conformance to this profile must be capable of handling this element." The capability obligation had two sides: a sending system must populate the element when the data is available; a receiving system must process the element without error when it is present in a received resource. Neither side required the element to appear in every resource instance.

Dev illustrated the distinction with Riverside General's allergy data. The US Core AllergyIntolerance Profile marked the reaction element as Must-Support but left its cardinality at 0..*. A patient with no known allergies would have an AllergyIntolerance resource with no reaction element, which is valid. But a patient with a documented allergic reaction to penicillin should have a resource including the reaction element with appropriate clinical detail. A system that routinely omitted the reaction element when reaction data was available in its source records was not conformant -- not because the instance was invalid, but because the system was failing its Must-Support capability obligation. Validation tooling would not catch this failure. Only a clinical data audit would.

Diagram 7.4: Required vs. Must-Support Obligation Comparison

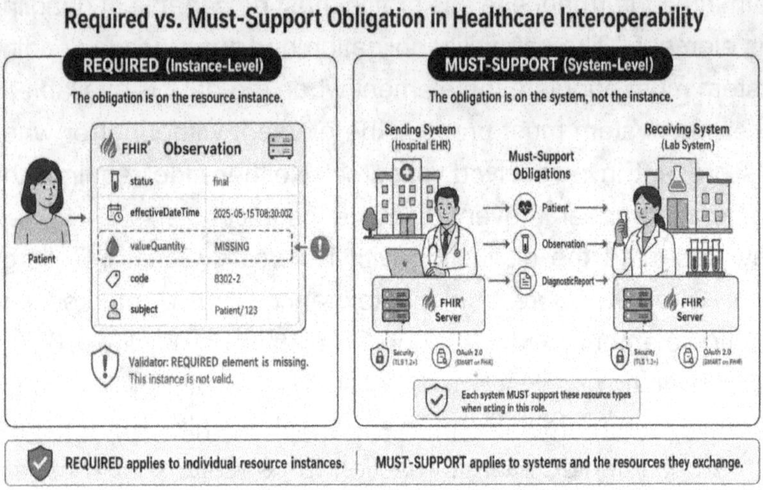

Patrick grasped the implication immediately: validation test results were not a sufficient check for conformance. A system could pass every validation test while still failing US Core conformance because it systematically failed to populate Must-Support elements when the data existed. The gap was invisible to automated testing and visible only through clinical data review, which was why Dev had insisted on a data completeness audit as part of the certification effort, over the objections of department heads who felt the audit was redundant given the clean validation results.

Sandra had her own perspective, less technical and more contractual. When Riverside General signed a data sharing agreement with a payer or the regional HIE, the agreement typically referenced US Core conformance as a technical requirement. If the hospital's systems were not populating Must-Support elements when the data was available, the hospital was technically in breach of those agreements even if every validation test passed. The conformance obligation in US Core was not only a regulatory standard; it was a contractual baseline incorporated into interoperability agreements across the industry. Sandra pulled the two most recent payer agreements from her files and put them on the table. Both contained a single sentence binding Riverside General to current US Core conformance, with no further definition. "This sentence," she said, tapping the page, "is what a plaintiff's counsel will read after a breach, and it is what an auditor will read during a complaint investigation. They will not be persuaded that we passed the validator if we did not populate the elements the standard requires us to support when we have the data. The contract incorporates the standard by reference. We do not get to interpret around it." Patrick listened with the expression he reserved for items that were about to move up the priority list.

US Core: The Federally Mandated Floor

US Core was developed by HL7 International through its consensus process, with active participation from ONC, CMS, major EHR vendors, health information exchanges, payers, and provider organizations. The result had both the technical rigor of an HL7 standard and the regulatory force of an ONC certification requirement: certified health IT products in the United States must support US Core, and that requirement was encoded in the ONC Health IT Certification Program criteria.

Dev explained the relationship between the FHIR base specification, US Core, and organization-specific profiles using a framework he called the "profile pyramid." The base FHIR specification formed the bottom layer: broadest and most flexible, applicable to healthcare systems in every country. US Core formed the middle layer: constrained for the US regulatory and clinical context, mandated for certified health IT.

Organization-specific profiles formed the top layer: further constrained for Riverside General's specific clinical workflows and integration patterns. Each layer was valid independently, because each profile layer was a proper subset of the layer below it.

The US Core R4 specification defined profiles for more than twenty-five resource types. The clinical coverage was broad: Patient and related demographic resources, Practitioner and PractitionerRole for provider identity, Organization and Location for facility information, Encounter for clinical visits, Condition for diagnoses, Observation with sub-profiles for laboratory results, vital signs, and smoking status, MedicationRequest for prescription data, AllergyIntolerance for allergy records, Immunization for vaccination history, DiagnosticReport for laboratory and imaging reports, Procedure for clinical interventions, DocumentReference for clinical notes, CarePlan and CareTeam for care coordination data, and Goal for patient-centered treatment objectives.

Diagram 7.5: US Core Profile Pyramid and Resource Coverage

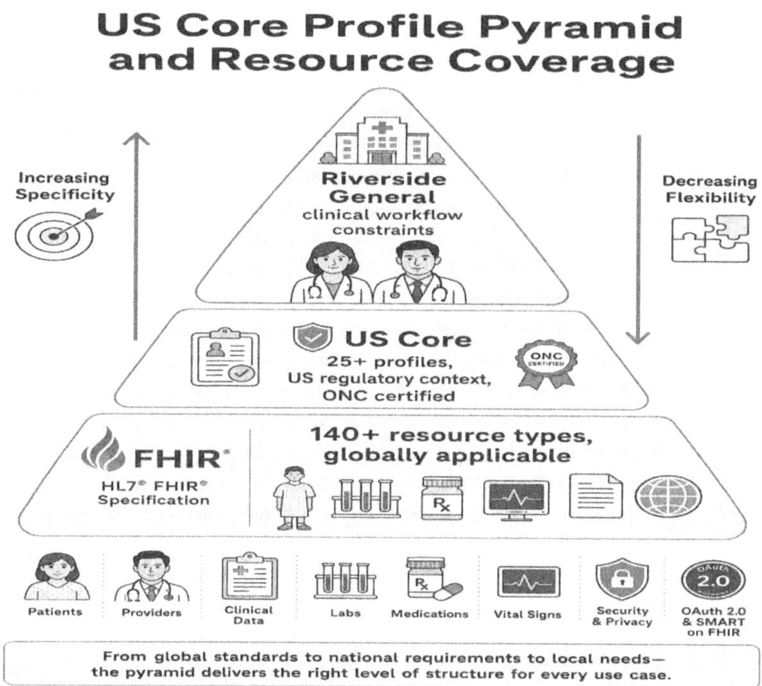

Beyond the profiles, US Core defined required search parameters: the specific combinations of search criteria a conformant server must support. A server claiming US Core conformance must be able to respond to a search for all Conditions for a given patient, all Observations of a given type for a given patient, all MedicationRequests for a given patient with a given status, and dozens of other standardized query patterns. Without these requirements, a client application could not reliably retrieve data from a US Core conformant server using standardized API calls.

Dev also explained the US Core version lifecycle. US Core 3.1.1 had been the stable R4-based version most certified products implemented first. US Core 4.0.0 and 5.0.0 were followed by incremental updates that added new profiles and clarified Must-Support obligations. US Core 6.0.0 introduced significant changes, including new profiles for blood pressure panels and a revised approach to medication data. Each version change required implementers to assess whether existing implementations remained conformant or needed updates. Dev had built a version-tracking item into the quarterly IT governance review to prevent surprise transitions, and Sandra tracked US Core version requirements with the same attention she gave to HIPAA updates and CMS reporting deadlines.

The StructureDefinition in Practice

Dev spent twenty minutes of the session opening the FHIR specification website and walking Patrick through what a StructureDefinition actually looked like. He pulled up the US Core Patient Profile page. He worked through the differential table, pointing out how each row corresponded to an element in the base Patient resource and how the profile's constraints appeared in the columns for cardinality, type, and conformance. The goal was not for Patrick to memorize the structure, but to give him a mental model so that when his team said a profile change was "just a cardinality update," he would understand what they meant and could ask an intelligent follow-up question. The differential was the most useful starting point for understanding what a profile required. It showed only the changes from the base resource: elements whose cardinality had been tightened, elements that had been marked Must-Support, value set bindings that had been added or changed, and extensions that had been introduced. For a developer implementing a profile for the first time, the differential was the specification of work. Every row in the differential table represented something that the base FHIR implementation would not handle correctly without modification.

The snapshot view was longer and less immediately readable, but it was the view validation tools used. A snapshot was a complete definition of every element in the resource as constrained by the profile, including elements the profile had not changed from the base. When a validation error message referenced an element path, the snapshot was where a developer looked to understand exactly which constraint had been violated and what the resource needed to satisfy it.

Patrick asked Dev to explain the slicing notation that appeared in validation error messages. Slicing allowed a profile to define multiple distinct constraints on a single repeating element. The US Core Patient Profile sliced the identifier element to define constraints on specific identifier types (medical record number, social security number, driver's license) while leaving the overall identifier element open to additional types. Each named slice had its own cardinality and constraint rules. Understanding slicing was necessary for Dev's team to implement profiles correctly and for Sandra to interpret validation reports.

Diagram 7.6: Why Dev Reads the Snapshot and Not the Differential When Debugging a Validation Failure — What Each View Shows and What It Hides

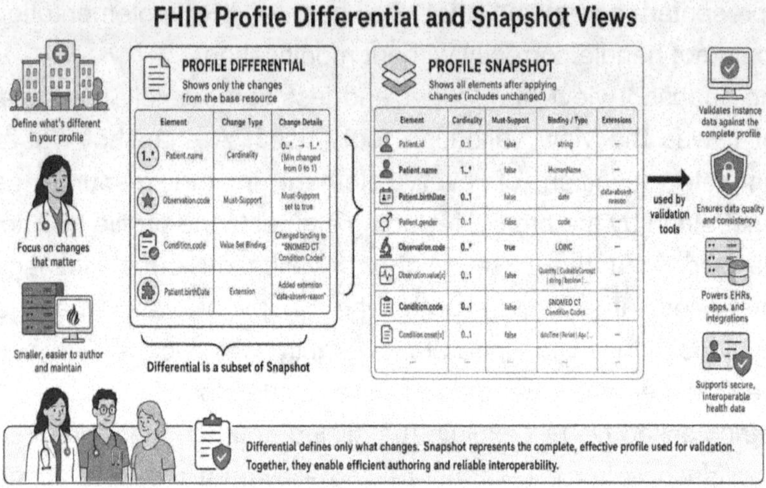

Slicing appeared in nearly every significant FHIR profile because the repeating element pattern was pervasive in FHIR resources. Coding elements, identifiers, names, addresses, and contact points were all modeled as arrays of objects, and profiles frequently needed to constrain specific members of those arrays without restricting the array as a whole. Developers who understood slicing could read profile specifications fluently; those who did not tended to implement by trial and error.

Da Vinci IGs and the Payer-Provider Frontier

Dev brought up the Da Vinci Project specifically for Patrick's benefit, because the use cases it addressed appeared directly in Patrick's operational budget. Prior authorization was the canonical example: every week, Riverside General's clinical staff spent hours on the phone with insurance companies, faxing clinical documentation and waiting for authorization decisions before they could schedule procedures. The manual burden was significant, the delays affected patient care, and the costs -- in staff time and delayed revenue -- showed up on Patrick's operational reports.

The HL7 Da Vinci Project was a collaborative initiative involving health insurance companies, providers, and EHR vendors working together to define FHIR-based solutions for administrative workflows at the boundary between clinical care and healthcare financing. Da Vinci had produced a suite of implementation guides, each addressing a specific administrative use case, all built on top of US Core as the clinical data foundation. The guides were published as HL7 standards that major payers and EHR vendors were actively implementing in response to CMS Interoperability Rules.

The Prior Authorization Support (PAS) Implementation Guide defined a FHIR-based workflow for submitting prior authorization requests from a provider's EHR directly to a payer's authorization system, receiving real-time or near-real-time decisions, and attaching clinical documentation through standardized FHIR document bundles. For Patrick, the most significant implication was the reduction in staff time: a PAS-based prior authorization workflow could reduce a process that took 48 hours of manual effort to under an hour of automated processing, with clinical staff focused on cases requiring clinical judgment rather than on fax machines and hold queues. Patrick had a specific picture for this: the cardiology authorization queue where Dr. Chen's MRI orders sat for two days while staff faxed clinical notes that the payer's system could have read directly.

Diagram 7.7: Da Vinci Project Use Cases and IG Relationships

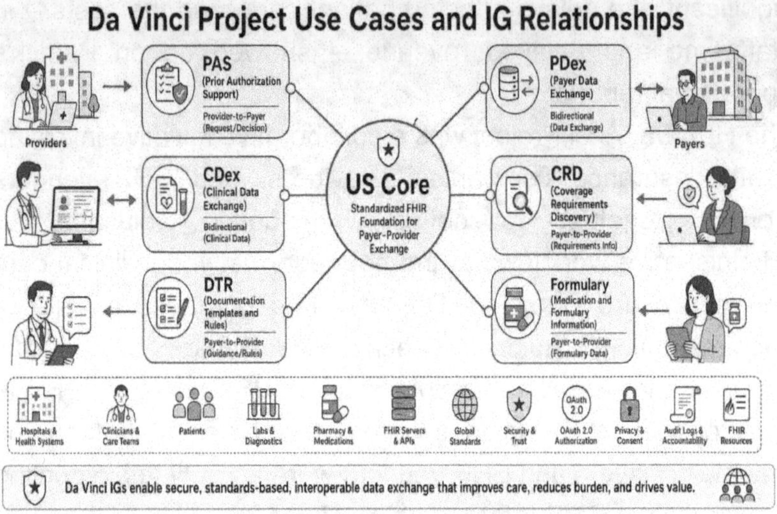

The Payer Data Exchange (PDex) Implementation Guide addressed the CMS requirement that payers support patient-requested data access and payer-to-payer exchange using FHIR. PDex defined how a payer exposed its claims data, clinical data, and prior authorization history via an FHIR API, enabling patients to retrieve their data through third-party applications and allowing new payers to receive prior coverage data when patients changed insurance. Sandra had been tracking PDex closely because the payer-to-payer exchange requirement had direct compliance implications for Riverside General's contracted payer relationships. The anchor for Patrick was the open-enrollment month every January, when his patient services line filled with calls from new members whose prior authorizations had not transferred and whose coverage histories had to be reconstructed by phone.

The Coverage Requirements Discovery (CRD) and Documentation Templates and Rules (DTR) guides worked together to address a problem Patrick recognized from Dr. Chen:

the burden of gathering clinical documentation for prior authorization at the time of care. CRD allowed the EHR to query the payer's system in real time when a clinician ordered a procedure, returning coverage requirements before the order was finalized. DTR enabled the payer's documentation requirements to be expressed as a structured questionnaire integrated directly into the EHR workflow, so clinical staff completed the required documentation without leaving their clinical application. The pain CRD and DTR addressed was the one Dr. Chen complained about most: discovering at the end of a visit that a procedure required documentation the EHR could have surfaced at the moment of ordering, and sending the patient home to wait while staff chased forms.

Dev set realistic expectations. Implementing Da Vinci IGs required that Riverside's EHR vendor support them, that its contracted payers support them, and that Riverside General's IT infrastructure could serve as the provider-side endpoint. All three conditions were at least partially met: the EHR vendor had announced CRD and DTR support in its next major release, two of the hospital's largest payers were participating in a CMS-sponsored Da Vinci pilot, and the US Core certification work would provide the technical foundation Da Vinci implementations required.

SMART on FHIR: Authorized App Access

Dev shifted to authorization when Patrick asked the question he had been building toward: "If we have all these FHIR APIs, how do we make sure the right people can access them and the wrong people cannot?" The answer was SMART on FHIR.

SMART on FHIR -- Substitutable Medical Applications and Reusable Technologies -- was developed at Boston Children's Hospital and later adopted by HL7 as a formal specification. It defined a standardized way for third-party applications to obtain authorized, scoped access to FHIR data using OAuth 2.0 as the underlying authorization protocol. The result was an architecture that allowed applications to connect securely to EHR data without requiring the EHR vendor to build every application themselves -- a design principle foundational to the competitive health app marketplace ONC's information blocking rules were intended to foster.

The authorization flow began when a user launched an application, either from within the EHR (an EHR launch) or as a standalone application. In the EHR launch scenario, the EHR passed context to the application at launch time: the current patient ID, the current encounter, and the identity of the launching user. The application requested specific FHIR data scopes (patient/Patient.read, patient/Observation.read, user/MedicationRequest.read). The EHR's authorization server issued an access token scoped to exactly what had been requested and authorized. Every subsequent FHIR API call from the application included the access token, which the FHIR server validated and used to enforce the scope boundaries.

Diagram 7.8: SMART on FHIR Authorization Flow

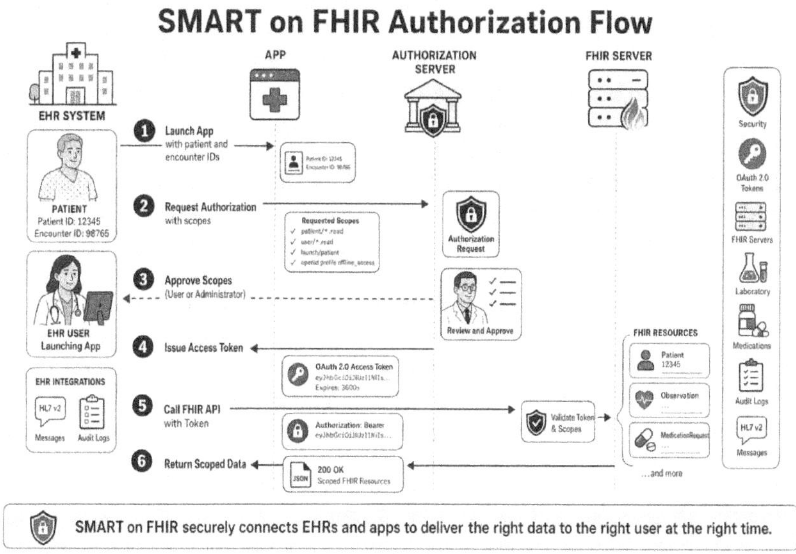

Dev explained the scope syntax: a FHIR scope like "patient/Observation.read" said, "this application is authorized to read Observation resources for the current patient context." A cardiology decision support application could be authorized to read cardiovascular-relevant observations and conditions without being authorized to access mental health records, medication lists, or other data outside its clinical scope. The principle of least privilege was built into the authorization architecture.

Sandra found the SMART on FHIR model reassuring from a privacy governance standpoint. The scope-based access control meant that every application's permissions were explicit, auditable, and limited to what the application had specifically requested and the authorization server had granted. When a new third-party application was being considered for deployment at Riverside General, Sandra could review its scope requests and evaluate whether they were proportionate to the application's stated purpose. An application requesting write access to medication records for a clinical reference tool had a clear scope mismatch. An application requesting read access to the current patient's medication list for a drug interaction checker had a proportionate scope request. The difference was legible in the SMART on FHIR scope declaration.

US Core required SMART on FHIR as the authorization mechanism for patient-facing applications. The ONC certification criteria for patient access API support specified that a certified system must support SMART on FHIR for standalone launch, allowing patients to authorize third-party applications to access their data. Without it, patient-facing app access required custom integration work for every application—the model that had prevailed before the standard existed and had effectively excluded third-party applications from accessing EHR data. SMART on FHIR was the technical prerequisite for a competitive, interoperable patient-facing health app ecosystem.

How to Read an Implementation Guide

Dev's final section was where the room stopped taking notes and started asking questions. Rather than describing what the implementation guides contained, he opened the US Core Implementation Guide homepage on hl7.org and began navigating — not as a tour, but as an investigation. The navigation structure was not the point; what it revealed was.

The home page of a published IG contained several navigation items every implementer needed to know. The Introduction or Background section described the use case, the regulatory context, the audience, and the relationship to other standards the guide depended on. For US Core, this section explained the ONC certification requirement, the relationship to the FHIR base specification, and the version history. The Introduction was the entry point for anyone approaching the guide for the first time.

The Artifacts Summary page was the most useful navigation destination for gap analysis work. It listed every profile, extension, value set, code system, capability statement, and search parameter defined in the guide, with links to individual artifact pages. For each profile in the list, the implementer asked whether their system produced and consumed resources conforming to that profile and what changes were needed. A quick scan of the artifact list also gave a reasonable estimate of the implementation effort required.

The individual profile pages had a consistent structure. At the top was a brief narrative describing the profile's purpose and Must-Support obligations. Below that was the differential view as a structured table. Below the differential was the snapshot view, and at the bottom the conformance page listing searches and operations a conformant server must support when serving that resource type.

Diagram 7.9: Navigating an IG on HL7.org

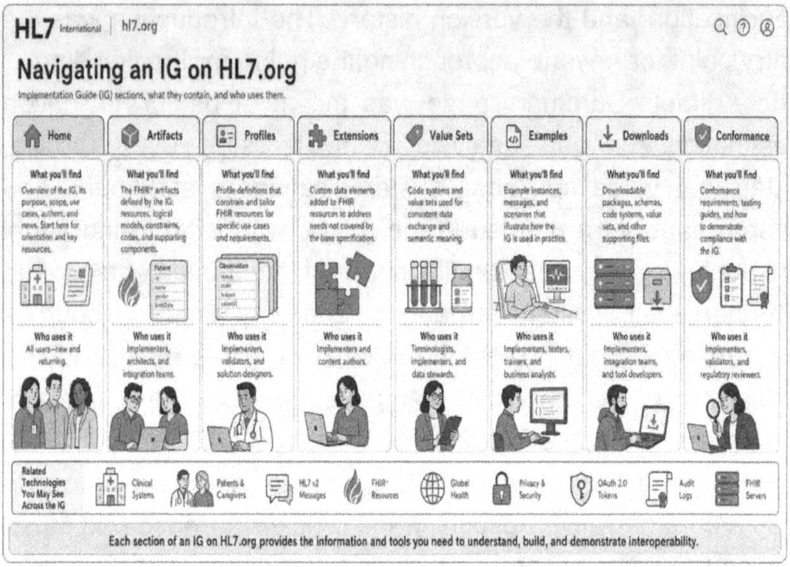

Dev pointed out the Examples section: sample resource instances illustrating how profiles should be populated for common clinical scenarios. Reading an example of a US Core Patient resource with race, ethnicity, and preferred language populated was more instructive than reading the differential table for those extensions. Dev clicked into the example, zoomed in on the extension block, then scrolled back to the differential table and pointed at the row marked "Must-Support". "Race and ethnicity are not optional decorations here," he said. "They are Must-Support, with specific OMB categories and granular detail values from the CDC code system. A system that drops the extension or substitutes its own coded list fails conformance."

Maya had been reading the same screen. “I assumed we were capturing race and ethnicity as part of the standard registration intake,” she said. “Demographic preference, clinical context — something the registration desk collects but clinical systems handle loosely.” Dev shook his head. “Must-Support means three things. The registration workflow has to collect it. The EHR has to store it. And a system that receives a Patient resource with the extension and silently drops it fails US Core conformance." Maya was quiet for a moment. "So this is a registration workflow change, not a clinical one." "Correct," Dev said. "The IG tells you that before you build anything." It was the first time in the project that the guide had corrected Maya rather than confirmed her. She updated her assessment without protest.

Patrick was also surprised — he had assumed demographic fields were a registration-screen choice the hospital could shape however it preferred. Sandra wrote a note to add demographic capture to the next compliance committee agenda. This was what reading an IG carefully revealed: obligations that looked administrative were actually architectural. Dev's team read the examples first, then the differential, then used the snapshot as a reference when validation errors appeared.

The Downloads section was where implementers found the NPM package: the machine-readable form of the guide's artifacts, in JSON, that validation tools used to verify resource instances against the guide's profiles. The dev team used the HL7 FHIR Validator CLI to accept an NPM package and validate a resource file against all profiles in the package. Before the gap analysis project, the team had not routinely run resources through the validator. After the project, running the validator before deployment was a required step in the development workflow.

Sandra asked about the Conformance section, which Dev explained was the most important section for organizations preparing for ONC certification. It described exactly what a system must support to claim conformance: which profiles, which search parameters, which capability claims. Dev had extracted the conformance requirements into a tracking document that mapped each requirement to the system at Riverside General responsible for meeting it, with a status column updated at each sprint review. The tracking document was Sandra's single reference for US Core certification status.

Putting It Together at Riverside General

The gap analysis Dev presented identified 11 resource types in which Riverside's existing FHIR output did not meet US Core profile requirements. Seven required only technical changes to the FHIR mapping layer: adding missing elements, correcting value set bindings, implementing required extensions. Four required workflow changes upstream of the FHIR layer: the registration workflow needed updating to capture race and ethnicity data; the medication reconciliation workflow needed updating to populate required MedicationRequest elements; and the clinical documentation workflow needed updating to ensure Condition resources carried the required clinical and verification status elements.

Patrick asked which of the four workflow-dependent gaps was most complex. Dev said the race-and-ethnicity capture workflow was the most politically sensitive, requiring staff training and patient communication about why the data was being collected. The medication reconciliation gap was the most technically complex, involving coordination between the EHR, the pharmacy system, and the FHIR mapping layer. Both required engagement from clinical leadership and from Sandra's compliance team. The US Core implementation was not an IT project in the narrow sense. It was an organizational data governance project with significant technical components.

Sandra had already begun drafting the staff communication for the race and ethnicity workflow change, drawing on Office of

Minority Health guidance on communicating the purpose of demographic data collection to patients. She and Dev agreed the communication needed to explain both the clinical rationale and the regulatory context without overwhelming front-desk staff with technical detail. It would be piloted at one registration desk before rollout, and patient responses monitored before hospital-wide deployment.

Dev closed the meeting by returning to the sentence Sandra had highlighted. "We support FHIR" was true, and would remain true after the project was complete. But after the project, it would also be true that the hospital supported US Core, that it produced and consumed US Core-conformant resources, that its FHIR server responded to US Core search parameters, and that its authorization framework supported SMART on FHIR for patient-facing applications. The difference between those two statements was the six months of work ahead. Patrick thanked Dev, scheduled the first project review for two weeks out, and sent Sandra a message confirming the certification work had executive sponsorship.

Takeaways

Dev was right from the moment he sent his two-minute reply: FHIR compliance and US Core compliance are not the same achievement. A system can pass every base-level validation test and still fail ONC certification requirements because it does not implement the profiles, search parameters, and conformance requirements that US Core defines. The distinction matters legally because the ONC certification criteria explicitly reference the US Core. It matters contractually because data-sharing agreements use US Core conformance as a technical baseline. And it matters clinically, because systems that omit US Core-required data elements fail to exchange the information that makes patient records portable, comparable, and actionable across organizational boundaries.

US Core is the floor of US healthcare interoperability, not the ceiling. Organizations implementing US Core are beginning their profile work rather than completing it: real-world clinical workflows, specialty requirements, and population health analytics programs require additional profiles, extensions, and capabilities beyond what US Core mandates. The path forward is an organization-specific profile library that builds on the US Core foundation and adapts it to the institution's particular data governance needs and integration patterns.

Must-Support is a conformance obligation on the system, not a presence requirement on individual resource instances. A US Core-conformant system must populate every Must-Support element when data is available in its source records. It must process every Must-Support element without error upon arrival in a received resource. Systems that treat Must-Support as equivalent to optional are not conformant, regardless of validation test results. The conformance gap between "passes validation" and "truly meets Must-Support obligations" is visible only through systematic data quality audits, which is why audits must be a required component of any certification preparation process.

Da Vinci implementation guides represent the operational frontier of FHIR adoption for organizations seeking to reduce administrative burden. Prior authorization automation through PAS, real-time coverage discovery through CRD, structured documentation collection through DTR, and payer-to-provider data exchange through CDex are not future capabilities. They are published HL7 standards with active implementations by payers and EHR vendors today. The organizations building US

Core infrastructure are building the foundation on which Da Vinci workflows will run in their next implementation cycle.

SMART on FHIR is the authorization infrastructure that enables the app ecosystem. Without it, every third-party application accessing EHR data requires a custom integration—a model that limits innovation, concentrates control, and makes patient-facing health app development prohibitively expensive for all but the largest vendors. SMART on FHIR decouples application development from EHR vendor decision-making, creating conditions for a competitive marketplace. For Sandra, it makes third-party app access auditable at the scope level, converting a broad access grant into a precise, reviewable permission record.

Implementation guides should be read before a single line of interface code is written. The guide specifies which elements are required, which terminologies are bound, which search parameters must be supported, and which conformance claims must be declared. A team that discovers profile requirements after building its data model will spend significant time rebuilding what should have been built the first time correctly. Reading the IG first -- Introduction, Artifacts, Profiles, Conformance, Examples, in that order -- is the practice that most reliably distinguishes efficient FHIR implementations from expensive ones.

The work of implementing profiles is organizational as much as it is technical. Data elements that profiles require must be captured in the clinical or administrative workflow before they can appear in a FHIR resource, and aligning those workflows involves clinical leadership, patient communication, staff training, and governance decisions no developer can make alone. The organizations that succeed with US Core implementations are those where the IT team, the clinical leadership, and the compliance function work as genuine partners from the first day of the project -- where Dev, Patrick, and Sandra are at the same table before the first gap analysis row is written.

Dev's private version of the chapter's lesson was the one he refused to put on a slide: most teams claiming FHIR support have not actually read the implementation guide they claim to conform to, and that is the failure mode he expects to encounter on every vendor call until the industry catches up. Patrick's version was operational and unsparing: "We support FHIR" is a sentence that should never appear in a contract Riverside General signs again without a profile name attached, because the version without the profile name is the one that produced the HIE remediation invoice on his desk. Sandra's version was the contractual one she now repeats whenever a department head suggests treating a Must-Support element as optional: the standard is in the contract, the contract is in the file, and the file is what the auditor reads. The three takeaways are different. The three people meant the same thing.

8 Regulations, Compliance, and Info Blocking

Sandra Torres read the letter from the Office of the National Coordinator for Health Information Technology twice before she called Patrick Osei. Three pages of measured federal prose, with its core allegation near the bottom of the second page: an anonymous complainant had alleged that Riverside General's patient portal was failing to provide timely access to laboratory results, conduct that potentially constituted information blocking under 45 CFR Part 171. Sandra set the letter face-down on her desk, looked at the ceiling for a moment, and picked up the phone.

"Patrick, we have thirty days to respond to ONC. I need two hours on your calendar today, and I need Dev in the room."

His voice had the carefully controlled quality of a man who understood that panic was not useful. "What kind of response are we talking about?"

"We either demonstrate we're compliant, invoke an applicable exception, or explain what we're doing to come into compliance. All three paths require documentation we may or may not already have." Sandra pulled up the patient portal release notes on her second monitor. "Clear your afternoon."

This chapter is Sandra's chapter. She is Riverside General's compliance and privacy officer. The regulatory landscape she navigates is not abstract policy theory. It is a live enforcement environment with published complaint statistics, penalty authority, and the weight of two federal agencies behind it. Every API endpoint Dev builds, and every portal feature Patrick approves, will pass through her review. Understanding why that review matters requires understanding the regulatory framework that created it.

Why Federal Interoperability Rules Changed Everything

For most of the history of electronic health records, the systems that stored patient data were designed with little regulatory obligation to share it. The outcome was a healthcare technology ecosystem characterized by deep fragmentation, proprietary interfaces, and active resistance to data portability. Vendors had commercial incentives to make their systems the center of a provider's data gravity; providers had institutional incentives to maintain control over patient population data. The patient's interest in accessing their own health information was a secondary consideration.

The HITECH Act of 2009 drove widespread EHR installation but produced a landscape of poorly connected systems rather than a coherent national health information infrastructure. By the mid-2010s, federal policymakers had concluded that EHR adoption alone would not produce interoperability. The problem was structural: organizations holding that data had incentives to keep it closed, and voluntary market incentives were insufficient to open it.

The 21st Century Cures Act, signed in December 2016, created an explicit federal prohibition on information blocking: a legal obligation that healthcare providers, health IT developers, and health information networks could not take actions that interfere with, prevent, or materially discourage access to, exchange of, or use of electronic health information. It also directed ONC to establish certification criteria requiring standardized FHIR-based APIs for patient access to health information, thereby linking the technical architecture of modern health data exchange directly to the regulatory enforcement framework.

The ONC's 21st Century Cures Act Final Rule, published in May 2020, translated the statutory provisions into operational requirements: a broad definition of electronic health information, eight exceptions that would not constitute information blocking, certification requirements for health IT developers covering FHIR R4, US Core profile conformance, SMART on FHIR authorization, and USCDI data access, and the enforcement and complaint intake mechanism through which Riverside General now found itself receiving a federal letter.

Diagram 8.1: The Regulatory Clock That Was Already Running Before Patrick's Team Noticed

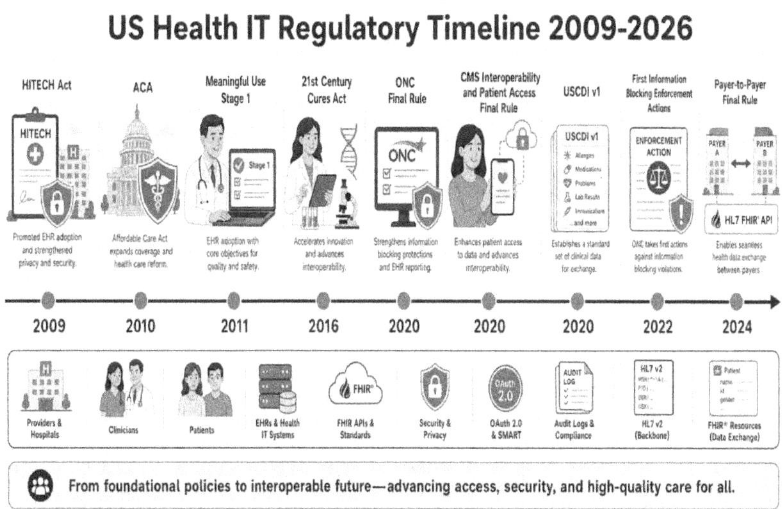

Sandra's first task was to establish a precise factual record of what the hospital's patient portal did and did not do with laboratory results. Healthcare providers have an obligation not to impose unreasonable delays on patient access to electronic health information, and laboratory results are within the explicit scope of EHI subject to that obligation. If the portal was holding lab results longer than clinically necessary, and no documented exception basis existed for that delay, the hospital had a potential finding on its hands.

Dev pulled the portal configuration logs and found the answer within an hour. The portal was configured with a 72-hour hold on laboratory results before release to the patient-facing view, implemented two years earlier by a physician committee concerned that patients might receive abnormal results without a clinician available to provide immediate context. The concern was clinically legitimate. Without documentation linking the configuration to the Preventing Harm Exception under 45 CFR Part 171.204, it may have constituted information blocking.

"We have a configuration decision that was made with good clinical reasoning and zero documentation of the regulatory basis," Sandra told Patrick and Dev at their afternoon meeting. "That's our gap. The question now is whether the clinical reasoning actually satisfies the exception criteria."

The 21st Century Cures Act: What Congress Required

The 21st Century Cures Act addresses dozens of healthcare and biomedical research topics. Its information blocking provisions appear in Sections 4004 and 4006, drafted with specificity reflecting documented evidence of how health IT vendors and healthcare organizations had resisted data sharing.

The information blocking prohibition covers three categories of actors: healthcare providers, health IT developers of certified health IT, and health information networks and exchanges. The law prohibits any practice the actor "knows, or should know, is likely to interfere with, prevent, or materially discourage" access to, exchange of, or use of electronic health information, unless the practice falls within a regulatory exception or is otherwise required by law. The knowledge standard matters: the prohibition is not limited to intentional obstruction, and ignorance of the regulatory implications is not a reliable shield.

The API requirement in Section 4006 directed ONC to establish criteria requiring health IT developers to provide API access to patient health information without special effort, without requiring proprietary APIs, and without fees that make access cost-prohibitive. A patient using a third-party app should be able to access their health information through standard FHIR APIs without navigating proprietary procedures or paying unreasonable fees.

The information blocking prohibition creates a legal obligation not to restrict access. The API requirement creates a technical obligation to provide access through a standardized FHIR-based mechanism. A hospital that builds a functional FHIR R4 server but configures it to prevent patient access may satisfy the letter of the API requirement while violating the information-blocking prohibition. Sandra's review process exists precisely to catch the gap between what the technical implementation makes possible and what the operational configuration actually delivers.

Diagram 8.2: Information Blocking: Actor Categories and Obligations

Information Blocking: Actor Categories and Obligations

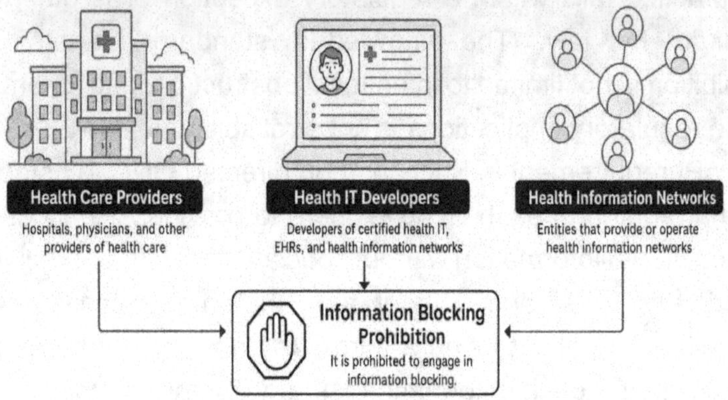

This diagram illustrates the three categories of actors subject to the information blocking prohibition and their shared obligation to enable access, exchange, and use of electronic health information.

The Act also addressed EHR vendor obligations directly, making information blocking by health IT developers a condition-violating act subject to certification consequences. The vendor's certification commitments include obligations to provide FHIR API access to all USCDI data elements, support SMART on FHIR authorization, and refrain from practices that interfere with third-party application access. One of Sandra's first questions in reviewing the complaint was whether the 72-hour lab result hold was a hospital configuration choice or a vendor-imposed restriction. Dev's investigation confirmed it was on the hospital side: the vendor provided the capability to release lab results immediately, along with a configuration option to apply a time delay of the hospital's choosing. Riverside General had chosen the 72-hour delay, and the regulatory consequence fell on the hospital.

The Eight Information Blocking Exceptions

The information blocking prohibition would be unworkable if it required immediate, unrestricted access to all electronic health information in all circumstances. A hospital responding to a cyberattack cannot be required to maintain full API access to patient data while actively remediating a breach. A physician committee that delays portal release of abnormal lab results for clinical safety reasons is making a decision that, with proper documentation, falls within the exception framework rather than outside it. The regulations address these realities through eight defined exceptions.

Sandra had a printed summary of all eight exceptions in her compliance binder, annotated with notes from every information blocking investigation she had reviewed since the regulations took effect. The Preventing Harm Exception at 45 CFR 171.204 permits practices that are likely to prevent harm to a patient or another person. The practice must be no broader than necessary, implemented consistently rather than selectively, and based on an objective risk rather than a generalized preference for caution. The 72-hour lab result hold could fit this exception if the hospital could demonstrate that releasing abnormal results without clinician review created a documented risk of harm, and that 72 hours was the minimum hold necessary. Sandra's problem was not that the exception was inapplicable. It was that no one had written down the analysis when the configuration was made.

The Privacy Exception at 45 CFR 171.202 permits practices necessary to comply with applicable privacy laws, including HIPAA. This exception is narrower than it appears: it permits compliance with actual legal obligations, not an organization's general preference for privacy protection. A hospital that refuses to release patient data to a patient-authorized app because it is worried about privacy, without a specific legal basis, has not satisfied this exception.

Diagram 8.3: The Eight Information Blocking Exceptions

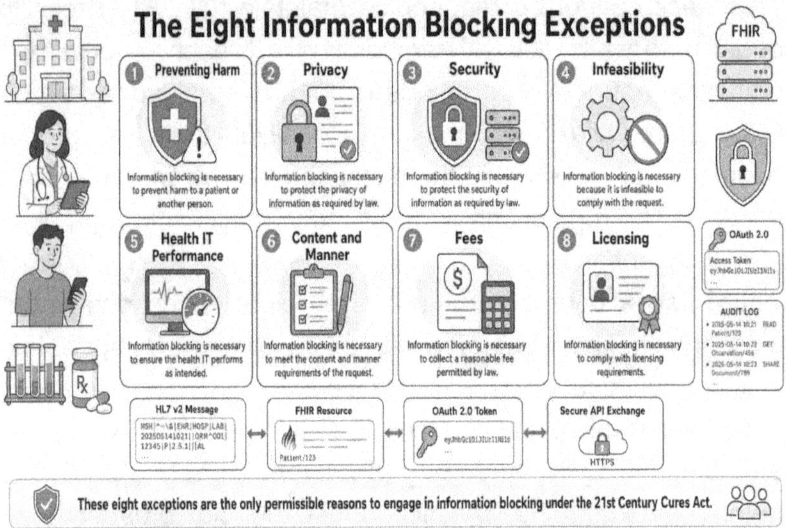

The Security Exception, at 45 CFR 171.203, permits practices necessary to protect the security of EHI, requiring that the practice be consistent, proportionate, and based on an objective security risk. Disabling FHIR API access in response to an active cyberattack falls within this exception; requiring all third-party apps to use a proprietary authorization protocol when SMART on FHIR is available does not. The Infeasibility Exception, at 45 CFR 171.205, permits non-compliance where compliance is technically or operationally impossible. "Technically impossible" is a higher bar than "would require significant development effort." The Health IT Performance Exception permits practices necessary to maintain or improve health IT system performance, including temporary service interruptions for maintenance, with advance notice to users where possible.

The Content and Manner Exception, at 45 CFR 171.301, permits fulfilling an access request using different content or manner than requested, preserving the obligation to fulfill it without permitting outright refusal. The Fees Exception permits reasonable, cost-based, non-discriminatory fees for access in defined circumstances. The Licensing Exception permits licensure requirements for certain commercial uses of EHI involving proprietary content, but does not allow such requirements to block access to standard clinical data.

After reviewing all eight exceptions, Sandra identified the Preventing Harm Exception as the most likely basis for Riverside General's defense of the 72-hour hold. The clinical reasoning was sound; the documentation problem was addressable. She assigned Dev to pull incident records for cases in which a patient contacted the hospital after receiving an abnormal portal result without prior physician contact, and she assigned herself to draft the exception documentation framework that should have been created when the configuration was first implemented.

ONC Certification and FHIR API Requirements

ONC Health IT Certification is the federal program through which health IT developers demonstrate that their products meet ONC's criteria for interoperability, privacy, security, and functionality. Certification is required to participate in Medicare and Medicaid quality programs, and a healthcare organization operating on an uncertified platform faces both payment and regulatory consequences. ONC's certification criteria, updated significantly by the 21st Century Cures Act Final Rule in 2020, require certified health IT to support a standardized patient-facing FHIR R4 API. That API must expose all data elements defined in the United States Core Data for Interoperability, support SMART on FHIR launch and authorization, and conform to the US Core Implementation Guide's profile requirements for every FHIR resource type within the USCDI scope.

Diagram 8.4: ONC Certification Criteria and FHIR Requirements

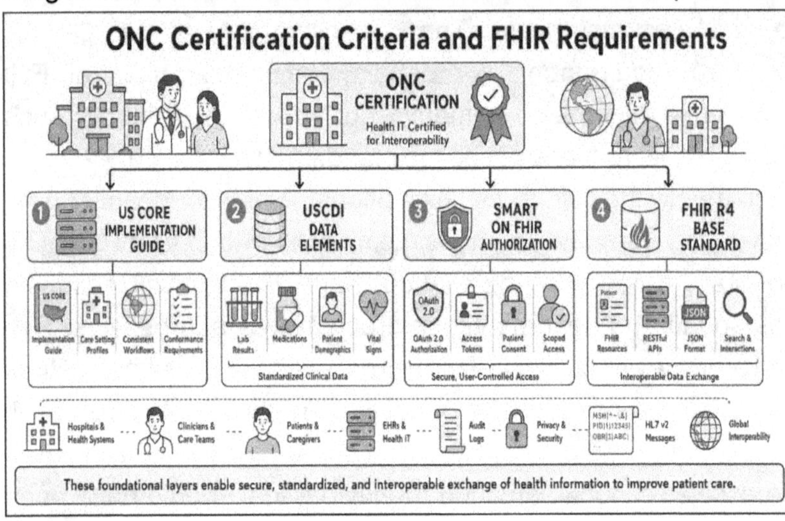

For Riverside General, the certification requirements create both an asset and an obligation. The asset is the contractual commitment their EHR vendor has made to ONC: the certified platform must support FHIR R4 APIs, expose USCDI data, and support SMART on FHIR. If the vendor fails to deliver those capabilities, the hospital has grounds for a contract dispute, and the vendor risks decertification. The obligation is less comfortable: certification requirements apply to the configuration and operation of certified health IT by the healthcare organization, not only to the vendor's development choices. A healthcare organization that uses its EHR's configuration options to restrict API access in violation of information blocking rules cannot shelter behind its vendor's certification. Riverside General's 72-hour lab result hold was exactly this kind of hospital-side choice.

Dev's review also surfaced a second issue Sandra had not anticipated. The hospital's SMART on FHIR authorization configuration allowed patients to authorize third-party apps to access demographic data and problem lists, but disabled patient authorization for laboratory results, citing a security scope restriction that was documented nowhere in the compliance record. If a patient had tried to access lab results through a third-party app and been denied, that denial would have been a second potential information-blocking issue, independent of the hold.

"We have two problems," Sandra told Dev. "The hold is defensible if we build the documentation. The SMART scope restriction is harder to explain without a written rationale." Dev recognized the particular discomfort of discovering that a technically reasonable-seeming decision made without regulatory context had produced regulatory exposure. Dev would restore the laboratory results scope in the SMART on FHIR authorization configuration; Sandra would draft the Preventing Harm Exception analysis for the hold. Outside counsel would review both documents before the ONC response was submitted.

HIPAA and FHIR: The Privacy Rule in the API Economy

HIPAA's Privacy Rule and Security Rule did not disappear when FHIR arrived. FHIR is a data format and exchange protocol; the data it carries, when it includes individually identifiable health information, remains protected health information subject to every Privacy Rule and Security Rule obligation. It does not change what that information may be exchanged for, with whom, or under what conditions.

The Privacy Rule's core obligation for API-based exchanges is the authorization and minimum necessary framework. PHI may be disclosed through a FHIR API only for a permitted purpose and only to the extent necessary for that purpose. An API endpoint that returns a complete patient record in response to a query for medication allergies is not minimum-necessary compliant, even if the query is authorized.

Sandra had incorporated minimum necessary analysis into her pre-launch review checklist long before FHIR was a factor at Riverside General. Adapting it to FHIR required translating the analysis into the language of resource types and authorization scopes. Sandra's review confirmed that the declared scopes matched the app's stated purpose and that the EHR's authorization server restricted the app to those scopes rather than granting broader access by default.

Diagram 8.5: HIPAA Privacy Rule and FHIR API Compliance

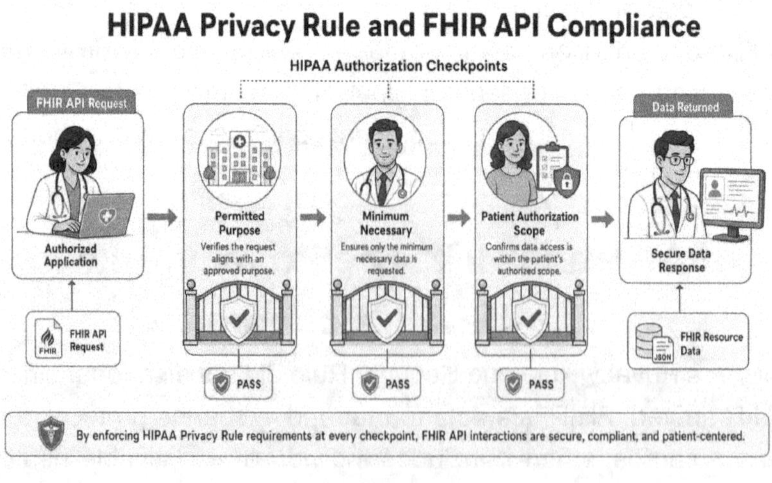

The Security Rule's technical safeguard requirements translate to specific implementation obligations: TLS 1.2 or higher for every API communication carrying PHI; access controls ensuring that unauthenticated requests cannot retrieve patient data; audit logging for every access event capturing user or app identity, patient identity, timestamp, data elements accessed, and purpose of access; and log integrity controls preventing audit records from being altered or deleted.

Dev had implemented these controls, but Sandra's Security Rule review process required documented evidence of configuration: screenshots confirming the TLS certificate configuration and audit log fields. It tested access control validation confirming that an unauthenticated request to the FHIR patient endpoint returned a 401 response. Quarterly audit log reviews verified that logs were generated, retained, and actively examined by a designated individual.

The audit log review requirement consistently surprised developers new to HIPAA compliance. HIPAA requires not only that logs exist but also that they be reviewed. An organization generating thousands of access log entries per day without a defined review process for anomalies and unauthorized access attempts has the logging infrastructure HIPAA requires but has not satisfied the Security Rule's underlying protective purpose.

Patrick had once asked Sandra whether the compliance overhead was really necessary for a technical data exchange standard. Sandra answered with the story of a hospital that had launched an elegant FHIR patient access API and discovered six months later that the API was returning data for any patient whose identifier was guessed, with no authentication required, because authentication middleware had been misconfigured during a server migration. HIPAA audit log reviews would have surfaced the anomaly within days. No one had been reviewing them. "The audit log review is not overhead," Sandra told Patrick. "It is the early warning system."

The CMS Interoperability and Patient Access Rules

The Centers for Medicare and Medicaid Services published its Interoperability and Patient Access Final Rule (CMS-9115-F) in May 2020. Where ONC's rule focused on health IT developers and their certification obligations, CMS-9115-F focused on payers: Medicare Advantage organizations, state Medicaid programs, CHIP programs, and qualified health plans in the federally facilitated marketplaces. The rule imposed FHIR API requirements on the payer side of the healthcare ecosystem for the first time, extending the interoperability mandate to the claims and coverage data held by health insurance organizations.

The Patient Access API requirement, effective July 2021 for most covered payers, required payers to implement a FHIR R4 API that members could use to access their own claims data, clinical data, and formulary and coverage information through authorized third-party apps. The API had to conform to US Core and to the CARIN Alliance Blue Button 2.0 Implementation Guide. Payers were also required to implement a Provider Directory API, making their network provider and facility information available in FHIR format to support consumer plan selection and care coordination.

Diagram 8.6: CMS Interoperability Rules: Payer API Requirements

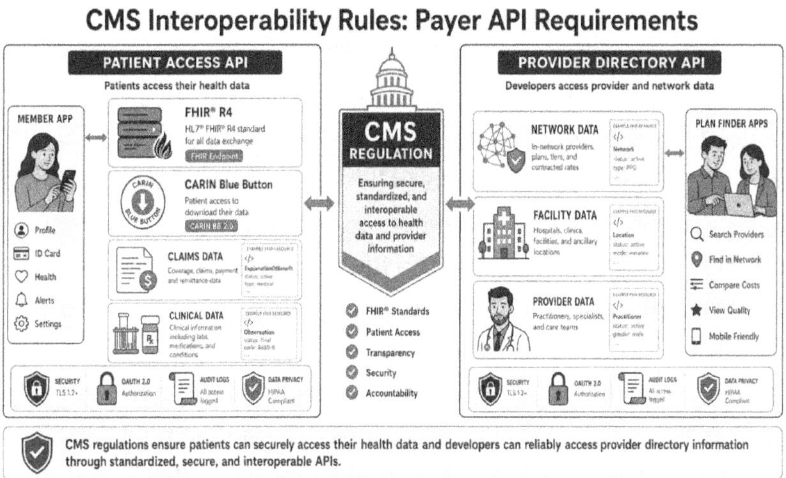

The patient is the constant. Every specification, regulation, architecture choice, and governance framework in this chapter and this book exists to serve a patient who deserves accurate, timely, accessible health data. Mrs. Rivera is not an exceptional case study. She stands for every patient whose care depends on information flowing accurately and accessibly across the systems that hold it. Organizations that keep her at the center of every interoperability decision will build systems that are not only technically conformant but truly worth building.

You understand that HL7 v2 is not a failure but a foundation that has carried healthcare data reliably for decades and will continue to do so in many hospitals. You understand that FHIR is not a destination but a direction: it creates the conditions for interoperability without guaranteeing it and requires continued investment to fulfill its promise. You understand that clinical terminology is not overhead but the semantic layer that makes technical connectivity meaningful, and that a FHIR Observation with a proprietary code is not truly interoperable.

You understand that regulations exist because patients were harmed when these behaviors were absent. That compliance with the Information Blocking Rule and the HIPAA Security Rule is not a burden on innovation but a prerequisite for trust. You understand that security in a FHIR environment is not a setting but an ongoing practice, and that AuditEvent, OAuth scopes, and the FHIR Consent resource are the concrete tools that protect or expose patient privacy with every API call.

Dev and Patrick created a four-part governance structure: a monthly migration leadership meeting as the main decision forum (with all four leads, a standing agenda, and a decision log), a quarterly steering committee review for escalations and major scope or schedule changes (including the CEO, CFO, and board technology committee chair), a weekly technical coordination meeting for Dev's operational work, and a daily 15-minute stand-up during hypercare. Formal change control applied to scope increases of more than five routes, schedule shifts of more than two weeks for any compliance-linked milestone, or budget variances above 10 percent in any phase; any of these required steering committee approval. The board chair reviewed and approved these thresholds before the project began.

You understand that regulations exist because patients were harmed when these behaviors were absent. Compliance with the Information Blocking Rule and the HIPAA Security Rule is not a burden on innovation but a prerequisite for trust. You understand that security in a FHIR environment is not a setting but an ongoing practice, and that AuditEvent, OAuth scopes, and the FHIR Consent resource are the concrete tools that protect or expose patient privacy with every API call.

USCDI: The Minimum Data Set for Interoperability

The United States Core Data for Interoperability is the federally maintained minimum set of health data classes and elements that must be accessible through ONC-certified health IT APIs. ONC publishes versioned releases with defined effective dates. USCDI v1, published with the 2020 Final Rule, covered clinical notes, diagnostic imaging, immunizations, laboratory data, medications, patient demographics, problems, procedures, and vital signs. Subsequent versions have expanded the set: USCDI v3 added health status assessments, sexual orientation and gender identity, and care team member attributes; USCDI v4 expanded laboratory data requirements. Each new version creates a corresponding update obligation for certified health IT developers and, in turn, for healthcare organizations dependent on their certified EHRs' API capabilities.

Sandra tracked USCDI version updates as part of her regulatory monitoring program, because each new version could create new API access obligations for Riverside General. The version-to-certification-to-enforcement timeline was not always intuitive, and Sandra had learned from a prior close call that "the certification requirement isn't effective yet" was not a safe assumption without confirming the effective date for the element at issue.

Diagram 8.7: USCDI Version Evolution and FHIR API Scope

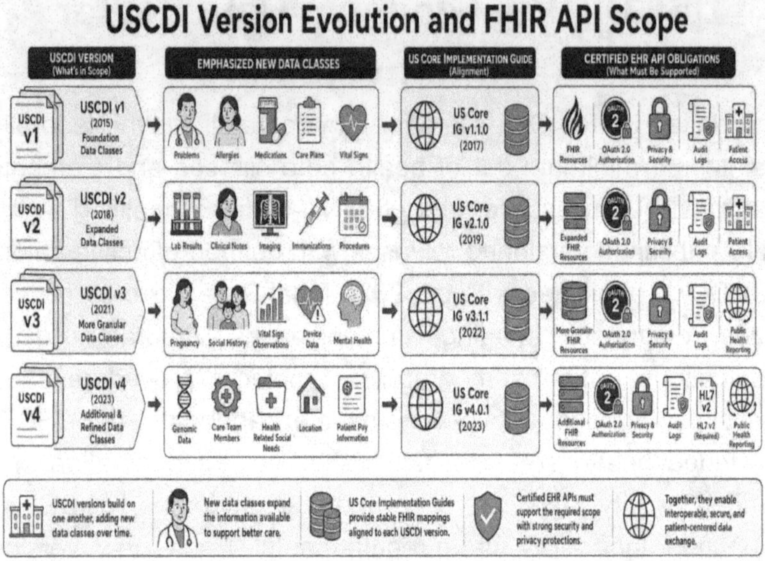

The USCDI also defines the scope of information blocking obligations for healthcare providers. As USCDI expands, the scope of information that providers must not restrict without an applicable exception broadens accordingly. Reaching compliance with the current USCDI version was a milestone, not a finish line.

Patrick had once described USCDI updates as "the regulatory treadmill." Sandra had responded that the treadmill was moving at a pace manageable with a continuous monitoring program, and that the organizations that fell off were the ones that had treated the 2020 compliance deadline as the last compliance event they would ever need to manage. "We're not those organizations," Sandra said. "Thanks to you," Patrick replied.

The Global FHIR Regulatory Landscape

FHIR adoption is not confined to the United States. The same technical standard ONC has mandated for US certification underpins national health data exchange programs in the United Kingdom, Australia, Canada, and across the European Health Data Space — each with regional regulatory variation in profile requirements, terminology bindings, and licensing rules. For Riverside General, those differences are context rather than a current project. Chapter 14 walks through the comparative international picture — NHS England, EHDS, AU Core, and CA Core+ — in operational depth. What Sandra's arc requires is the underlying point.

Diagram 8.8: Regional FHIR Regulatory Variation Overview

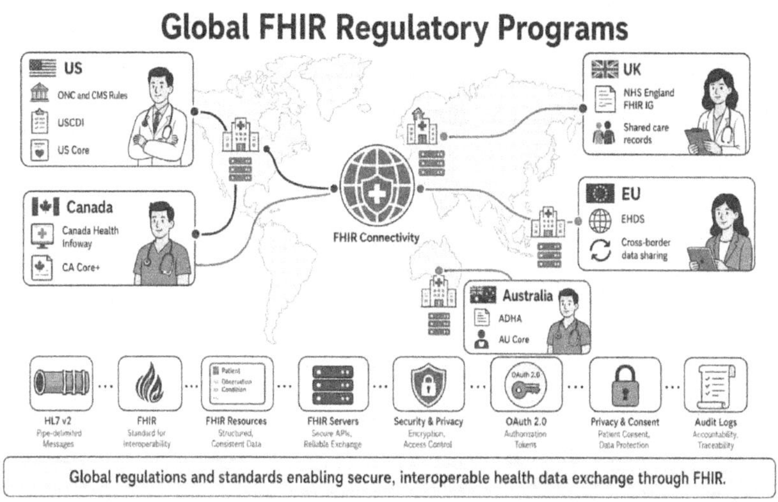

Interoperability by mandate is not an American invention — it is a convergent solution to a universal problem, and every major developed health system reached the same conclusion at roughly the same time. Sandra uses that framing when briefing new members of the governance committee: the ONC rule is not a US peculiarity but the US instance of a global pattern. The regional details belong in Chapter 14.

Sandra's Pre-Launch Compliance Review

Three weeks after the ONC complaint letter arrived, Sandra formalized the pre-launch compliance review process she had been running informally for two years. As Riverside General's FHIR-based capabilities expanded, the surface area for potential information blocking and HIPAA exposure was growing faster than an informal process could keep up with. The formalized process had five phases, each with defined inputs, outputs, and sign-off requirements.

The first phase was access scope verification: Sandra identified the specific EHI data classes involved in any new feature, confirmed the feature did not restrict access without a documented exception, and verified that any restriction fell within one of the eight exceptions with written analysis. A one-paragraph memo documenting the clinical rationale for a lab result hold time, citing 45 CFR 171.204 and the conditions for the Preventing Harm Exception, was sufficient. An undocumented configuration setting was not.

The second phase was HIPAA authorization basis confirmation. Sandra confirmed the legal basis for every data disclosure: patient authorization through SMART on FHIR, treatment purpose, payment purpose, or healthcare operations purpose. For patient-authorized disclosures, she reviewed the declared authorization scope against the data the feature would disclose and confirmed the minimum necessary standard applied.

Diagram 8.9: Sandra's Pre-Launch Compliance Review Framework

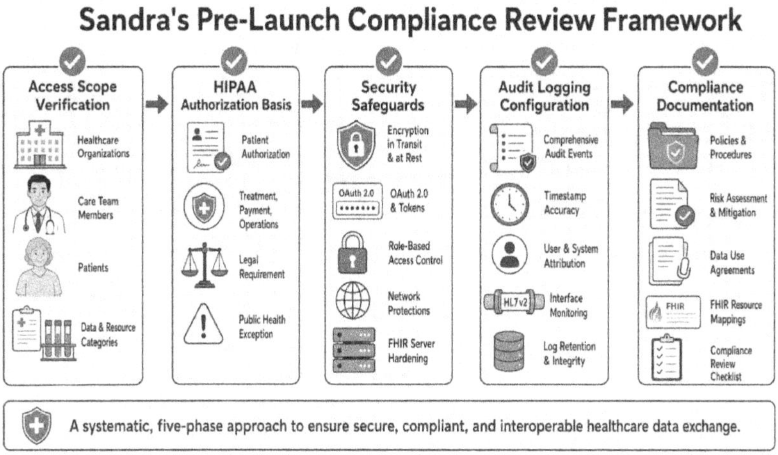

The third phase was security safeguard verification. Sandra confirmed from Dev's security control checklist that TLS encryption was configured for the new endpoint, authentication and authorization controls were active, access controls restricted the endpoint to authorized users and apps, and the audit logging configuration captured the required fields for any new data elements or user/app categories introduced by the feature.

The fourth phase was audit log configuration review. Sandra confirmed that the new feature's access events would be captured in the audit log with all required fields (user or app identity, patient identity, timestamp, data elements accessed, purpose), that logs were retained for the required HIPAA minimum period, and that the logs were within scope for the hospital's quarterly review process.

The fifth phase was compliance documentation: a review record for every feature, documenting the date, data classes assessed, exception analyses performed, HIPAA authorization basis, security controls verified, and issues remediated. The record was maintained in a dedicated compliance folder within the hospital's document management system and was readily retrievable in response to an ONC or Office for Civil Rights inquiry.

The formalized process added, on average, two to three days to the launch timeline for any new data access feature. Patrick accepted this without resistance after the ONC letter. Dev came to appreciate it as a quality control layer that caught configuration problems before they became production incidents.

The ONC complaint investigation concluded eight weeks after Sandra submitted Riverside General's response. The Preventing Harm Exception analysis for the 72-hour lab result hold, supported by incident data compiled by Dev and reviewed by outside counsel, was found sufficient to establish the exception. The SMART scope restriction had been remediated before the response was submitted. ONC closed the investigation with a determination of no information blocking, a finding Sandra filed in the compliance record with the same care she would have given to an adverse finding.

"The outcome doesn't change the process," she told Dev when she shared the closing letter. "The next complaint goes through the same review. The next feature launch goes through the same pre-launch check. The exception documentation framework we built for this one becomes the template for the next configuration that needs one."

Dev printed the ONC closing letter and taped it above his workstation, not as a trophy but as a reminder of the distance between a technical decision made without regulatory context and a regulatory response built on documented analysis. That distance was navigable, and bridging it required Sandra's review, Dev's technical transparency, and Patrick's willingness to accept that compliance is not overhead.

Enforcement, Complaint Intake, and the Disincentives Program

The information blocking prohibition has teeth. Enforcement mechanisms, operational since the first formal information-blocking determinations in 2022, are structured to impose consequences at multiple levels of severity.

Health IT developers and health information networks face financial penalties of up to $1 million per violation. The penalty is assessed per violation, not per incident, so a systematic practice of information blocking across multiple products or customer relationships can generate multiple violations with compounding consequences. ONC weighs the nature and extent of the information blocking, the harm to patients and providers, the financial benefit to the actor, and whether the actor took prompt corrective action.

Healthcare providers face a distinct enforcement mechanism: the Disincentives Program, established by CMS to create programmatic consequences for provider information blocking. The Disincentives Program links information-blocking determinations to providers' participation in and payment for federal quality programs, thereby affecting value-based care incentive payments and quality reporting outcomes. For a hospital like Riverside General that participates in multiple CMS quality programs, an adverse information blocking determination is a material financial event even without a direct per-violation penalty.

Both mechanisms share a complaint intake process managed by ONC. Any person can file an information blocking complaint through ONC's complaint portal. Complaints are reviewed for jurisdictional sufficiency and then investigated through requests for documentation, interviews, and technical review of the systems and configurations at issue. The complaint that Riverside General received was anonymous, which is explicitly permitted under the intake rules. ONC does not require complainants to identify themselves, a design choice reflecting a judgment about the power imbalance between individual patients or small third-party developers and the large organizations against whom complaints are most often filed.

Sandra's post-investigation analysis identified the most likely source: a patient who had waited 72 hours for an abnormal lab result and, during that wait, had filed the complaint directly with ONC without ever calling the hospital. This sequence had a practical implication for the hospital's patient-facing communications: patients experiencing access delays, even delays within a valid exception framework, needed to understand why the delay existed and how long it would last. An unexplained delay, visible through the portal's "results pending" status indicator, was an invitation to file a complaint. A portal notification documenting the clinical review process and the expected timeline was a manageable transparency measure.

Patrick assigned the portal notification improvement to Dev's sprint backlog and added a standing agenda item to the patient experience committee for portal access transparency. Sandra documented the patient communication design improvement in the compliance record as a risk mitigation measure against future information-blocking complaints.

Designing for Compliance from the First Line of Code

The regulatory framework in this chapter is relevant to every developer who writes code that touches patient data, every system architect who designs an API, and every administrator who approves a configuration. Information blocking liability follows the configuration decision, not the intent behind it. HIPAA Security Rule liability follows the access log that was never reviewed. The gap between technical implementation and regulatory requirement is almost always a documentation and process gap rather than a technical capability gap.

Dev had come to understand this through experience. Riverside General's FHIR server was technically capable of everything the regulations required. The information blocking complaint was not about technical capability. It was about a configuration choice and a documentation failure. The SMART scope restriction was a separate configuration choice, also technically reversible, made without regulatory analysis. Neither issue required rebuilding the platform; both required building the process around it.

The principle Sandra had articulated to every new member of Dev's team was simple in statement and demanding in practice: compliance begins in the design review, not in the post-launch audit. A data access feature designed with HIPAA minimum necessary in mind, with exception documentation for any access restriction, with audit logging in the design specification, and with a defined authorization basis for every data disclosure is no harder to build than a feature designed without those considerations and is substantially cheaper to defend.

The regulatory environment for FHIR and health data interoperability is not static. New ONC certification criteria, new CMS interoperability rules, new versions of USCDI, and new information blocking enforcement guidance are published on a rolling basis. Organizations that treat compliance as a one-time achievement to be maintained without active monitoring will find themselves out of compliance sooner than they expect. The organizations that fare best in this environment run compliance as a continuous program, not a periodic project.

Patrick had learned this lesson through the ONC complaint and the two months of internal disruption that followed. His instruction to Sandra after the closing letter arrived was simple: "Whatever budget you need to run the monitoring program, it's approved." Sandra wrote the budget request the next morning: a regulatory monitoring subscription service, a half-time compliance analyst position to support the pre-launch review program, and a training program for Dev's development team covering information blocking, HIPAA Security Rule technical safeguards, and FHIR API compliance design. The total cost was less than the outside counsel fees incurred in the ONC complaint investigation and considerably less than the cost of an adverse information-blocking determination.

The compliance program Sandra built was the organizational immune system that kept a technically capable, clinically well-intentioned hospital from becoming a cautionary example in someone else's training materials. Every developer who understood the eight information-blocking exceptions, every configuration choice accompanied by written exception documentation, and every audit log review that surfaced an access anomaly before it became a breach was evidence that the investment had been worth it.

The regulations will continue to evolve. New versions of USCDI will expand the scope of electronic health information subject to information blocking obligations. New CMS rules will extend interoperability requirements to new participants in the healthcare payment system. Riverside General's compliance program, built on Sandra's pre-launch review process and the organizational learning that followed the ONC complaint, is designed to evolve with that landscape rather than lag behind it.

The letter from ONC was not the beginning of Riverside General's regulatory exposure. It was the moment when the exposure became visible. Sandra's work in the months that followed transformed that visibility into capability: the capability to build, operate, and defend a health data access program that served patients, supported providers, and met the obligations Congress, ONC, and CMS had created to ensure that the promise of health data interoperability was not defeated by the institutions responsible for delivering it.

9 Data Exchange Patterns

The Whiteboard Dev Drew First

The morning meeting had run long. Patrick Osei stood at the fourth-floor conference room window watching an ambulance round the corner toward the emergency bay, its lights cutting through November drizzle. Behind him, Dev Okafor had already uncapped a dry-erase marker and turned to the whiteboard before anyone else found a seat.

"Five boxes," Dev said, drawing them with the practiced confidence of someone who had made this diagram in a dozen different rooms. He labeled them as REST, Messaging, Documents, Subscriptions, and Bulk Data. "These are the five ways FHIR moves data. They are not interchangeable. Pick the wrong one, and you will spend six months fixing it in production."

Patrick turned from the window and studied the whiteboard. He had sat through enough vendor demonstrations over the past year to distrust the phrase "fully FHIR-compliant" with the same skepticism he applied to "fully integrated" in any EHR sales presentation. But the whiteboard in front of him was different from a vendor slide. Dev was not selling anything. He was organizing a problem.

"Which one do we need for our problem?" Patrick asked.

"That depends entirely on what the problem is. That is the point of this whole conversation."

Dr. Maya Chen leaned forward from her seat to Patrick's left. "Can you give me the version a clinician can follow? I understand why a lab result needs to be fast and why sending a million patient records is a different situation. But the mechanics in between are blurry."

"That," Dev said, picking up the marker and underlining each box in turn, "is exactly what we are going to walk through."

Why Exchange Pattern Choice Is Architecture

Maya's question for this section is: Why does it matter which technical method we use to send data if the data arrives correctly either way?

The answer lives in the gap between a working test environment and a working production system. In a test environment, differences between exchange patterns are mostly invisible. A small patient population, a cooperative vendor sandbox, and a team that controls the timing and volume of interactions will produce successful results with almost any approach that technically moves data from one system to another. Production is different.

In production, Riverside General generates tens of thousands of clinical events per day. A potassium value of 6.8 mEq/L must be communicated to the ordering physician within a tight, clinically defined window. Admission, discharge, and transfer events fire at all hours and must reach every downstream system that depends on patient movement. At the same time, the quality reporting team submits annual data extracts to CMS covering hundreds of thousands of encounters, and the population health analytics program requires quarterly snapshots of the entire active patient panel.

No single exchange pattern handles all of these use cases well. A pull-based REST pattern, ideal for on-demand record retrieval, becomes an architectural liability when pressed into real-time alerting: the alerting system must poll the server continuously, generating hundreds of requests per minute, most of which return empty results, thereby imposing load on the server and introducing latency measured in polling intervals rather than in seconds. A topic-based Subscription pattern that handles real-time alerting elegantly becomes absurd for exporting five years of patient observations: the Subscription fires once per matching event, which would produce millions of individual notification deliveries over a multi-year dataset rather than a single coordinated export.

The principle Dev was communicating is simple to state and surprisingly difficult to internalize: exchange patterns are tools, and the right tool depends on the job. Understanding why each pattern was designed and what its failure modes look like when applied to the wrong use case separates architects who design integrations that work in production from those who require emergency remediation after go-live.

Diagram 9.1: Five Patterns, One Decision: Choosing How FHIR Data Moves

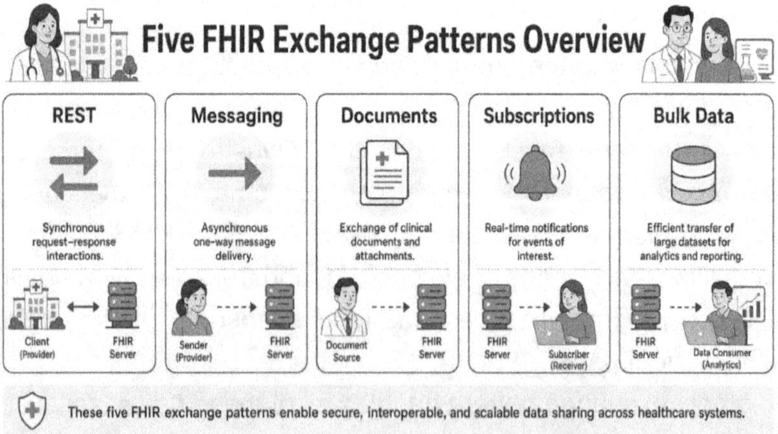

The decision to use a particular exchange pattern is made at the design layer, not at the implementation layer. An organization that defers pattern selection to the implementation team, letting

each developer choose the pattern they know best rather than the pattern the use case requires, ends up with an incoherent integration architecture at scale. The architecture review Dev conducts before any new integration work begins at Riverside starts with a single question: which pattern does this use case require?

RESTful FHIR: On-Demand and Client-Controlled

Maya's question for this section is: When I open a patient's chart, how does the system determine which data to show me, and is that the same as when a third-party app pulls a patient's records?

RESTful FHIR is the exchange pattern that most developers encounter first, and most vendors implement first. Its dominance in the FHIR ecosystem is not accidental: REST is the architectural style underpinning the modern web, and the engineers who built the first FHIR implementations were web engineers who naturally gravitated toward the familiar request-response model enabled by HTTP. A FHIR REST interaction is an HTTP request to a well-defined URL governed by a small set of verbs (GET, POST, PUT, DELETE, PATCH), returning a FHIR resource or Bundle in JSON or XML.

When Maya opens a patient chart and the EHR displays the current medication list, a REST interaction occurs: the EHR client sends an HTTP GET request to the FHIR server to retrieve all active MedicationRequest resources for that patient. The server processes the query, assembles a Bundle containing matching resources, and returns it. The interaction is synchronous: the client waits for the server to respond before displaying anything. Each tab navigation fires an independent, bounded REST interaction.

The same fundamental mechanics are at work when a third-party patient-facing application, authorized through SMART on FHIR, retrieves a patient's records. The authorization layer differs (the patient's own access token rather than a provider session token), and the FHIR profiles may vary depending on whether the endpoint is a provider-facing EHR API or a Patient Access API. But the HTTP GET request, the FHIR server query, the Bundle response, and the client-side rendering are structurally identical.

FHIR's RESTful interaction model defines a standard vocabulary of operations any compliant server must support. The basic CRUD operations form the foundation: POST creates a new resource; GET with a resource ID reads a specific resource; PUT updates a resource; DELETE removes it. Above the CRUD level, FHIR defines search operations allowing clients to query for resources matching complex criteria: all Observations for a patient with a specific LOINC code, all Encounters for a given date range, all active MedicationRequests for a specific care team. A Bundle transaction groups multiple resource creates or updates into a single atomic HTTP request. The $everything operation retrieves all resources associated with a patient in a single Bundle, useful for care transitions but unwieldy for patients with long medical histories.

Diagram 9.2: What Travels Over the Wire When Dev Calls a FHIR Endpoint — Headers, Status Codes, and the OperationOutcome the Server Returns When Something Is Wrong

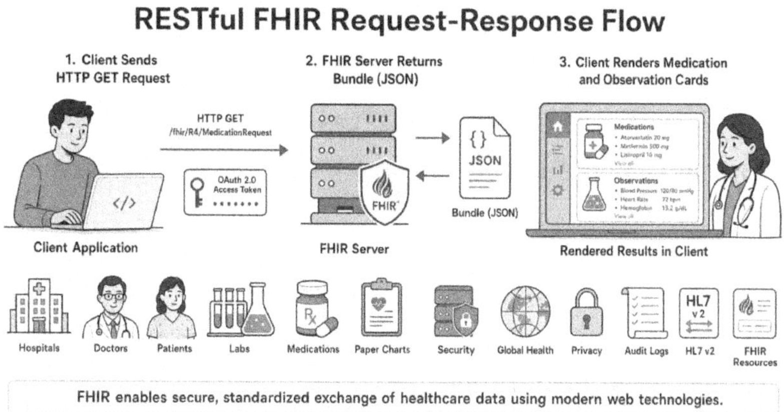

REST's failure mode, when applied to the wrong use case, is most visible in polling-based alerting implementations. An organization without Subscription infrastructure that wants real-time critical lab value alerting will reach for REST as a substitute: the alerting system polls the FHIR server every 30 seconds to request new critical Observations. Three alerting systems, each polling every 30 seconds, generate more than 8,600 requests per day, most of which return empty results. The server processes every one.

REST is a synchronous, client-controlled, on-demand pattern. Those three words define the boundaries of its appropriate use. When the consuming application controls the timing of its data requests, when the data is needed at a specific moment rather than when an event fires, and when the application can wait for the server to respond before proceeding, REST is the right choice. When any of those conditions do not hold, a different pattern deserves serious evaluation.

FHIR Messaging: The Event-Driven Pattern

Maya's question for this section is: When a patient is admitted to our hospital, how does every system that needs to know about the admission find out?

The question Maya is asking is one that HL7 version 2 was built to answer thirty years before FHIR existed. An ADT A01 message fires from the admitting system the moment a patient is registered and pushes notification of that admission to every downstream system subscribed to that message type: the pharmacy, the care team communication platform, the clinical decision support tool, the billing system. The sending system pushes the notification the moment the event occurs; the receiving system need not ask. This event-driven, push-based model has been the architectural backbone of hospital operations for decades.

FHIR Messaging brings that event-driven model into the FHIR-native world. A FHIR message is a Bundle of type "message" whose first entry is a MessageHeader resource that carries the event code, the identities of the sending and receiving systems, a unique message identifier, a timestamp, and a response element indicating whether this is an original notification or a reply. The remaining Bundle entries are the clinical resources relevant to the event: the Patient resource, the Encounter resource, the Location resource, and any other clinical resources the receiving system needs to process the event.

At Riverside General, Dev's team maintains FHIR Messaging for three specific use cases. The first is the admission, discharge, and transfer workflow: when a patient is admitted, transferred between units, or discharged, an FHIR message is sent immediately to all downstream systems with a workflow dependency on that event. The pharmacy needs the patient's current location to route medication delivery. The care team communication platform needs the care team assignment for the new encounter. None of these systems can afford to wait thirty seconds for a polling interval to expire.

The second use case is referral management across organizational boundaries. When a Riverside physician generates a referral to an external specialist, the referral workflow produces a FHIR message containing the MessageHeader, the Patient resource, the ServiceRequest representing the referral, and the relevant clinical resources. The receiving specialist's system processes the message and sends a response message confirming receipt.

The third use case is the legacy HL7 v2 migration path. Riverside inherited downstream systems designed to receive HL7 v2 messages through workflow handlers deeply embedded in their application logic. In the interim before those systems are replaced, the integration engine translates HL7 v2 ADT and ORU messages into equivalent FHIR messages, allowing upstream systems to migrate to FHIR-native event production while the downstream modernization program proceeds at its own pace.

Diagram 9.3: How a FHIR Message Bundle Differs From a REST Call — Why the MessageHeader Defines Routing When There Is No Endpoint to Query

FHIR Messaging Bundle Structure

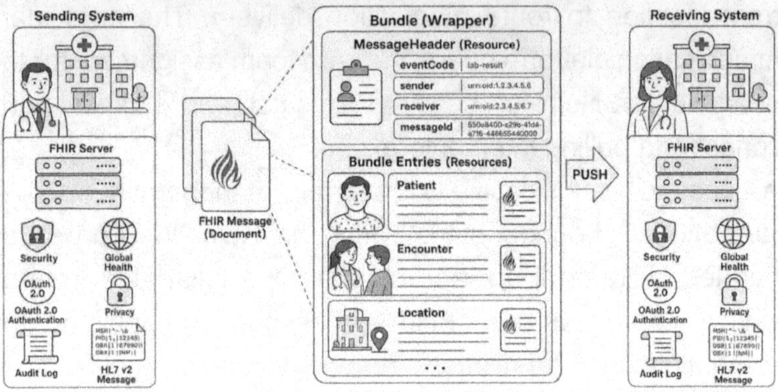

A FHIR message is wrapped in a Bundle, with a **MessageHeader** that provides context and routing, followed by related FHIR resources delivered securely from the sending to the receiving system.

FHIR Messaging transfers responsibility for the timing of data delivery from the consuming system to the producing system. In a REST model, the consuming system decides when to ask for data. In a Messaging model, the producing system pushes data when an event occurs, and the consuming system must be ready to receive it. This requires message queuing, delivery-retry logic, dead-letter handling, and monitoring dashboards that surface when a downstream system stops receiving messages it should be receiving. When a receiving system processes a FHIR message, it sends a response Bundle containing a MessageHeader referencing the original message ID and a status indicating success or failure, a model directly analogous to the MSA segment in HL7 v2 ACK messages.

FHIR Documents: Snapshots Across Boundaries

Maya's question for this section is: When one of our patients sees a specialist outside Riverside, how does that specialist know what medications the patient is on and what conditions we have documented?

The clinical document exchange problem predates digital health records. Long before HL7 or FHIR, clinicians solved it with paper: a discharge summary mailed to the referring physician, a consultation letter filed in the patient's chart. The paper document had one property that proved architecturally important even in the digital era: it was complete and self-contained when it was created.

FHIR Documents preserve this property in a digital format. A FHIR Document is a Bundle of type "document" whose first entry is a Composition resource serving as the document's table of contents and metadata: the document type, the subject patient, the author and attesting clinician, the date of composition, the document sections, and references to the entries that populate each section. Every resource referenced by the Composition (every Condition, every MedicationStatement, every AllergyIntolerance, every relevant Observation) appears as an additional Bundle entry, making the document self-contained: a receiving system processing the Bundle needs no additional REST calls to retrieve referenced resources.

This self-containment is architecturally meaningful for cross-organizational exchange. When Riverside sends a care transitions document to an external specialist, that specialist's system may not be connected to Riverside's FHIR server. Even if it does, a live query when the specialist opens the document might return different data than the document was intended to convey, because live FHIR resources reflect the current state of the record rather than the state at the time of composition. The FHIR Document pattern resolves this by making the document a point-in-time snapshot fixed at the time of composition, regardless of subsequent changes in the source system.

Diagram 9.4: What a FHIR Document Bundle Must Contain to Be Legally Persistent — and How the Composition Resource Ties Every Section to a Signed Whole

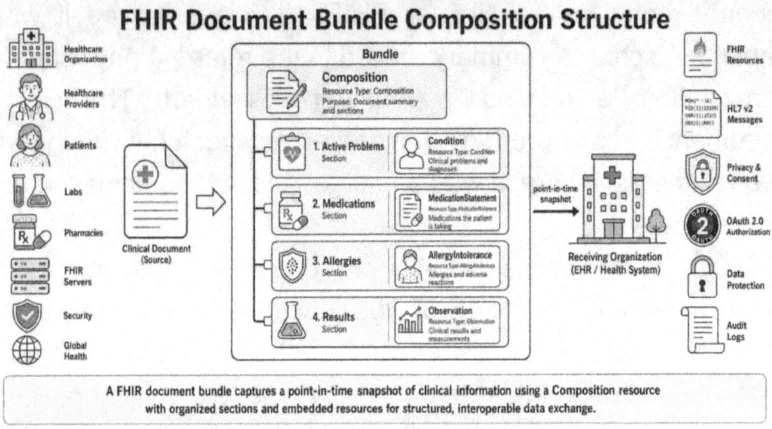

At Riverside, the FHIR Document pattern governs three workflow categories. First, the care transitions summary: when a patient is discharged, the discharge workflow automatically generates a FHIR Document with sections for active problems, current medications, reconciled allergies, pending results, and follow-up instructions. Second, the referral summary: when a physician generates an outbound referral, the referral workflow compiles an FHIR Document that captures the clinical context and transmits it as the clinical attachment. Third, the patient-requested summary: when a patient requests their medical record for a new provider, Riverside generates an FHIR Document in the International Patient Summary format, which is available through the patient portal or the Patient Access API.

The US Core Implementation Guide defines FHIR Document profiles that Riverside must support for regulatory compliance. The Continuity of Care Document profile maps the USCDI's clinical information requirements to FHIR Document sections and resource types, ensuring that another can correctly interpret a FHIR Document generated by one certified health IT system. A receiving specialist's system needs no custom parser for Riverside's proprietary format, because the document conforms to a published standard the specialist's system was built to understand. One operational consideration deserves attention: snapshots become stale. A medication list accurate at the time of discharge may be partially outdated by the time a receiving facility processes it. FHIR addresses this through standardized document provenance mechanisms (the Composition. date and Composition. attester elements) and a standardized approach to document replacement (a Composition can reference a prior Composition as a replacement), making the risk transparent and auditable rather than hidden.

FHIR Subscriptions: Replacing the Polling Loop

Maya's question for this section is: Instead of having our alerting system constantly check for new lab results, can the lab system tell us when something important comes in?

Maya's question is the founding motivation for the FHIR Subscription pattern. The polling loop she is describing is not a design choice; it is a workaround for the absence of a push notification mechanism. Every organization that has implemented clinical alerting on top of a REST-only FHIR endpoint has written some version of a polling loop, and every one has eventually had the same conversation with its infrastructure team about whether the polling frequency can be reduced because the polling load is impacting server performance.

A FHIR Subscription is a resource stored on the FHIR server representing a registered interest in a specific type of clinical event. In FHIR R4B and R5, Subscriptions operate through a topic-based model: the server defines Subscription Topics as formal specifications of the clinical events it can notify about (e.g., a new laboratory result, a patient admission, a high-priority medication order). A client application creates a Subscription resource that specifies the topic, filter criteria, delivery channel (a REST endpoint, WebSocket connection, or messaging channel), and the notification content format it expects.

When the server processes a clinical event matching a registered Subscription's topic and filter criteria, it assembles a notification and delivers it to the subscriber's specified channel. A "full resource" notification includes the complete FHIR resource that triggered the event. An "id-only" notification includes only the resource type and ID, with a follow-up REST call to retrieve the full resource. An "empty" notification indicates a matching event occurred without identifying the triggering resource.

The architectural difference between the polling loop and the Subscription model lies in the degree of system coupling. In a polling model, the consuming system repeatedly queries the producing system, often receiving no new data. In a Subscription model, the producing system notifies the consuming system when an event actually occurs, and the server load profile is proportional to the rate of actual clinical events rather than to the polling frequency of all subscribing systems.

Diagram 9.5: FHIR Subscription Notification Flow

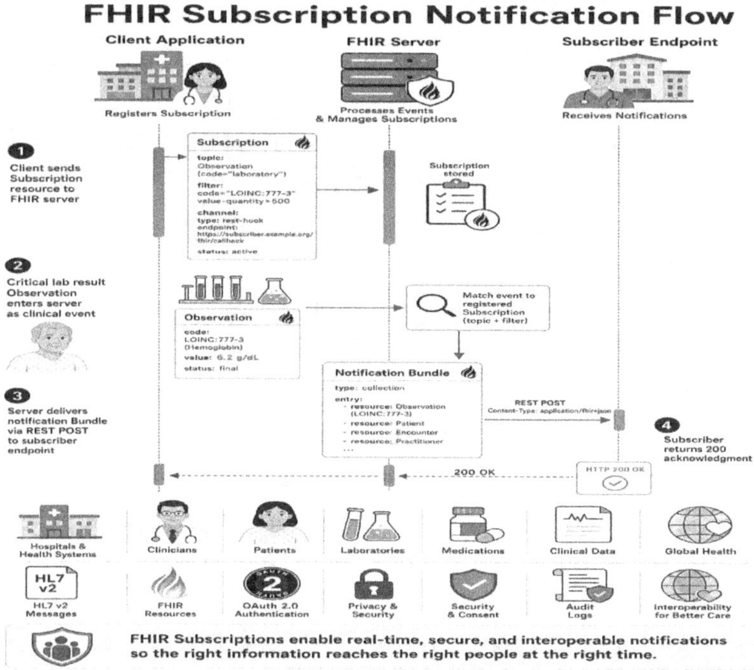

Dev's team at Riverside implemented the Subscription infrastructure in two phases. The first addressed critical lab value alerting, which had been running on a polling loop for eighteen months. That loop generated approximately 4,200 REST queries per day, 94 percent returning empty result sets. After the Subscription implementation, server query load from the alerting subsystems dropped by more than 99 percent, and latency between the return of a critical result and the notification reaching the ordering physician dropped from an average of 28 seconds to under 3 seconds.

The second phase extended Subscription-based notifications to the care management platform for high-risk ED admissions. Before implementation, care managers manually checked the ED census list every 30 minutes. After implementation, they received a Subscription notification within four seconds of a high-risk patient's ED registration, including encounter details and the patient's risk score.

The following pseudocode illustrates the notification handler Dev's team implemented for the critical lab alerting use case:

on_notification_received(bundle): for each entry in bundle.entries: resource = entry.resource if resource.type is "Observation": if has_critical_interpretation(resource): clinician = lookup_ordering_clinician(resource) deliver_alert(clinician, resource, priority = "STAT") log_audit_event(resource, clinician, "CRITICAL_ALERT_DELIVERED") else if resource.type is "Encounter": if matches_high_risk_flag(resource.subject): care_mgr = lookup_assigned_care_manager(resource.subject) notify_care_manager(care_mgr, resource, protocol = "HIGH_RISK_ADMISSION")

The Subscription filter capability is where the topic-based model's operational power becomes apparent. Without filters, a subscribing system receives a notification for every event of the registered type and must evaluate each for relevance. With filters, a care management platform subscribed to the patient-admitted topic receives notifications only for patients flagged with a specific care management status code, assigned to a specific care program, or above a defined risk threshold.

Bulk FHIR: Population Scale Without Disruption

Maya's question for this section is: When we need to send all of our diabetic patients' records to the quality reporting program, how do we do that without shutting down the EHR?

The concern embedded in Maya's question is a real operational risk. A FHIR REST endpoint responding to individual resource queries in 200 milliseconds will not respond to a query for all Observations for 18,000 patients in any acceptable timeframe: the server will time out, return a partial result indistinguishable from a complete one, or compete for resources with the clinical transactions that must continue running. The Bulk FHIR specification exists precisely because the REST pattern cannot scale to population-level data volumes without unacceptable operational trade-offs.

Bulk FHIR defines an asynchronous, server-side export mechanism. A bulk export request specifies the export type (system-level for all patients, group-level for a defined cohort, or patient-level for a single patient), the resource types to include, and an optional date range filter. The client receives a 202 Accepted response, not the data itself, and the server begins processing asynchronously. The client periodically checks a status endpoint; when the export completes, the server provides a manifest listing download URLs, and the client downloads those files at its own pace.

The output format is NDJSON (Newline-Delimited JSON): each line of a plain-text file is a complete JSON object representing a single FHIR resource. NDJSON suits large-scale data processing because it can be read line by line without loading the entire file into memory, split into chunks for parallel processing, and directly ingested by analytics tools and data pipeline frameworks.

Diagram 9.6: Bulk FHIR Asynchronous Export Workflow

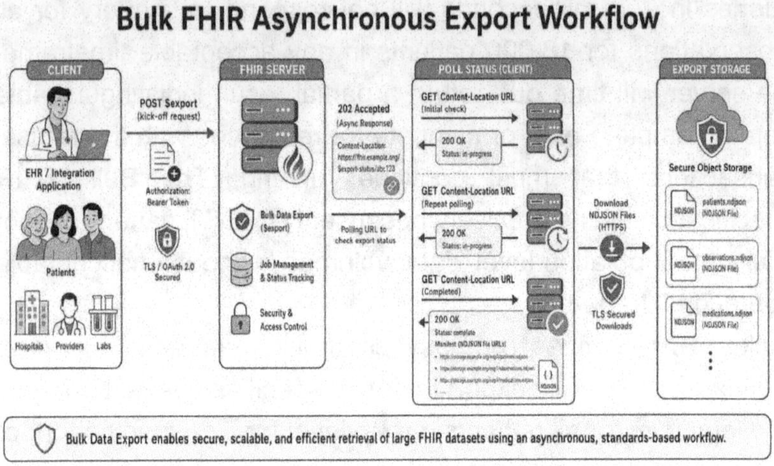

At Riverside General, the Bulk FHIR export infrastructure supports four operational programs. First, annual CMS quality reporting: the quality analytics team runs weekend-night exports that produce approximately 2.3 gigabytes of NDJSON files; running the same extraction via synchronous REST interactions was estimated to require 14 hours of continuous querying. Second, the population health analytics platform, which requires quarterly snapshots of the complete active patient panel. Third, the research data repository, receiving monthly extracts under an IRB-approved protocol. Fourth, the payer data exchange program under the CMS Payer-to-Payer API requirements, using Bulk FHIR as the transport for population-level data elements.

The following pseudocode illustrates the client-side bulk export workflow:

```
initiate_bulk_export(server, types, since, group_id): resp =
http_post(server + "/Group/" + group_id + "/$export", {_type:
types, _since: since}) assert resp.status == 202 poll_url =
resp.headers["Content-Location"] while true: s =
http_get(poll_url) if s.code == 200: manifest = s.body; break
```

else if s.code == 202: wait(s.headers["Retry-After"]) else: raise ExportError(s.body) for each url in manifest.output: download_and_process_ndjson(url)

Dev established three operational disciplines for Riverside's Bulk FHIR program. First, all exports run during defined maintenance windows. While the asynchronous design minimizes clinical system impact, a server processing a large export generates nonzero disk I/O, CPU load, and memory pressure. Second, every export is logged in the audit framework, including the requesting system, the patient population scope, the resource types exported, and the initiation and completion timestamps, thereby satisfying HIPAA's audit control requirement. Third, download URLs in the export manifest are treated as sensitive credentials and expire within 4 hours, as they provide unauthenticated access to exported files during their validity period.

Diagram 9.7: What Riverside's Bulk Export Actually Produces — How NDJSON Files Are Named, Separated by Resource Type, and Linked in the Manifest

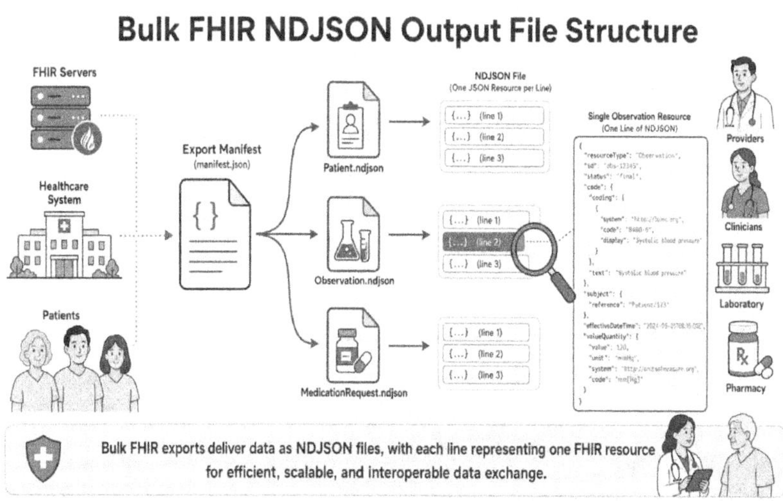

The Four-Pattern Decision Framework

Maya's question for this section is: When our team is planning a new integration, what is the first question we should ask to figure out which pattern to use?

Dev's answer to this question has not changed since he first drew the boxes on the whiteboard: three questions, in sequence, determine the pattern. The questions require honest answers about the use case rather than assumptions about what the vendor's platform prefers.

The first question is: Who controls the timing? If the consuming application decides when to request data (e.g., when a user opens a chart or a workflow step reaches a decision point), REST is the natural starting point. If the producing system controls the timing (a laboratory reports a result, a patient is registered, a medication is dispensed), an event-driven pattern is required: FHIR Messaging for workflow-continuity scenarios and migration-from-v2 contexts, Subscriptions when the consumer needs to register interest in a class of events and receive asynchronous notifications over time.

The second question is: How much data needs to move? A single resource or small Bundle is REST territory. A complete clinical document belongs to the Document pattern, where self-containment ensures the recipient has everything needed without follow-up queries. Thousands or hundreds of thousands of resources belong to Bulk FHIR, the only pattern that handles that volume without imposing unacceptable operational risk on the source system.

The third question is: Does the timing require real-time response, near-real-time notification, or batch processing? Real-time response (under one second) is REST's domain. Near-real-time notification (under ten seconds) is the domain of Subscriptions.

Batch processing (minutes to hours) is the domain of Bulk FHIR. FHIR Messaging occupies a middle position: push-based and event-driven like Subscriptions, but used for workflow-continuity scenarios where the acknowledgment model provides reliability guarantees that Subscription channels may not offer.

Applying these three questions to Riverside's integration backlog produces a decision mapping Dev keeps posted in the integration team's work area. On-demand chart data retrieval maps to REST: consumer-controlled, small volume, real-time. Critical lab value alerting maps to Subscriptions: event-driven, small notification payload, near-real-time. Patient admission notifications map to FHIR Messaging: event-driven, complete event payload, workflow-continuity requirement. Discharge summary transmission maps to FHIR Documents: point-in-time snapshot, complete self-contained record, no live connectivity assumption. Annual CMS quality measure extraction and quarterly population health analytics refresh both map to Bulk FHIR: batch, large volume, time-windowed.

Diagram 9.8: Three-Question Pattern Decision Framework

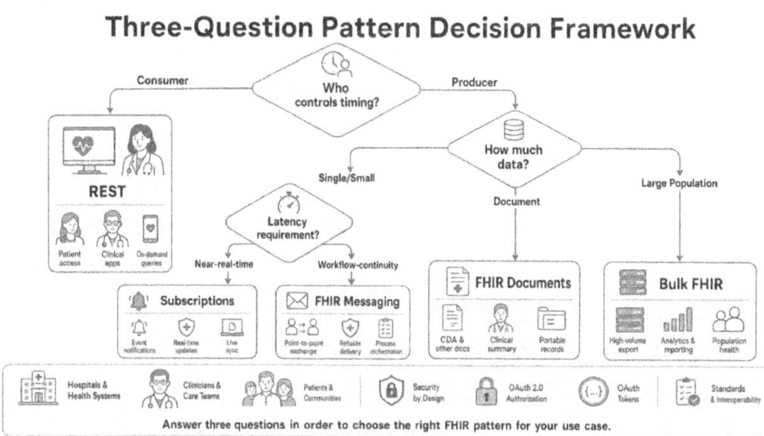

The framework does not produce surprising answers when applied to well-characterized use cases. Its value is making the architectural decision explicit, documented, and defensible, preventing the most common failure mode in FHIR integration planning: never asking which pattern is appropriate in the first place.

Combining Patterns: What Production Looks Like

Maya's question for this section is: Do most hospitals use just one of these patterns, or do they use all of them at the same time?

Nearly every mature FHIR implementation uses multiple patterns simultaneously. The architecture Dev has built is a routing infrastructure that directs each class of interaction to the appropriate pattern. The integration engine at the center dispatches each FHIR interaction to the appropriate pattern handler based on the interaction type, use case, and the agreed-upon integration contract with each partner.

The REST handler processes synchronous read requests from clinical applications, patient portal queries, and SMART-on-FHIR-authorized third-party apps, managing caching of frequently accessed reference data and enforcing the search parameter profiles that each application is authorized to use. The Messaging handler processes inbound HL7 v2 messages from legacy systems not yet migrated to FHIR-native event production, translating each message type to its FHIR equivalent and managing the acknowledgment lifecycle. The Subscription handler manages the registry of active Subscriptions, routes incoming clinical events to matching subscription filter evaluations, assembles and delivers notifications, and queues notifications for retry when a subscriber endpoint is unreachable, escalating to a manual exception queue if the endpoint remains unavailable beyond the maximum retry window. The Bulk FHIR handler manages the asynchronous export job queue, schedules exports within defined maintenance windows, enforces export scope authorization, and provides the operations dashboard through which Dev's team monitors active and completed exports.

Patrick came to fully understand the multi-pattern architecture through a retrospective on an incident that Dev ran six weeks after the initial pattern discussion. A care manager complained that she was not receiving timely notifications when high-risk patients were admitted to the emergency department. The investigation revealed that the ED admissions notification had been implemented using a polling loop against the REST endpoint because the developer assigned to it was more familiar with REST than with the Subscription infrastructure. The polling interval was five minutes: for a use case where near-real-time notification was a care-quality requirement, that interval represented a gap in patient care.

Dev ran the retrospective not to assign blame but to institutionalize the pattern decision process. From that point forward, every new integration at Riverside begins with a one-page Integration Design Brief that documents the answers to the three framework questions, the selected pattern, the alternatives considered, and the rationale for why the selected pattern best fits the use case. Patrick signed the first three Integration Design Briefs himself, not to approve the technical decision but because his signature communicated to the organization that pattern selection was a governance decision, not a developer preference.

Diagram 9.9: Multi-Pattern Integration Architecture at Riverside General

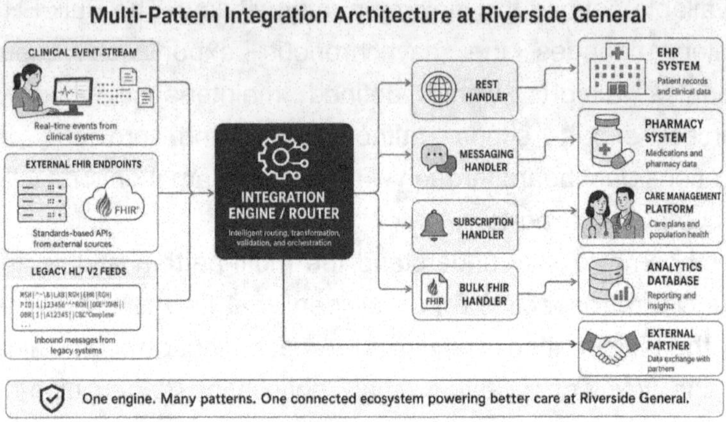

Pattern Documentation and Institutional Memory

Maya's question for this section is: Once we decide which pattern to use, how do we make sure the next team that touches the integration knows why we made that choice?

The answer is a documentation and governance discipline, and it is the most neglected aspect of FHIR integration architecture in teams under delivery pressure.

Pattern documentation serves three functions that become visible only when they are absent. The first is onboarding. A developer joining Riverside's integration team in the third year of the FHIR implementation inherits a system built by engineers who made dozens of architectural decisions not obvious from the code. Why does the critical lab alerting system use a Subscription rather than a polling loop? Because a five-minute polling interval caused a care quality gap, documented in the incident retrospective from two years ago. Without that documentation, the new developer might simplify the implementation back to a polling loop, recreating the problem the Subscription was designed to solve.

The second function is change management: the Integration Design Brief provides the baseline for evaluating any proposed change to an integration's pattern, turning a potentially uninformed vendor negotiation into a structured technical discussion with clear acceptance criteria. The third function is regulatory defensibility. HIPAA's audit control requirement (164.312(b)) is satisfied by demonstrating that the audit log design was intentional and covers the interactions that need to be audited. Sandra's compliance reviews focus as much on documentation of architectural decisions as on technical controls. She had drafted the audit-log section of the Integration Design Brief template herself, after a quarter of reviewing AuditEvent configurations that no one could explain. Her standing requirement was simple: every integration record must state, in one sentence, which AuditEvent subtypes are emitted, which actor and entity fields are populated, and what retention applies. "If we cannot say that on the day we deploy," she told Dev when they negotiated the template, "we will not be able to say it under deposition either." Dev had pushed back on the prose burden and lost; he later admitted the field had caught two integrations whose audit coverage looked complete in code but quietly omitted the patient identifier on read events.

Dev's Integration Design Brief template is a one-page document with six fields: the use case description in plain language, the answer to each of the three framework questions with a one-sentence justification, the selected pattern and alternatives considered, the integration partner and agreed contract, and the compliance and audit requirements applicable to this integration. The documentation discipline it enforces is organizational as much as technical. Patrick can read an Integration Design Brief and understand why a Subscription was chosen over a polling loop, but not the mechanics of a Subscription. Sandra can identify the compliance requirements that need to be validated. Maya can confirm that the use case description matches the clinical requirement she articulated. The technical decision and its implications become transparent to the entire governance team.

The integration architecture at Riverside General handles more than 14,000 FHIR interactions per day across all five patterns. The only significant architectural remediation in twenty-two months of production was the care manager notification incident that established the Integration Design Brief process. Every subsequent integration has been designed using the three-question framework, documented using the template, and reviewed by the governance committee before implementation begins.

Patrick's answer to the question he asked on the whiteboard day has not changed in its substance, only in its precision. The pattern they need for any given problem is the one that correctly answers who controls the timing, how much data needs to move, and how quickly it must arrive. The whiteboard is still in the conference room. The five boxes are still on it. Dev has never erased them.

Takeaways

Choosing the wrong exchange pattern is an architectural error that is expensive to reverse in production. A polling-based alerting implementation that generates thousands of empty REST queries per day; a REST-based export that competes with clinical transactions while attempting to extract a population-scale dataset; and a Document pattern applied to a real-time event notification use case: these are the specific failure modes that motivated each alternative pattern's design. The pattern vocabulary exists because the wrong pattern, applied at scale, creates real operational and clinical consequences.

REST is not a universal answer to FHIR data exchange. It is the right answer for a well-defined class of problems: synchronous, client-controlled, on-demand data access at manageable per-interaction volumes. Treating every FHIR integration as a REST integration is the most common architectural error in FHIR implementation projects, and it compounds with each polling-based workaround built to compensate for REST's limitations outside its design domain.

FHIR Subscriptions in R4B and R5 represent a significant maturation over the filter-based subscription model in earlier FHIR versions. The topic-based model is more expressive, more interoperable, and more operationally manageable at scale. Organizations planning real-time notification use cases with flexibility in their FHIR version selection should evaluate R4B or R5 Subscription capabilities as a strong argument for moving to the more recent version, particularly if the alerting use cases involve filter criteria more complex than the base FHIR R4 filter model supports.

Bulk FHIR is not optional for organizations with population health, quality reporting, research, or payer data exchange programs. The data volumes these programs require are genuinely incompatible with synchronous REST interactions, and organizations that discover this late will face an unplanned infrastructure project at a point in their implementation timeline when they have the least capacity for it. Planning for Bulk FHIR from the earliest stages of a FHIR program is the decision that prevents that crisis.

The integration engine is the infrastructure that makes multi-pattern architectures operationally manageable. An integration engine that can route REST interactions, handle Subscription notifications, process and generate FHIR Messages, package FHIR Documents, and initiate and monitor Bulk FHIR exports through a unified management and monitoring framework is the operational foundation that prevents the multi-pattern architecture from becoming a multi-tool maintenance problem. Pattern support capabilities should be a primary evaluation criterion in any integration engine selection process.

Pattern documentation is institutional memory, and institutional memory is the infrastructure that makes an integration architecture maintainable over a five-year horizon as teams change, vendors release new versions, and regulatory requirements evolve. The Integration Design Brief is a one-page document. The discipline of producing it for every integration, every time, before implementation begins prevents the organization from having to rebuild the same architectural understanding from scratch with every team transition. It is not a bureaucratic requirement; it is a professional practice, the same way that a structural engineer documents the load calculations for every beam in a building, not because anyone expects to inspect those calculations tomorrow, but because the building will stand for decades. The people who maintain it deserve to understand why it was built the way it was.

10 The Patient at the Center

The Call That Changed Patrick's Priorities

The message landed in Patrick Osei's inbox on a Tuesday morning, forwarded from the patient services line with a note that the coordinator did not know how to resolve it. Mrs. Elena Rivera, a sixty-three-year-old patient who had received care at Riverside General for nearly a decade, wanted to know why she could not see her own laboratory results in the health app her insurance company had recommended. Her insurer had told her that the hospital was required by law to share those results through a standardized electronic interface. Mrs. Rivera was methodical, informed, and accustomed to following up when she did not receive a satisfactory answer.

Patrick called her himself. He listened to her recount each step she had taken: downloading the app, creating an account, searching for Riverside General, finding it listed, tapping the connect button, and arriving at a page that said the connection was unavailable. She had called patient services, been directed to medical records, been told to use the patient portal, explained that she wanted the app rather than the portal, and been told the portal was the approved mechanism. She had called her insurer again, who confirmed that what the hospital told her was inconsistent with federal regulations.

By the time Patrick ended the call, he had made two commitments he did not yet know how to keep: fix Mrs. Rivera's problem before the end of the quarter, and never again be uninformed about patient access rights when a patient asked. He walked to the clinical wing to find Maya Chen, who had been pressing questions about patient-facing technology for months while Patrick was absorbed in the budget cycle. Maya was between meetings, reviewing annotated conference printouts. He told her about Mrs. Rivera. She slid a stapled document across the desk, the ONC information-blocking regulations covered in red-ink margin notes, and said, "I have been waiting for us to have this conversation."

What followed was a series of conversations across several weeks, as Maya and Patrick worked through the regulatory landscape, the technical architecture, and the operational implications of a patient access framework that Riverside General had installed but never fully configured. This chapter is the distillation of those conversations: what they needed to learn, what they discovered about the hospital's current state, and what they committed to changing.

Why Patient Access Is the Visible Face of FHIR

Patient access is the dimension of FHIR adoption that patients actually experience. Every other aspect of interoperability in this book operates in the background: provider directories synchronize without the patient watching, clinical documents move between systems inside administrative workflows, payer data exchange happens in pipelines the patient never sees. Patient access lands in a patient's hands, on their phone, in an application they chose themselves, and it either works or it does not. When it does not work, patients call. When patients call, and staff cannot explain why, the organization is failing to get right something it is legally required to get right.

The 21st Century Cures Act, signed in 2016 and implemented through ONC's Cures Act Final Rule in 2020, established enforceable rights regarding access to electronic health information. Patients have the right to access their health information through third-party applications of their choosing. Healthcare organizations have the obligation to enable that access through standardized FHIR-based APIs. Information blocking, defined as a practice that interferes with, prevents, or materially discourages access to electronic health information, became an offense with regulatory consequences. The gap between that regulatory framework and the operational reality Mrs. Rivera encountered is the most common FHIR compliance failure in the US healthcare system, caused almost always by incomplete implementation and inadequate staff education rather than deliberate obstruction.

Maya had understood the clinical stakes before she understood the regulatory ones. As chief medical officer, she reviewed patient engagement metrics and had noticed for years that Riverside General's portal adoption lagged industry benchmarks. Patients who did not engage with their health information between visits were less likely to adhere to treatment plans and more likely to present in the emergency department with conditions that better engagement might have caught. She had attributed the gap to poor portal design; it had not occurred to her that patients might be attempting to access records through other applications, failing, and giving up.

Patrick's concern was more immediate. If Mrs. Rivera's complaint represented a pattern, and both he and Maya quickly concluded it did, Riverside General was generating regulatory exposure with every failed app connection. The OIG had the authority to investigate information-blocking complaints and to impose civil monetary penalties. When Patrick pulled three months of call center logs, he found fourteen complaints matching Mrs. Rivera's description: patients who had tried to connect a third-party health app, received an error, and called for help. In every case, staff had directed the patient to the portal, and in no case had anyone investigated whether the failure reflected a problem with the FHIR infrastructure configuration.

Diagram 10.1: Patient Access Gap: Regulatory Promise vs. Operational Reality

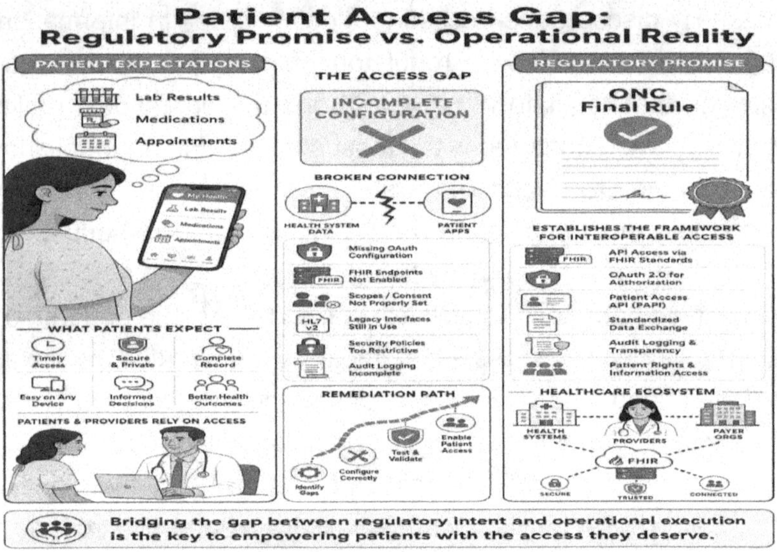

This is why patient access is the visible face of FHIR: its failures are experienced personally by patients, reported directly by patients, and enforceable by regulators at the individual patient's instigation. An administrator who does not understand the mechanics of patient access is not equipped to manage the consequences when it fails.

Mrs. Rivera's Rights: What the Law Guarantees

Maya's first task in educating Patrick was to establish what the law actually required. The 21st Century Cures Act does not simply say that patients should have access to their records; it specifies what records, through what mechanism, at what cost, and with what restrictions on the healthcare organization's ability to limit access.

Under the Cures Act framework and the ONC Final Rule, Mrs. Rivera's rights are concrete and enumerable. She has the right to access her electronic health information as defined by the United States Core Data for Interoperability. This federally maintained dataset specifies which data elements are subject to the access mandate. USCDI includes laboratory results, medication lists, medication allergies, encounter diagnoses, smoking status, procedures, vital signs, immunizations, demographics, care team members, clinical notes, and health insurance information. She has the right to access that information through a third-party application of her choosing, not only through the hospital's patient portal, without incurring cost-prohibitive fees, and without delays or requirements that go beyond what the information-blocking exceptions permit.

The enforcement mechanism is direct. Patients who believe their access has been improperly blocked can file information blocking complaints through ONC's complaint portal. Those complaints can be investigated by the OIG, which has authority to impose civil monetary penalties on healthcare providers, health IT developers, and health information exchanges. Riverside General's fourteen unresolved call center complaints were exactly the kind of pattern an OIG investigation would find significant.

Patrick's initial instinct was to understand patient access rights in terms of what the hospital was required to do. Maya's instinct, shaped by clinical practice, was to understand them in terms of what a patient in Mrs. Rivera's situation should actually be able to accomplish. A compliance frame asks whether the hospital has checked the right boxes. A patient-centered frame asks whether the patient can meet the law's requirements. Organizations that stop at the compliance frame will routinely pass internal audits while generating the kind of patient complaint that eventually lands in an OIG investigation.

Diagram 10.2: Patient Rights Under the 21st Century Cures Act

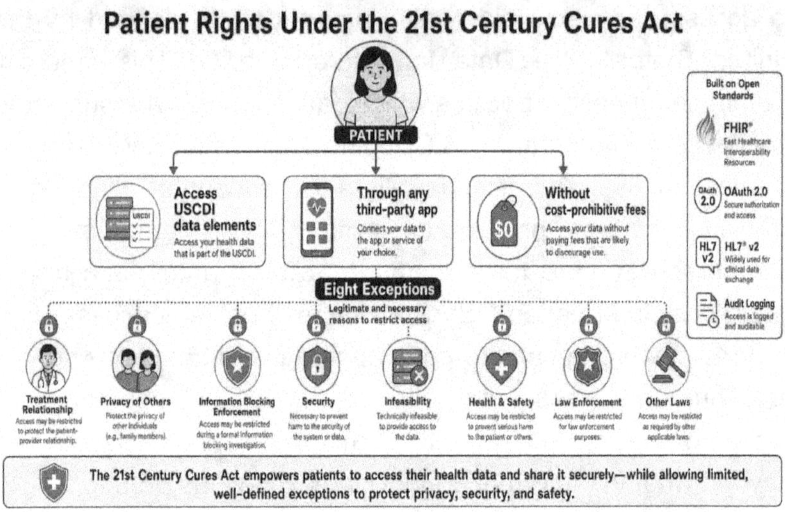

Maya made a point of explaining that information blocking rights extend beyond the institutional contexts in which they are most often discussed. Most conversations about information blocking focus on provider-to-provider data sharing or payer access to clinical information. Patient access is the third category, and the most personal: when a healthcare organization blocks a patient's access to their own health information, it is not a dispute between institutions but an institution failing one of the individuals it exists to serve.

The Patient Access API: Architecture for Consumers

The technical mechanism enabling Mrs. Rivera to connect her health app to Riverside General is the Patient Access API: an FHIR R4 endpoint that a certified health IT product must expose, secured with SMART on FHIR authorization, and populated with data formatted according to US Core profiles that cover all USCDI data elements. Dev Okafor helped Maya understand the architecture, and Maya translated Dev's explanation into terms that let Patrick make informed configuration decisions without needing to master the underlying engineering.

The Patient Access API is distinct from other FHIR interfaces a health system maintains. Provider-facing APIs exchange structured data between clinical institutions; payer APIs support claims and prior authorization. The Patient Access API is designed specifically for consumer-facing applications: the endpoint that a health app calls when the patient has authorized it to retrieve data. That consumer orientation shapes how the API is configured and secured.

From Mrs. Rivera's perspective, the API is invisible. What she sees is the connect button in her app, followed by the authorization screen. Behind that screen, a sequence of steps unfolds: the app discovers Riverside General's FHIR server URL and associated authorization server; the app redirects her browser to the authorization server with the data scopes it is requesting; the server authenticates her identity and presents a consent screen; she approves; the server issues a scoped access token; the app retrieves her data. Each step must work correctly. A failure at any step produces the dead end she encountered.

Dev identified the failure within a few hours. The authorization server required patients to have an active portal account with a verified email address before completing the SMART on FHIR flow. Mrs. Rivera had a portal account, but the email address had changed after a provider transition and had never been updated. The server could not match her credentials, failed silently, and the app displayed a generic error that told her nothing about which step had failed. The implementation had passed ONC certification tests because those tests used correctly configured accounts; real patients with drifted configurations generated real failures.

Diagram 10.3: The Four Layers a Patient Request Passes Through Before Returning a Record — and Where Riverside's API Failed the 21st Century Cures Act Audit

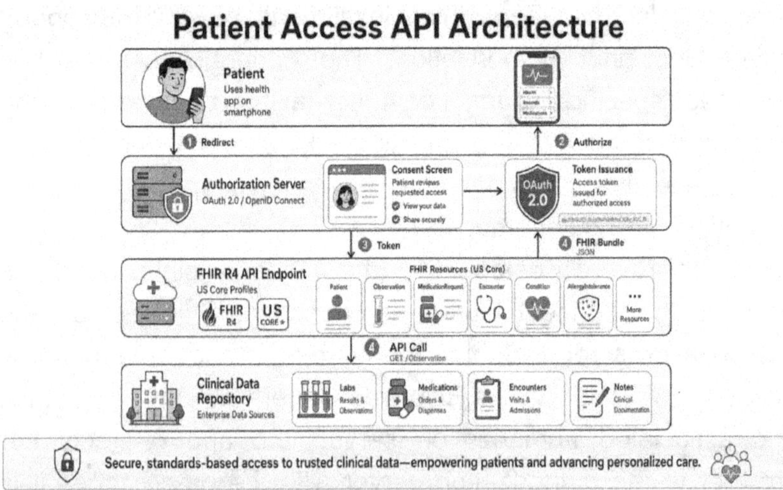

When Mrs. Rivera's app requests her laboratory results, the FHIR server returns Observation resources formatted to the US Core Laboratory Result Observation profile: structured resources with coded fields for test name, result value, reference range, observation status, and observation date. Medication data comes as US Core-profiled MedicationRequest resources.

The profiles ensure sufficient consistency that any FHIR-capable app can interpret the data regardless of which EHR vendor generated it.

Dev's review also revealed data completeness gaps: provenance information was missing from several resource types, and the clinical notes implementation was returning notes for some document types while silently omitting others. Riverside General had not been aware of these gaps because the missing elements were not the most commonly accessed ones, but in an environment where regulators were actively investigating complaints, incremental distance from full compliance was not a comfortable position to occupy.

SMART on FHIR: Auth Without Shared Passwords

The SMART on FHIR authorization framework is the security layer that enables patient-controlled, consumer-facing access to health data at scale. Maya explained it to Patrick with a useful comparison: SMART on FHIR is to health data what the Google login button is to everyday consumer applications. It allows a patient to authorize a third-party app to access their health information without sharing their portal password, using a delegation mechanism that is revocable, scoped, and auditable. The underlying protocol is OAuth 2.0, extended with healthcare-specific conventions. When Mrs. Rivera taps Connect, the app redirects her browser to Riverside General's authorization server with the FHIR data scopes it is requesting. The server authenticates her identity and presents a consent screen. The consent screen is the element Maya cared about most, because it is where the patient exercises genuine informed consent. A well-designed screen tells Mrs. Rivera which specific data categories the app is requesting, how long the token will remain valid, and what she should understand about data that leaves the hospital's control. A poorly designed screen presents FHIR scope identifiers that a non-technical patient cannot interpret. Both designs may satisfy certification requirements; only one serves Mrs. Rivera's interests.

After she approves, the authorization server issues a short-lived access token, scoped to the data elements she approved and to this specific app. If she later wants to revoke the app's access, she can do so through the portal's consent management interface.

Diagram 10.4: SMART on FHIR Authorization Flow for Patients

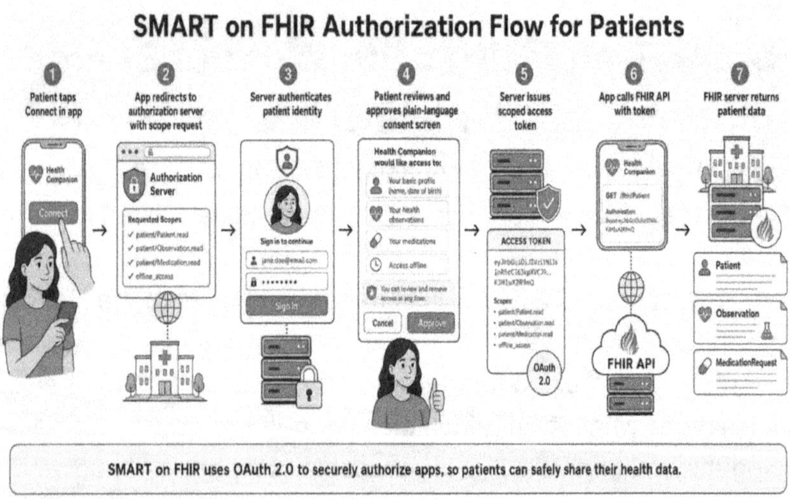

The token lifecycle carries important operational implications. Access tokens expire, and refresh tokens, which allow an app to obtain new tokens without requiring the patient to re-authorize, may or may not be issued depending on server configuration. Dev had configured Riverside General's server to issue refresh tokens with a rolling validity window, but this had not been communicated to patients, so several called the patient services line asking why their apps showed stale data. The fix was partly technical and partly communicational: updating the patient-facing documentation to explain the refresh mechanism, and configuring apps to display a clear re-authorization prompt rather than silently showing outdated information.

Consumer App Ecosystem: A World Beyond the Portal

One of the most significant shifts in Patrick's understanding came from a conversation Maya arranged with Riverside General's patient advisory committee, which consisted of 12 patients who met quarterly to provide feedback on services. Maya explained the patient access project and asked whether any committee members had tried to connect a health app to their records. Seven of twelve had. Four had tried Riverside General and failed. Three had successfully connected to other health systems they also used. Two had given up and accepted that Riverside General was not accessible through their preferred app. Not one had filed a formal complaint.

That conversation illustrated something operationally important. The fourteen call center complaints were not the full scope of the problem; they were the fraction of patients who experienced a failure and chose to call. The majority had stopped trying. The gap between the complaint rate and the actual failure rate meant the compliance exposure from information blocking was considerably larger than the complaint log suggested.

The consumer health app ecosystem is more diverse and widely used than most healthcare administrators recognize. Apple Health and Google Health are the two platforms with the largest patient-facing footprints: pre-installed on their respective operating systems, implementing SMART on FHIR connections to hundreds of health systems, and presenting health record data in formats patients find more accessible than traditional patient portals. A patient with an iPhone can open the Apple Health app, navigate to Health Records, search for their provider, and complete a connection in about two minutes without any technical knowledge, provided the health system's Patient Access API is correctly configured and discoverable.

Diagram 10.5: Consumer Health App Ecosystem

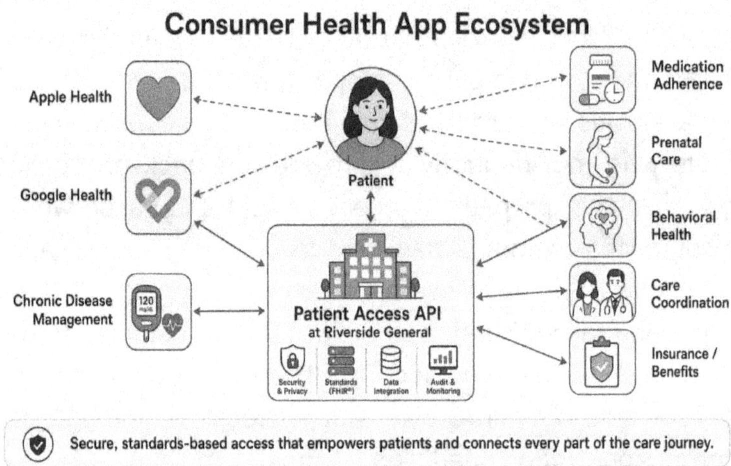

The third-party developer ecosystem beyond the major platform players is large and growing: diabetes management apps that connect to laboratory and pharmacy data; prenatal apps that pull encounter data and immunization records; medication adherence tools that use medication request data to generate reminder schedules; and care coordination platforms that aggregate records from multiple providers. All depend on the same Patient Access API infrastructure Mrs. Rivera was trying to use.

For Riverside General, ecosystem diversity means the hospital cannot screen every third-party app a patient might want to connect. The regulatory framework does not permit it to do so as a condition of access, except within the narrow boundaries of the Security Exception, which permits security requirements that are reasonable and proportionate to actual risk. Dev offered a framework that clarified the distinction: the Patient Access API is a patient-authorized interface. The patient initiates the connection and consents to the data sharing. The hospital's role is to verify the patient's identity, present a clear consent screen, and issue a scoped access token. The hospital is not the gatekeeper of which apps a patient may choose; the patient is.

Permissible Restrictions: The 8 Exceptions

Sandra Torres prepared a briefing on the information blocking exceptions, organized around when Riverside General could legitimately restrict third-party app access. Her analysis was more nuanced than the simple framing Patrick had initially taken from his conversations with Maya.

The eight exceptions define the circumstances under which a healthcare organization may lawfully restrict access to electronic health information: Preventing Harm, Privacy, Security, Infeasibility, Health IT Performance, Content and Manner, Fees, and Licensing. Each has specific conditions that must be satisfied; none operates as a general permission to restrict access when the organization finds it inconvenient.

The Security Exception is most relevant to third-party app connectivity concerns. A healthcare organization may implement security practices that restrict access if those practices are consistent with applicable security frameworks, applied nondiscriminatorily, and reasonably tailored to the security risk they address. The critical phrase is reasonably tailored. A security measure so burdensome that it effectively prevents any app from connecting is not reasonably tailored to a security risk; it is a de facto access prohibition with a security rationale attached. Sandra worked with Dev to audit every security control applied to the Patient Access API against this proportionality standard.

Diagram 10.6: Information Blocking Exceptions and the Security Exception

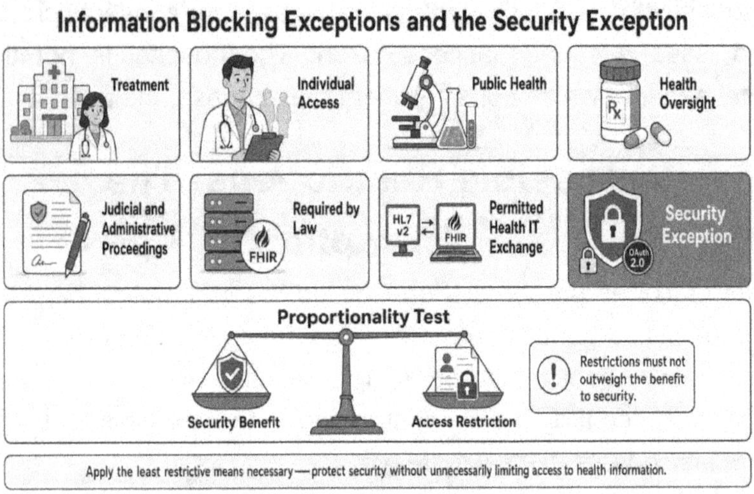

The Privacy Exception permits restrictions necessary to comply with applicable privacy law, but it does not permit restricting all patient data because of a concern about some data; it requires restrictions specifically tailored to the information and the law protecting it. The Preventing Harm Exception permits restrictions designed to prevent physical harm to the patient or another person; it is not a broad safety valve that permits restrictions whenever a clinician is concerned about a patient's reaction. Maya had been attentive to this exception because she had encountered clinicians who wanted to apply it as a general override for releasing clinical notes to patients with mental health conditions, a use inconsistent with both the letter and spirit of the exception.

Patrick's key takeaway was that the exceptions were not a flexible toolkit. They were a narrow set of defined circumstances that, if misapplied, could themselves serve as the basis for an information-blocking finding. The safest posture was to treat patient access as the default, document every restriction with reference to a specific exception, and audit those restrictions regularly.

Information Blocking Trap: Security as Access Barrier

The eight information-blocking exceptions Sandra walked through in the ONC response (Chapter 8) now reappear in a consumer-app context — the same regulation, a different operational frame. The most operationally dangerous information blocking risk Patrick discovered during the remediation was not a deliberate restriction. It was a well-intentioned security program that had quietly evolved into an access barrier without anyone noticing. Riverside General's security team had established a third-party application review process several years before the FHIR patient access mandate was finalized, designed to evaluate clinical applications that connect to hospital systems, including EHR integrations, radiology applications, and laboratory systems. When the Patient Access API went live, the security team extended the same review process to consumer apps attempting to connect to the FHIR endpoint.

The review took three to four months on average and required a security questionnaire, documentation of security practices, penetration testing, and a business associate agreement. For enterprise clinical applications with dedicated compliance teams, the process was manageable if slow. For consumer health app developers, many of them small organizations or individual developers, the requirements were effectively prohibitive. The practical effect was that no consumer health application had completed Riverside General's third-party security review in the eighteen months since the Patient Access API went live.

Dev discovered the broader problem while tracing Mrs. Rivera's specific failure. Examining the logs revealed a clear pattern: dozens of applications had reached the authorization server, failed at the app registration check the security team required as a precondition, and been turned away with an error that did not explain the security review requirement. Patients who followed up were told to have their app developer contact the security team. Developers who contacted the team received the requirements and, in most cases, never responded again.

Diagram 10.7: The Information Blocking Trap: Over-Restrictive Security Review

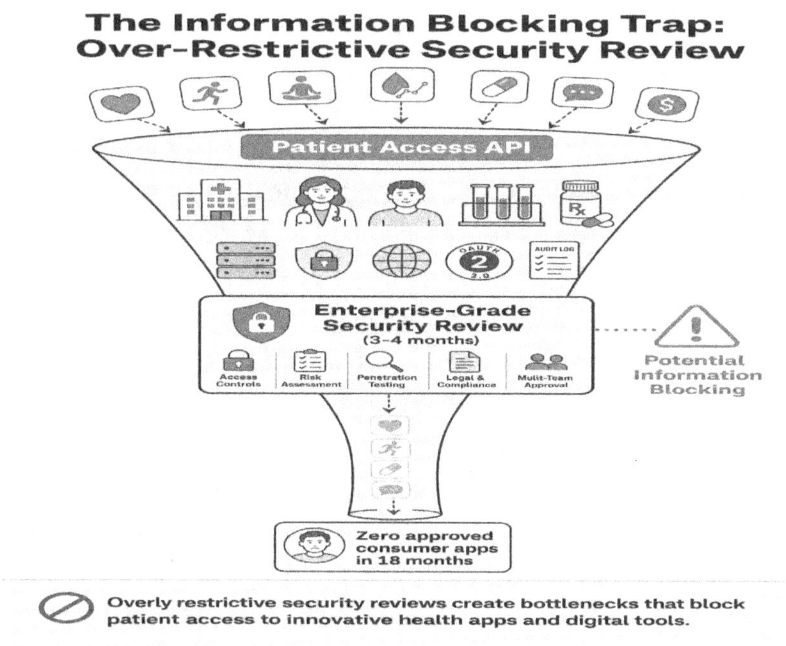

Sandra's analysis was direct. The security review, as applied to consumer apps connecting through the patient-authorized Patient Access API, did not satisfy the Security Exception's proportionality requirement. The requirements were designed for enterprise systems with direct network access to clinical data, not for a consumer app that could only access data a specific patient had authorized via a scoped token for the duration of its validity. The SMART on FHIR model already provides substantial protection: the app has no direct database access, stores no persistent credentials, and logs every interaction. Requiring penetration testing before accessing this already-scoped interface was disproportionate.

The remediation involved a tiered security review framework: a streamlined registration process for consumer apps accessing only through the patient-authorized API, calibrated to the

SMART on FHIR model, and a rigorous process reserved for applications requiring direct system integration. The streamlined process reduced the average review to approximately two weeks. Within the first month, six consumer health apps had successfully registered and connected to Riverside General's Patient Access API. The organizational lesson was about the dynamics that produce well-intentioned information blocking: the security team had applied an existing process to a new context without recognizing that it required a different approach. The failure was not malice; it was a gap in accountability in which no one had explicit responsibility to evaluate the patient-access implications of security policies designed for a different purpose. The pattern recurs in Chapter 11, where a third-party analytics application is approved through a process that satisfies each step in isolation but never asks whether the steps, taken together, actually constitute durable governance. Both failures share a structure: a well-intentioned process applied without a clear owner asking whether the process still fits the context it now serves.

Building a Compliant Patient Portal: Feature Checklist

Patrick and Maya agreed early in the remediation that they needed a concrete operational picture of what a fully compliant Patient Access API implementation required. Dev prepared a technical assessment, and Patrick spent several hours translating it into terms he could use to prioritize resources and communicate with the board.

The first requirement is a SMART-on-FHIR-compliant authorization server capable of verifying patient identities, issuing scoped access and refresh tokens, and logging all authorization events. Riverside General had this capability in its certified EHR platform, but the identity verification workflow had a configuration gap that had affected Mrs. Rivera. Fixing it required a configuration change and a process update for patient services to help patients whose portal credentials had drifted from their current identity.

The second requirement is an FHIR R4 API endpoint that returns US Core-profiled resources for all current USCDI data elements. Riverside General had the endpoint, but the data completeness audit identified gaps in provenance metadata and clinical notes coverage. The third requirement is a patient consent management interface where patients can see which applications they have authorized, what data scopes each covers, and a mechanism for revoking authorizations they no longer want active. Riverside General's portal included a basic version buried under three menu levels, formatted in a way most patients found confusing. A usability improvement project was added to the remediation list.

Diagram 10.8: Compliant Patient Access API Feature Set

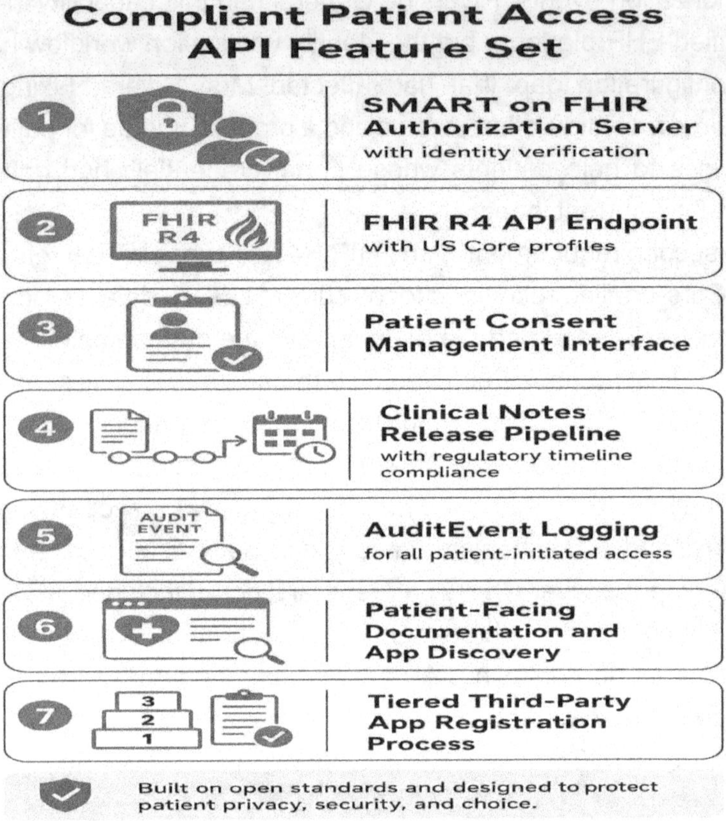

The fourth requirement is a clinical notes release pipeline complying with ONC's information blocking timeline requirements. Notes must be released at the time of finalization, without the delays historically justified on grounds that patients would be confused by reading them before a provider reviewed them. Riverside General had implemented note release for certain note types and main campus locations. Still, the audit revealed that notes generated in affiliated outpatient practices were not being routed through the same pipeline.

The fifth requirement is comprehensive AuditEvent logging for every patient-initiated access interaction: every authorization request, token issuance, API call by a patient-authorized application, and revocation. Riverside General's logging was partial; events were captured but not aggregated in a way that enabled systematic analysis of access failure trends. The sixth requirement is patient-facing documentation explaining what the Patient Access API is, how to connect a health app, what data will be shared, and what the patient's rights are. Riverside General's portal had a single paragraph written in technical language; the remediation produced a plain-language rewrite developed in collaboration with the patient advisory committee. The seventh requirement, generated by the remediation rather than inherited from the regulations, is the tiered third-party app registration process Sandra and Dev had designed to replace the enterprise security review. Working through this feature list component by component with Maya, Patrick found it clarifying. Each requirement was concrete, had a clear owner, and had a measurable success criterion—the gap between where Riverside General was and where it needed to be had become a project plan.

Consent, Revocation, and Ongoing Patient Control

Maya's thinking on patient consent had been shaped by years of clinical practice in informed consent for procedures. She recognized that consent for health app data access shared the requirement for meaningful patient understanding: consent given without comprehension is not really consent, and a consent screen a patient cannot interpret is not doing the work consent is supposed to do. The key difference from clinical informed consent was the ongoing nature of authorization: unlike a one-time procedure, a data access authorization persists, can be extended via refresh tokens, and requires a practical mechanism for the patient to revisit and withdraw it.

A consumer health app must request specific scopes corresponding to the FHIR resource types it needs before connecting to Riverside General's FHIR API. The authorization server presents those scopes to the patient. If displayed as FHIR identifiers (patient/Observation.read, patient/MedicationRequest.read), most patients cannot interpret what they are consenting to share. If displayed in plain language ("Your laboratory results," "Your prescription medication list"), the consent is meaningfully informed. Regulatory requirements do not prescribe the exact language of the consent screen, but patient rights under the Cures Act imply that consent must be genuinely meaningful, not merely technically present.

Riverside General's consent screen had been implemented using the EHR vendor's default configuration, which displayed FHIR scope identifiers directly. During a walkthrough Dev organized, Maya asked a passing nurse to interpret the screen as a patient would; the nurse identified "your lab results" correctly but was uncertain about three of the seven scope categories. If a healthcare professional could not reliably interpret the screen, a non-clinical patient was essentially authorizing access to an unspecified dataset. The EHR vendor provided a configuration option for plain-language scope descriptions, and Dev deployed it within a week. The patient advisory committee confirmed that all seven categories were clearly understandable; one member also noted that the screen needed to explain what would happen to the data once it was shared with the app. That addition was incorporated.

Diagram 10.9: Patient Consent and Revocation Lifecycle

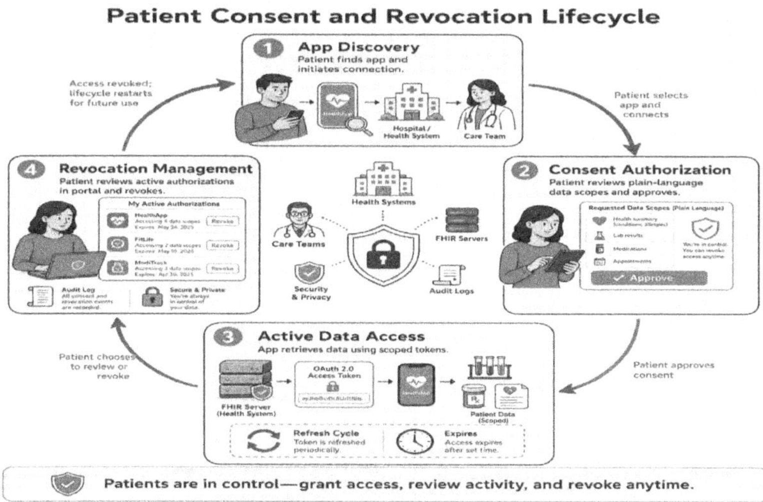

The revocation mechanism is as important as the consent mechanism, and both need to be symmetrically designed. If connecting an app takes two minutes and three taps, but revoking an authorization requires navigating an obscure settings menu and waiting for a staff member to act, the system is not treating consent and revocation as equivalent patient rights. The Cures Act framework treats patient control over data authorizations as a continuous right, not a one-time grant.

After the usability improvement, Riverside General's revocation interface displayed all active app authorizations in a single view: the app name, the authorization date, the data scopes covered, and a revocation button for each entry. Revoking an authorization immediately invalidated the associated refresh tokens, preventing the app from obtaining new access tokens afterward. Active tokens already issued remained valid until their natural expiration, configured at one hour; this behavior, effective for future access but not instantaneous for currently valid tokens, was disclosed in patient-facing documentation. Three committee members tested the full connect-review-revoke flow before the updated interface was deployed to the entire patient population; all three completed revocation. One member expressed interest in a notification when an app accessed her data; feedback Dev recorded as a future enhancement.

From Complaint to Framework: Mrs. Rivera's Outcome

Patrick called Mrs. Rivera eight weeks after their initial conversation to report that the problem had been investigated, diagnosed, and corrected. The specific issue, the outdated email address on her portal account, had been resolved by the patient services team three weeks after her initial call. By the time Patrick called back, Mrs. Rivera had been viewing her laboratory results in her app for nearly a month. She said it worked exactly as her insurer had told her it should.

The fourteen patients whose call center complaints had been closed without resolution were all followed up by the patient services team. Seven successfully connected their preferred app. Four had already moved to other applications or accessed their data through the portal. Three could not be reached. The systemic remediation, including the authorization server fix, data completeness corrections, consent management interface improvement, plain-language documentation, and tiered security review framework, was substantially complete within the quarter Patrick had promised. Dev built a monitoring dashboard that tracks authorization success and failure rates by application and error type, enabling detection of new failure patterns before they generate significant call volume.

The patient access call center volume declined by roughly forty percent in the quarter after remediation was complete. The complaints that still came in were requests for help navigating a specific step in a process the patient understood, rather than reports of complete failure. That was the operational signature of an implementation that had finally begun to work. Riverside General had passed its ONC certification for the Patient Access API. It had failed Mrs. Rivera. Those two facts were not contradictory; they described exactly the gap that operational diligence is required to close, and that no certification process alone can close.

Takeaways: Patient Access as a Standard of Care

The argument Patrick made to Riverside General's board at the end of the remediation was not primarily a compliance argument. He made the compliance case clearly, because the regulatory exposure was real. But the argument he pressed most forcefully was a patient service argument. Patients who could access their health information through applications of their choosing were better equipped to manage their own health: more likely to follow through on referrals, more likely to bring complete medication lists to specialist appointments, and more likely to stay engaged between clinical visits. A Patient Access API that worked was not a regulatory checkbox; it was a contribution to the health of the patients Riverside General existed to serve.

The regulatory obligation is real, enforceable, and growing. ONC continues to update the USCDI data elements subject to the access mandate, and the regulatory environment is moving toward greater patient data rights, not fewer. Organizations without a robust, well-monitored Patient Access API accumulate technical debt that will grow more expensive as requirements expand. Riverside General's fourteen call center complaints in a single quarter were not an anomaly; they were an early indicator of a systemic failure that would have produced many more complaints and, eventually, regulatory scrutiny if it had not been addressed.

SMART on FHIR patient authorization is worth understanding in operational depth. Organizations that understand it can configure it correctly, monitor it effectively, audit it for compliance with information-blocking regulations, and explain it to patients in plain language. Dev's diagnostic work on Mrs. Rivera's failure had taken hours, not because the problem was technically complex, but because no one had ever closely examined how the authorization server behaved with real patient accounts. That operational visibility is not a luxury; it is a basic competency for any organization with meaningful Patient Access API traffic.

Information blocking risk concentrates at the intersection of legitimate concern and disproportionate response. The security team's concerns were not unfounded; real security considerations exist in a consumer-facing API. But the response, an enterprise review process applied without modification to consumer apps operating through a scoped patient-authorized token model, was disproportionate to the actual risk. Designing security controls that are proportionate, documented, and defensible under the Security Exception is not a concession to regulatory pressure; it is a matter of getting the risk assessment right.

Consent and revocation deserve equal operational investment. A framework that makes authorization easy and revocation difficult is not neutral; it prioritizes data access over patient control. The consent screen Mrs. Rivera finally saw was designed to help her understand what she was sharing. The revocation interface was designed to make withdrawing that sharing as straightforward as granting it. Both choices illustrate what patient-centered care means when patients have access to their health data.

The call center volume is a leading indicator worth watching. An organization monitoring its Patient Access API only through technical checks will catch configuration failures but will miss the category Mrs. Rivera experienced: a technically present implementation that fails real patients because real account configurations do not match test configurations. Monitoring at the experience level, through call center patterns, patient advisory feedback, and proactive outreach to patients with failed connections, complements technical monitoring, making an API genuinely functional rather than merely certified. Patient access is not a project with a completion date; it is an ongoing operational capability Riverside General will need to sustain for as long as it cares for patients.

Patrick had kept the annotated ONC regulations printout Maya had slid across her desk at the start of their conversations. It was in the top drawer, slightly more worn than when she had handed it to him. He had not become a regulatory expert, but he was no longer uninformed. The next time a patient called about a failed app connection, he would know exactly which question to ask first.

11 Security, Privacy, and Access Control

At 2:47 a.m. on a Wednesday in late October, Sandra Torres's phone lit up with a priority alert from Riverside General's security operations platform. The message was terse: unusual outbound network traffic detected from the clinical data repository server cluster. Sandra had spent eleven years in healthcare compliance, and she had learned that 2:47 a.m. alerts did not resolve themselves by morning. She was logged in to the monitoring dashboard before her feet touched the bedroom floor. By 3:15 a.m., she had Dev Okafor on a conference call. What they saw was deeply uncomfortable: automated, systematic outbound connections to an external IP address, originating from a process with no legitimate reason to initiate outbound traffic. Dev isolated the affected server segment from the network within twenty-two minutes of Sandra's call. That decision would prove consequential.

By 6:00 a.m., the incident response team had assembled in what Sandra had started calling the War Room. Patrick Osei arrived at 6:30 a.m. carrying two cups of coffee and the expression of a man who understood he was about to have a very difficult day. Dr. Maya Chen joined by video from her home office, still in surgical scrubs from a case the night before.

The preliminary analysis confirmed the worst-case category but not the worst-case outcome: ransomware. The malicious process had gained access through a vulnerability in a third-party analytics application integrated with the FHIR server. It had begun its reconnaissance phase before Dev's isolation action cut off its command-and-control channel. No data had been encrypted. No patient records had been exfiltrated. But the forensic team's first question was one they could not immediately answer: had the attacker accessed or viewed any patient data during the reconnaissance phase?

That question could not be answered with certainty, not because the logs did not exist, but because the existing logs were not structured to allow definitive reconstruction of every resource access during the attacker's window. Sandra wrote four words in her incident notebook: "AuditEvent. Consent. Scopes. Now."

The ransomware attacker had found three unlocked doors. The first was insufficient audit logging that made it impossible to reconstruct the complete access history. The second was overly broad OAuth scopes on the analytics application, granted far more access than its stated purpose required. The third was the absence of a Consent resource implementation, leaving the FHIR server without a mechanism for enforcing patient-directed access restrictions at the resource level. Each gap had grown from time pressure, incomplete awareness of the FHIR security model, and the tendency to configure for convenience when the threat feels abstract. The threat was no longer abstract.

This chapter follows Sandra and Dev through the work of understanding and closing those gaps: the three-layer security baseline, OAuth 2.0 and SMART on FHIR authorization, the scope model and least privilege, the Consent resource for patient-directed access control, AuditEvent logging, break-glass access patterns, third-party app governance, HIPAA Security Rule mapping, and Sandra's completed post-incident checklist.

The Three-Layer Security Baseline

Maya's question for this section is: What are the absolute minimum security requirements that any FHIR system must meet before it handles real patient data?

The irreducible security baseline for a FHIR endpoint handling protected health information has three layers. The first is transport security. The second is authentication and token validation. The third is endpoint trust and registration. Each addresses a distinct attack surface, and each is necessary because the other two are insufficient without it.

Transport Layer Security encrypts every byte of data in transit between a FHIR client and a FHIR server. Without TLS, every network device along the path can read and modify both the request and the response without detection. With TLS, data is encrypted end-to-end, and the server's identity is cryptographically verified before any data is exchanged. The minimum acceptable version for a FHIR endpoint handling PHI is TLS 1.2; version 1.3 is strongly preferred. TLS 1.0 and 1.1 are deprecated and must not be used. The configuration covers not just the version but also cipher suites, the certificate authority chain, and certificate validity. An expired certificate turns off the identity verification component of TLS, regardless of the encryption in place.

Diagram 11.1: What the Ransomware Incident Exposed About Riverside's FHIR Security Assumptions

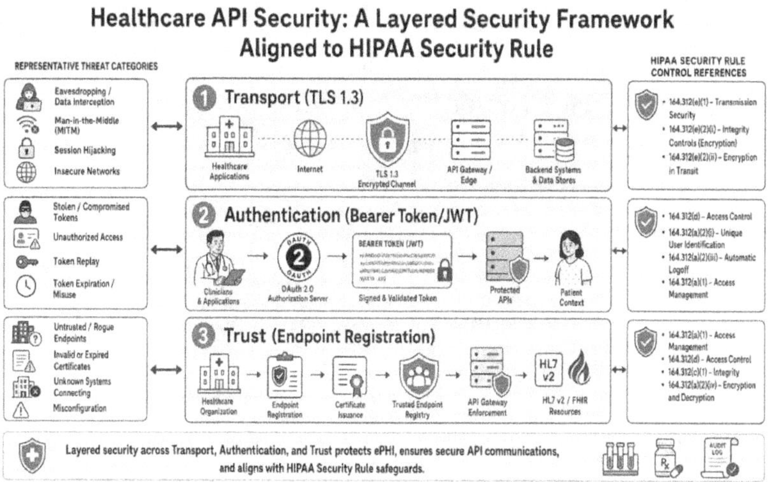

Dev pulled the TLS configuration report for every FHIR API endpoint on the first morning of the incident response. The primary FHIR server was running TLS 1.3 with a current certificate. The internal-facing API gateway used by the analytics application was running TLS 1.2 with a cipher suite configuration set two years earlier and never reviewed. That was not the ransomware's attack vector, but it was a gap requiring remediation.

The second layer is bearer token authentication. A FHIR server that requires an encrypted connection but not authenticated credentials for every request is protected only against passive eavesdropping, not against active unauthorized access. Bearer token authentication requires that every request carry a valid access token issued by a trusted authorization server, and that the server validate the token before processing any request or returning any data. In the SMART on FHIR model, these tokens are JSON Web Tokens: cryptographically signed credentials that carry claims about the authenticated subject, the authorized scopes, and the token's expiration time.

The third layer is endpoint trust and registration. A client application sending a valid bearer token to a server that is not the legitimate FHIR server has not been protected by the first two layers. Endpoint spoofing, in which a malicious system impersonates a legitimate FHIR server to intercept token-bearing requests, is a real attack class in healthcare IT environments. The SMART on FHIR application registration model provides the framework for mutual registration between clients and servers.

These three layers map directly to the HIPAA Security Rule Technical Safeguards: Access Control at 164.312(a)(1) maps to bearer token authentication and scope enforcement; Audit Controls at 164.312(b) maps to the logging of every token validation event and access decision; Transmission Security at 164.312(e)(1) maps to the TLS configuration. An organization that can demonstrate these layers are implemented and monitored on every FHIR endpoint handling PHI has established the technical foundation for its HIPAA Security Rule compliance argument.

OAuth 2.0 and SMART on FHIR

Maya's question for this section is: how does a third-party app get permission to access a patient's health records, and how does the system know what the app is and is not allowed to see? The authorization framework governing FHIR data access is built on OAuth 2.0, an open standard for delegated authorization designed to solve the problem healthcare data sharing precisely creates: how to allow a client application to access resources on behalf of a resource owner without the resource owner sharing their password with the client. In healthcare, the resource owner is a patient or provider, the client is an EHR or analytics tool, and the resources are FHIR resources containing PHI.

OAuth 2.0 introduces a trusted intermediary, the authorization server, which authenticates the resource owner, obtains their explicit consent to the client's request, and issues a time-limited access token that the client presents to the FHIR server to prove authorization. The FHIR server never sees the resource owner's credentials. The authorization server handles authentication; the FHIR server handles the data.

SMART on FHIR profiles OAuth 2.0 specifically for healthcare use cases. It defines how FHIR-aware applications register with authorization servers; the scope syntax for requesting access to specific FHIR resource types; the token response format and validation requirements; and the appropriate authorization flows for different application types: browser-based patient-facing applications, server-side clinical applications, and backend services without user interaction. SMART on FHIR is the reason a patient-facing application certified by one EHR vendor can be expected to work with any other SMART on FHIR-compliant EHR.

Diagram 11.2: Where the Third-Party App Gets Its Token and Where Dev Blocks It — How SMART on FHIR Authorization Stops an Overpermissioned Request Before It Reaches Patient Data

The SMART on FHIR Authorization Flow

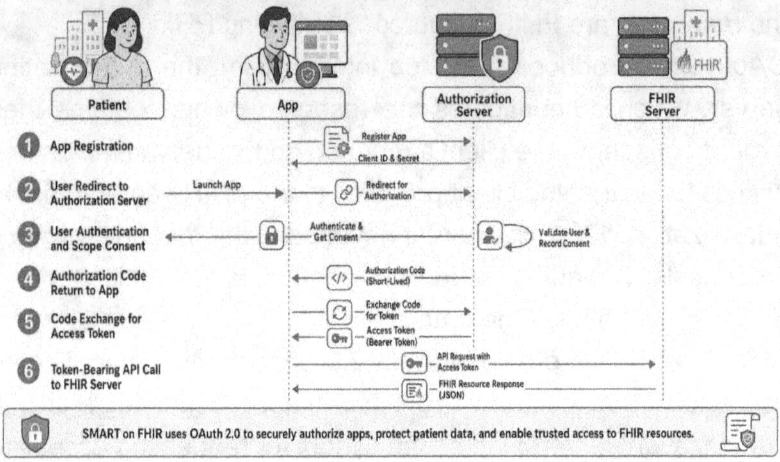

The flow begins with application registration before any patient launches the application. The developer registers with the authorization server, providing the application's name, redirect URI, required scopes, and privacy policy URL. Only registered applications can request access tokens. When a user launches the application, it redirects them to the authorization server with the client identifier, requested scopes, and a state parameter for integrity verification. After the user authenticates and consents, the authorization server redirects back with an authorization code. The application exchanges the code for an access token via a direct server-to-server request to the token endpoint, including the code, client credentials, and redirect URI.

The FHIR server validates every bearer token before processing any request: confirming the signature is valid, the token has not expired, the issuer is a trusted authorization server, the audience claim matches this FHIR server, and the scopes are sufficient for the operation. If any validation fails, the server returns a 401 Unauthorized and must not return any PHI. The analytics application had completed registration, obtained an access token, and used it to make API requests. The scope audit revealed not a flaw in the authorization framework itself, but a flaw in how the framework had been configured for that specific application. The framework had functioned exactly as designed. The design had been wrong.

Scopes and the Principle of Least Privilege

Maya's question for this section is: if a third-party app is connected to our FHIR system, how do we ensure it can see only the data it needs and nothing more?

Sandra's discovery of the overly broad OAuth scopes was not dramatic. There was no flashing alert, no anomaly detection algorithm flagging the scope configuration as dangerous. She found the problem by carefully reading the registration record, understanding what the application was supposed to do, and asking whether the granted scopes were consistent with that purpose.

The analytics application had been integrated eighteen months earlier to provide population-level reporting on care gaps. Its stated purpose required read access to Condition, Procedure, certain Observation, and Patient demographic resources. It did not need write access to anything. It did not need access to MedicationRequest, DiagnosticReport, or ClinicalImpression resources.

The scope grant told a different story. The application had been granted system-level read access to eight resource types, covering the majority of clinically significant resources in the FHIR server. It needed four. The overgrant had not resulted from a deliberate decision. The integration developer had erred on the side of inclusion to avoid problems. The review had focused on whether the application worked, not whether its scope grant was minimally sufficient. No one had reviewed the scope grant again in eighteen months.

Diagram 11.3: Scope Least-Privilege Comparison

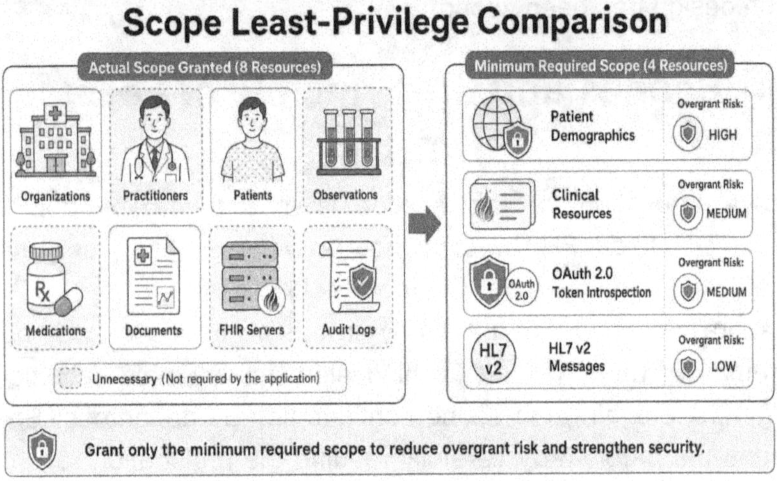

The principle of least privilege, applied to FHIR scopes, requires that every application be granted only the scopes necessary for its stated purpose, with a documented justification for each scope granted, and a review process that re-evaluates scope grants when the application's purpose changes, when it undergoes significant updates, or on a defined periodic schedule. This is the practical implication of the HIPAA minimum necessary standard, applied to API-level access controls.

The SMART on FHIR scope syntax makes the principle of least privilege operable at a high level of granularity. A scope of "patient/MedicationRequest.read" grants read access for the specific authenticated patient. A scope of "user/MedicationRequest.read" grants read access for patients within the authorized user's clinical purview. A scope of "system/MedicationRequest.read" grants read access for any patient in the system. The difference reflects distinctions among a patient-facing application, a provider-facing application, and a backend service that can access any patient's data, each of which requires a correspondingly higher level of justification. SMART on FHIR also defines operation-level scope modifiers specifying read, write, or both; an application that only needs to read data should never be granted write access, because a compromised application with write access can corrupt or delete clinical data in addition to viewing it.

Dev implemented the scope remediation within seventy-two hours of Sandra's audit findings. The analytics application's registration was updated to remove the four unnecessary scopes. Existing access tokens were revoked, forcing re-authentication with new tokens that carry only the revised scope. A scope review was scheduled for every other registered third-party application. Sandra drafted a new policy requiring dual sign-off from the IT lead and the compliance officer before any scope grant could be activated in production.

The FHIR Consent Resource

Maya's question for this section is: when a patient has said they do not want certain providers to see certain parts of their record, how does the system actually enforce that?

The FHIR Consent resource is the mechanism by which a patient's documented agreement to or refusal of specific uses of their health information becomes a machine-executable access control policy that the FHIR server can enforce in real time. Its absence from Riverside General's FHIR environment was not a deliberate choice but a consequence of design specifics that are easy to defer to competing priorities.

A Consent resource represents a patient's agreement across four dimensions. The scope defines which type of data use is authorized or denied: treatment, payment, research, or third-party disclosure. The category defines the specific data type covered: a patient might consent to sharing general medical records for treatment while specifically refusing to allow behavioral health records to be shared outside the treating facility. The patient reference ties the Consent to the specific patient. The provision element articulates the policy's structure, specifying which actors, resource types, time periods, and purposes are permitted or denied coverage.

The forensic team's analysis showed that during the attacker's window on the network, API requests had retrieved patient resources using valid bearer tokens with valid scopes. From the FHIR server's perspective, those requests had been authorized. The fact that a malicious process had used the tokens was not something the FHIR server could have known from the token alone.

Diagram 11.4: What Sandra Must Populate Before a FHIR Consent Resource Is Legally Enforceable — and Which Empty Fields Left Riverside Exposed During the Audit

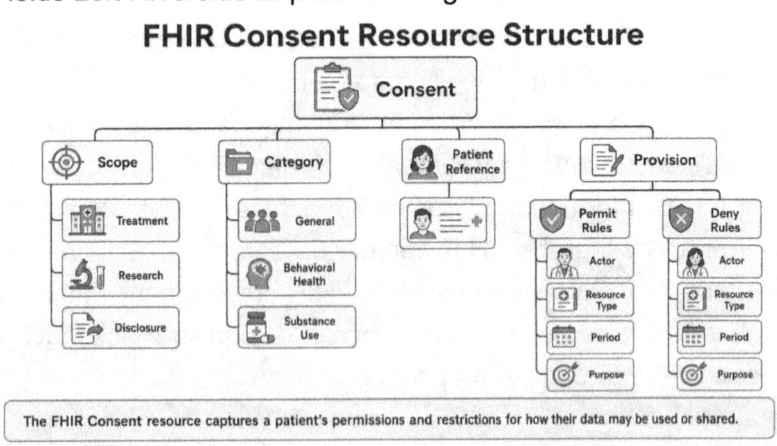

What a properly implemented Consent resource framework provides is an additional layer of access restriction operating independently of the OAuth scope model. The scope model determines what the application is authorized to access. The Consent resource determines what the patient has authorized to be shared. Both checks must pass before a resource is returned. An application with a valid scope for MedicationRequest resources that attempts to access a patient with a restricted medication record-sharing status would be denied by the Consent enforcement layer, even if scope validation passes. The Consent resource is particularly important for sensitive data categories that carry heightened regulatory protections: behavioral health records, substance use treatment records, HIV status information, and genetic data are all subject to state and federal protections that are more restrictive than the general HIPAA framework.

Implementing the Consent resource correctly requires decisions beyond the FHIR specification: which consent categories the organization will support, the default policy for data categories without an explicit consent record, how consent records will be created and updated, and how the access control engine will evaluate the intersection of scope-level and consent-level controls. These decisions require collaboration between the compliance officer, clinical informatics, legal counsel, and the patient experience team. Dev and Sandra worked through the Consent resource design over four weeks. They started with a default-permit model in which all patient data is shareable unless a specific Consent resource with a deny provision exists. This is not the most restrictive possible approach. Still, it is defensible for an organization without a mature consent management workflow and establishes the infrastructure for a more granular model later. Moving Sandra's two years of paper-based patient restriction records into Consent resources changed the organization's ability to enforce those restrictions, not just document them.

Maya, who had joined the consent design sessions expecting to argue for a fully patient-driven granular model from day one, revised her position by the third week. She put it on the record at the end of the fourth session. "I came into this thinking the right clinical framing was that the patient is the decision-maker and the system enforces whatever the patient decides," she said. "That is still the right framing. But I was wrong about the sequence. If we expose a granular consent UI to patients before we have a default-permit baseline that we can audit and reason about, we will produce inconsistent enforcement, clinicians will lose confidence in the access display, and patients will lose confidence in the consent record. The clinical framing I should have started with is: enforce a coherent default first, then earn the patient's trust to layer specificity on top of it." Sandra recognized the shift and underlined the line in her notes; Dev said nothing, but later told Patrick it was the most useful sentence anyone had said about consent in six months.

AuditEvent: The Structured Access Log

Maya's question for this section is: if something goes wrong with our patient data, how do we prove exactly what happened and which patients were affected?

The question consuming Sandra in her first week after the incident was a data question: given the available evidence, what could be said with confidence about which patients' data had been accessed during the attacker's window? The FHIR server, the API gateway, and the network infrastructure all produced logs. What was absent was a unified, structured, queryable representation of every FHIR resource access event in a format that could answer specific questions: which patients' resources were retrieved during a specific window, by which authenticated subject, using which access token, from which network address, with what outcome? Answering those questions required manual correlation across multiple formats with different schemas and different timestamp resolutions. That correlation took days. In a HIPAA breach analysis, days translate into ambiguity, and ambiguity translates into the presumption of a breach.

The FHIR AuditEvent resource exists precisely to prevent that situation. It is the standardized FHIR mechanism for recording every access interaction in a structured format that is consistent across implementations, queryable through the FHIR API, and sufficient to support post-incident forensic analysis, compliance auditing, and patient-facing access transparency.

Diagram 11.5: What the AuditEvent Captures When a Clinician Accesses a Record — and How Sandra Uses the Agent, Entity, and Outcome Fields to Reconstruct a Breach Timeline

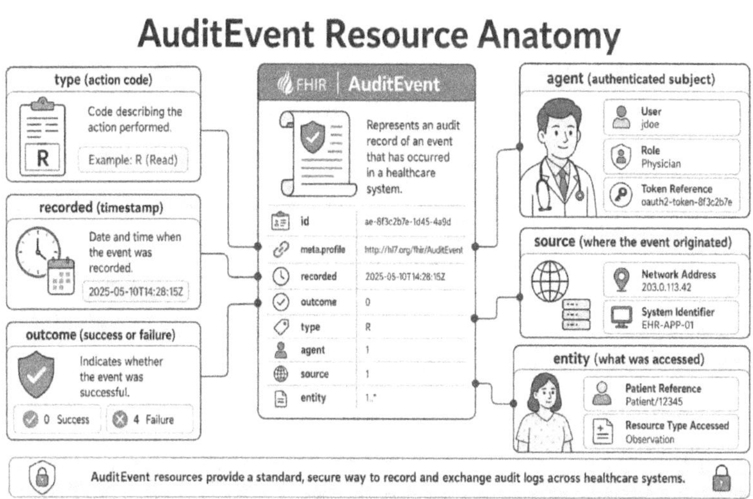

Every AuditEvent resource carries a minimum set of information. The type field encodes the category of action: an FHIR read, search, create, login, or consent override event. The recorded field captures the precise timestamp. The outcome field records whether access was permitted or denied. The agent array records the identity of every party involved: the human user, the client application, and any intermediary systems. The source element records the requesting system's network address and identifier. The entity array records the resources accessed, with references to the specific patients whose data was involved.

The gap Sandra identified was not a failure to log, but a failure to log in the AuditEvent format. The FHIR server, the analytics application, and the API gateway each produced logs in different formats. Consolidating these three streams required ad hoc parsing scripts built under incident pressure, introducing both delays and the risk of parsing errors.

Dev designed the AuditEvent logging implementation in the three weeks following the incident. The FHIR server was configured to generate a well-structured AuditEvent resource for every successful and failed API request and to write those resources to a dedicated AuditEvent repository. The API gateway was configured to emit structured event data compatible with the AuditEvent schema. A log correlation service converted the third-party application's proprietary log format to AuditEvent resources, providing unified query coverage.

Sandra ran the first comprehensive AuditEvent query exactly thirty days after the incident, querying for all access events involving the validation test patient population and verifying that every independently confirmed access was reflected with the correct agent, entity, timestamp, and outcome. The query returned complete results in under three seconds. Sandra wrote a single sentence in her incident log: "This is what we should have had on the night of the incident."

Break-Glass Access in Clinical Emergencies

Maya's question for this section is: what happens when a doctor in the emergency room needs to see a patient's full history immediately, but the system controls say they are not authorized?

The security architecture Dev and Sandra were building was designed to be restrictive by default. Every access request would require a valid token with appropriate scopes, every token would be validated, and Consent resource restrictions would be enforced at the resource level. That architecture was right for the vast majority of use cases. But Maya, joining post-incident technical reviews via video from the ICU whenever she could, raised the question every clinician asks when encountering a security-by-default architecture: what happens when an unconscious patient arrives in the emergency department, and the attending physician has no established clinical relationship with that patient in the FHIR server? The patient's allergy history, active medication list, and chronic condition diagnoses may be stored on a FHIR server to which the attending has no established access rights. The patient cannot consent. The normal authorization flow will deny access. But those records may be directly relevant to clinical decisions being made in the next fifteen minutes.

Break-glass is the security pattern designed to address exactly this scenario. It allows an authorized user to override normal access controls in a defined emergency, with three invariant requirements: the override must be documented at the moment it occurs, every resource accessed during the session must be individually logged with a specific indicator that normal authorization was overridden, and an immediate notification must be sent to the patient's primary provider and the compliance team for after-the-fact review.

Diagram 11.6: Break-Glass Access Pattern and Audit Trail

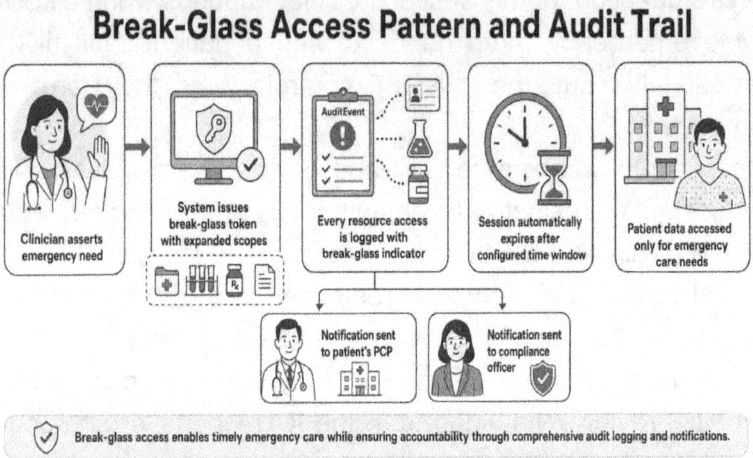

In a FHIR implementation, break-glass access is implemented through a dedicated authorization flow that bypasses normal scope restriction logic while maintaining a complete audit trail. The clinician activates it through a specific EHR workflow, confirms their identity and the clinical emergency assertion, and acknowledges that their access will be reviewed. The authorization server issues a special-purpose token with expanded scopes, marked with a break-glass claim that the FHIR server recognizes, and generates AuditEvent records indicating each access was made under break-glass authorization.

The time limit on break-glass sessions is a critical design parameter. A session that does not expire creates a persistent security risk. A session duration of thirty minutes is a common starting point, with the ability to request a single extension. Sandra defined the notification workflow for Riverside General as part of the post-incident remediation: notifying the patient's designated primary care provider within 15 minutes, including the activating clinician's identity, the session time window, and the resource types accessed. A break-glass notification sent to a general compliance inbox that is reviewed monthly does not satisfy the oversight requirement.

Maya's concern was not that break-glass would be too easy to activate, but that it would be too difficult under pressure. The design that Dev and Sandra arrived at after consulting Maya was a two-step workflow: a single affirmative button activation triggering an immediate confirmation dialog, then immediate token issuance. The clinician could complete detailed documentation of the justification within two hours, not at the moment of activation.

Third-Party App Security Governance

The information-blocking exceptions from Chapter 10 set the regulatory floor; the third-party app governance practices in this chapter build the operational ceiling above it. Maya's question for this section is: if we allow third-party apps to connect to our FHIR system, how can we ensure those apps do not put our patients at risk?

The analytics application that served as the ransomware's entry point had been approved for integration through a process that, at the time, had seemed adequate. Documentation had been provided. The application had been tested in a sandbox. The scope grant had been briefly reviewed. The application had worked correctly for eighteen months. The process had failed not because each step was wrong, but because the steps collectively did not constitute a durable security governance process. The failure mode mirrors the information-blocking trap examined in Chapter 10: a security review designed for one context applied to another without anyone explicitly owning the question of whether it still fits. There, the process was too heavy for consumer apps; here, it was too light for ongoing third-party access. The shape of the gap is the same.

A durable third-party application security governance process requires four components. The first is an initial security assessment with a defined scope and documented outcomes. The second is a scope-justification requirement that treats every requested scope as a decision that must be documented with a reason. The third is an ongoing monitoring posture watching access patterns for anomalies after go-live. The fourth is a periodic re-review process that does not wait for a security incident to prompt a second look at the application's configuration.

The initial security assessment should cover the application's authentication architecture, data retention practices, subprocessor relationships, incident response history, vulnerability disclosure program, and data processing agreement with the healthcare organization. All findings should be retained as part of the integration record.

Diagram 11.7: Third-Party App Security Review Process

The scope justification requirement transforms the scope grant from an administrative configuration step into a deliberate decision with documented accountability. For every scope the application requests, the integration record should answer three questions: which specific feature requires this scope, what data would be accessible if this scope were misused, and whether the risk of that misuse is acceptable given the application's demonstrated security controls.

Ongoing monitoring is where the analytics application's compromise might have been detected earlier. A properly configured anomaly detection system would have flagged the unusual access pattern, time of day, and request volumes generated by the ransomware process as deviations from the application's established baseline, potentially hours before the outbound traffic alert triggered. Dev implemented behavioral baseline monitoring for all registered third-party applications using the AuditEvent repository as the data source. The system established a rolling thirty-day baseline for each application and generated an alert when current behavior deviated by a configurable threshold.

The periodic re-review process addressed the assumption that a third-party application's security posture, once assessed and approved, remained static. Applications are updated. Vendors are acquired. Backend infrastructure changes. A vendor that was a two-person startup at the time of integration may have been acquired by a larger company with different data practices eighteen months later. A quarterly lightweight re-review is meaningfully more secure than no re-review at all. For high-risk applications with system-level scopes or access to sensitive data categories, Sandra specified an annual comprehensive re-assessment.

HIPAA Security Rule Mapping for FHIR

Maya's question for this section is: how do all these technical security controls connect back to what HIPAA actually requires us to have?

The ransomware incident was the first time Sandra had sat down with a FHIR server configuration and the HIPAA Security Rule text side by side, mapping each technical control to the specific regulatory standard it addressed. The HIPAA Security Rule Technical Safeguards are organized under five standards, each with implementation specifications designated as required or addressable. Required specifications must be implemented; addressable specifications must be implemented if reasonable and appropriate, or the organization must document why an alternative measure was used or why the specification is not applicable.

The Access Control standard at 164.312(a)(1) requires technical policies allowing access only to persons or software programs granted access rights. The FHIR implementation centers on the OAuth scope model: each access token carries specific, reviewed, documented scope grants. The SMART on FHIR registration process grants access rights; token validation controls access at each request.

Diagram 11.8: HIPAA Security Rule to FHIR Control Mapping

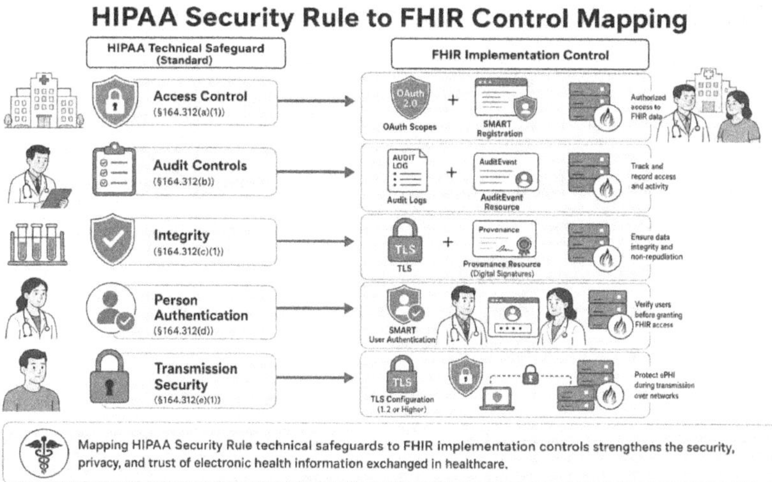

The Audit Controls standard at 164.312(b) requires mechanisms to record and examine activity in information systems that contain or use electronic PHI. The practical requirement to reconstruct access events during a breach investigation means that comprehensive AuditEvent logging for every FHIR resource interaction and failed access attempt is the only defensible implementation posture.

The Integrity standard at 164.312(c)(1) requires measures to ensure electronically transmitted PHI is not improperly modified without detection. TLS addresses this through cryptographic integrity checks embedded in the protocol: any in-transit modification is detected and the connection terminated. For data at rest, FHIR implementations may also use digital signatures on sensitive resources through the FHIR Provenance resource.

The Person Authentication standard at 164.312(d) requires procedures to verify that a person seeking access to electronic PHI is the one claimed. In a SMART on FHIR context, this is addressed through the authentication step of the OAuth flow. Multi-factor authentication, while not required by the HIPAA standard, is strongly recommended for any user whose credentials, if compromised, would provide access to large volumes of PHI. The Transmission Security standard at 164.312(e)(1) requires technical security measures guarding against unauthorized access to PHI transmitted over an electronic communications network; a FHIR endpoint using TLS 1.3 with current certificates and strong cipher suites satisfies this standard, while an endpoint using deprecated TLS versions or expired certificates does not.

The mapping exercise was the first time Riverside General's FHIR security architecture had been documented in terms that directly referenced the HIPAA Security Rule requirements it addressed. That documentation gave Dev a framework for evaluating future security decisions in regulatory terms, gave Patrick a basis for explaining the security investment to the board, and gave Sandra a structured checklist for the quarterly security reviews she was planning to implement.

Sandra's Post-Incident Security Checklist

Maya's question for this section is: after everything that happened, what is the actual list of items that need to be fixed and verified to ensure this cannot happen again?

Thirty days after the incident, Sandra presented the board with a seventy-three-page post-incident report. The section the board spent the most time discussing was the eight-item security checklist representing the minimum security posture Riverside General needed to achieve and maintain. Sandra framed it explicitly as the minimum rather than a comprehensive security program. The checklist was the floor, not the ceiling, and the floor needed to be established first.

The first item was a TLS version and certificate configuration audit covering every FHIR API endpoint, with quarterly repetition and immediate remediation for any endpoint that did not meet the minimum standard. The second item was an OAuth scope review for every registered third-party application, with a documented justification for each retained scope, removal of unnecessary scopes, dual sign-off for new scope grants, and annual re-review.

Diagram 11.9: Sandra's Eight-Point Post-Incident Security Checklist

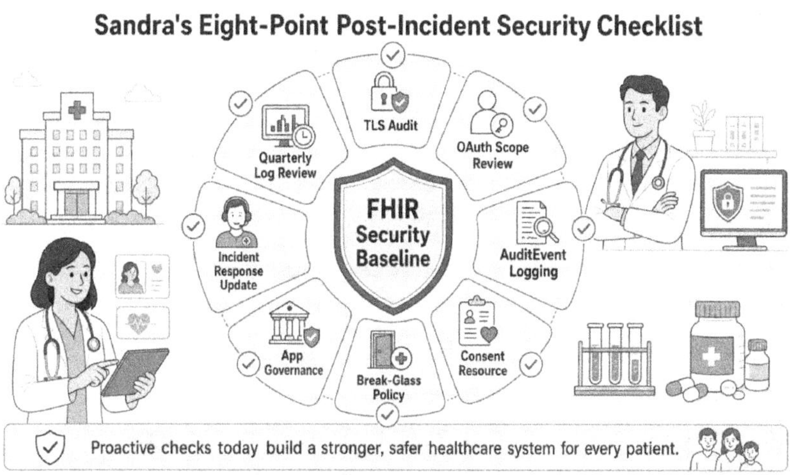

The third item was an AuditEvent logging configuration review confirming that every successful and failed access event generated a properly structured AuditEvent resource, with the required retention and a forensically validated query interface. The fourth item was Consent resource implementation status: a review of which patient consent restrictions were represented as Consent resources and which remained in paper or legacy formats, with a migration timeline. The fifth item was break-glass access procedure documentation and testing, specifying the activation workflow, token parameters, AuditEvent logging requirements, notification routing, and post-event documentation, which was tested quarterly in a sandbox environment.

The sixth item was third-party application security attestation and re-review: all currently registered applications reviewed against the new governance framework within sixty days. The seventh item was an incident response plan update to include FHIR-specific forensic procedures: AuditEvent repository access, token revocation, FHIR server isolation, and the HIPAA breach analysis workflow for FHIR access events, to be tested in a tabletop exercise within 90 days. The eighth item was a quarterly access log review for anomaly detection, looking for patterns suggesting unauthorized access, scope creep, or unusual access volumes.

The board approved the checklist and authorized the resources necessary to complete it. The ransomware near-miss had made it clear to every member of Riverside General's leadership team that FHIR security was not a technical implementation detail. It was an organizational obligation that touched every patient whose data lived on those servers.

Sandra closed her laptop at 9:47 p.m. the night of the board presentation. The conference room was empty except for her and the security checklist printout on the table in front of her. Eight items. A ninety-day clock. Three gaps that had been visible for months to anyone who had looked for them, and that had not been visible because no one had looked. The attacker had not succeeded—this time. The question Sandra was already turning over in her mind was what the next gap was: the one not on any checklist yet, the one that would only become visible after the next incident, or, if she did her job well, the one she would find before an attacker did.

The ninety-day remediation had begun.

Takeaways

Security in a FHIR-based system is a design requirement from the first day, not a layer added after the architecture is built. The ransomware incident at Riverside General did not result in a data breach, primarily because Dev's rapid network isolation cut off the attacker's command-and-control channel before the encryption phase could begin. That outcome had more to do with response speed and a measure of fortune than with a security architecture designed to prevent exactly that type of attack. The remediation work that followed was the construction of a security foundation that should have been in place before the first line of patient data was written to the FHIR server.

Overly broad OAuth scopes are a security failure that presents as a convenience. The analytics application's scope grant had been configured broadly because determining the minimum required scopes takes time and care, and the developer was under pressure to finish. That convenience created a potential attack surface in which a compromised application could access the majority of clinically significant data in the FHIR server. Scope hygiene must be addressed correctly at integration time and reviewed on a defined schedule thereafter.

AuditEvent logging is the difference between a contained security incident and an unquantifiable potential breach. The HIPAA Breach Notification Rule presumes a breach occurred if an organization cannot confidently demonstrate that PHI was not accessed during a security incident. A complete, structured AuditEvent log is the evidence that makes that demonstration possible; its absence creates a patient notification obligation, a legal exposure, and a reputational risk that properly implemented logging would have made unnecessary.

The FHIR Consent resource is a machine-executable access control policy, not a documentation tool. Its value lies in instructing the FHIR server to enforce patient consent at the resource level in real time. An implementation that stores consent documentation in a filing system but does not implement the Consent resource has documentation without enforcement, and documentation without enforcement does not protect patients. Break-glass access is only a security feature if the audit trail it generates is complete, specific, and independently reviewed; a break-glass implementation that does not log every resource accessed with a specific authorization-override indicator, and does not route a meaningful notification to reviewing parties, is an authorization bypass without oversight.

Third-party application security governance must be ongoing because the security landscape is not static. An application approved eighteen months ago may have changed its data access practices, backend infrastructure, or ownership in ways that materially affect its risk profile. Periodic re-review, behavioral monitoring, and a clear trigger framework for off-cycle security reassessments are the operational controls that prevent the overly broad scope problem from reappearing with a different application after the current one has been remediated.

The HIPAA Security Rule, mapped specifically to FHIR technical controls, is a checklist that can be addressed item by item by a FHIR security architecture built around TLS configuration, OAuth scope governance, AuditEvent logging, Consent resource enforcement, and the authentication controls embedded in SMART on FHIR. Organizations that approach HIPAA compliance as a documentation exercise will find the regulations easy to satisfy on paper but difficult in practice; organizations that approach it as a technical architecture requirement will find that a well-designed, rigorously implemented FHIR security architecture is also a well-documented compliance argument. The ninety days that followed Sandra's board presentation were the most demanding and the most clarifying of her time at Riverside General, and what that work produced set the foundation for everything that came next.

12 Interlude: Ninety Days at Riverside

The morning after the ransomware attack stalled at the perimeter, Patrick Osei did not celebrate. He sat in his office with a printed timeline of the previous eighteen hours and read it twice, looking for the part where Riverside General had been lucky rather than prepared. He found several such parts. The encrypted payload had reached a network segment that hosted legacy scheduling software, and only an automated alert, flagged by Dev Okafor at 2:14 a.m., had prevented the lateral spread that would have drawn patient records into the blast radius. Luck and a sleepless engineer had saved them. Patrick understood that neither could be counted on again.

The board meeting three days later was the hardest presentation of Patrick's career. He did not soften the findings. He walked the trustees through a reconstruction of the attack path, described the unpatched interfaces that had served as the initial entry vector, and named the number of patient records that could have been exposed if Dev had not caught the alert. Then he asked for something the board had never fully granted him: a mandate. Not a budget line for one project, but institutional authority to rebuild Riverside's data architecture from the access layer outward, with interoperability standards as the structural backbone rather than an afterthought bolted onto systems already in production. The board voted in forty minutes. Patrick left with a signed remediation charter and ninety days to show measurable progress.

He spent the first two weeks on personnel. Dev had been running the IT integration function with two contractors and a part-time analyst, a staffing model that worked in maintenance mode and failed under pressure. Patrick approved three new hires: a dedicated security engineer, a FHIR integration specialist recruited from a regional health information exchange, and a data governance coordinator who reported jointly to Dev and Sandra Torres. Sandra, for her part, used the remediation window to rebuild Riverside's privacy impact assessments from scratch, working backward from every interface that had touched the compromised segment to document exactly what data had moved, when, and under what authorization. The exercise was painful and clarifying in equal measure. By the end of week three, Riverside had a map of its own data flows that its compliance office had never possessed before.

The moment that mattered most during those ninety days did not appear in any board report. It happened in the fourth week, during a working session in which Dev was walking the new FHIR integration specialist through Riverside's existing HL7 v2 feeds. Dev had spent years treating FHIR as an external requirement, a compliance checkbox handed down from regulators who did not understand the operational complexity of running a mid-size community hospital. He had implemented what was required and defended what was already built. In that session, watching the specialist demonstrate how a properly structured FHIR resource bundle could be validated, versioned, and audited at the transaction level, something shifted. Dev asked a question about access control scopes, then another about audit logging granularity, then a third about how consent directives could be enforced at the API layer rather than enforced manually by staff. The specialist answered each one. By the end of the hour, Dev was at the whiteboard, sketching a proposed architecture for Riverside's patient data access layer that used FHIR not because a regulation said so but because it solved problems Dev had been managing around for five years. Patrick, who had stopped by to check on the session, stood in the doorway long enough to hear Dev say, "This is actually the right way to build it." He did not interrupt.

By day ninety, Riverside had closed the unpatched interfaces, established a formal incident response protocol, completed the privacy impact reassessments, and produced a twelve-month architecture roadmap that Dev and the new integration specialist had co-authored. The roadmap placed FHIR R4 at the center of every new integration and established a deprecation schedule for the point-to-point feeds that had grown up without documentation or oversight. It was not a finished building; it was a foundation plan, specific enough to execute and honest about what remained. Patrick presented it to the board as the beginning of a second phase of work, one that would require sustained investment and organizational patience. The board approved the roadmap without a vote against it. The ninety days had not made Riverside secure. They had made Riverside ready to become secure, which was the more honest claim, and the one worth making.

13 Building Interoperable Systems

The Whiteboard Meeting

Dev Okafor arrived in the conference room twenty minutes early. He uncapped a blue marker and wrote three words across the whiteboard: NATIVE. FACADE. EVENT-DRIVEN. Below them: performance, conformance, migration, cost, timeline.

The ninety days since the ransomware incident had been the most clarifying period of his career at Riverside General. The near-miss had turned architecture from a technical preference into a board-level directive. Patrick had returned from the emergency session with a single instruction: the hospital needed a defensible, documented FHIR architecture, with decisions made, recorded, and owned by someone who understood the implications. That person was Dev.

He had spent the ninety days running three proof-of-concept environments, one for each pattern, benchmarking them, breaking them, and rebuilding them. He had traced the root cause of ONC enforcement actions involving FHIR conformance failures back to a recurring theme: teams had chosen an architecture without understanding what it would cost to maintain.

Dr. Maya Chen arrived with a printed list carrying three items. Dev recognized the format; Maya's lists always contained exactly as many items as the things she refused to negotiate on. Sandra Torres followed with her compliance checklist already open. Patrick arrived last, slightly out of breath from a finance committee meeting.

"Two hours," Patrick said. "Walk us through it."

Dev turned to the whiteboard. "Three decisions," he said. "Everything else follows."

The choice in the room was a commitment that would shape Riverside's integration infrastructure over the next five to ten years. Each pattern represented a fundamentally different answer to the same question: given what Riverside currently had, what it was legally required to expose, and what it could realistically build in the available time, what was the right way to become a FHIR-capable institution? Dev had the answer, and the ninety-day sprint had been built around producing it with evidence rather than opinion.

Why Architecture Decisions Are Expensive to Undo

Maya's question for this section is: why does the choice of FHIR architecture matter so much at the beginning, before you have built anything yet?

In a healthcare IT system, the most expensive decisions are not the ones that cost the most money upfront. They are the ones cheapest to reverse before implementation and most expensive to reverse after go-live. A team that chooses the wrong FHIR pattern, implements it across six months of development, goes live with clinical users depending on it, and then discovers during a US Core upgrade that the chosen pattern cannot be updated without a near-total rebuild, does not experience that cost on the day the decision was made. They experience it eighteen months later.

FHIR architecture decisions are particularly prone to this pattern because the three major patterns make fundamentally different assumptions about where the source of truth lives, how data is stored, and how validation is enforced. A facade assumes that the legacy EHR remains authoritative and that FHIR is a translation layer in front of it. A native store assumes that the FHIR format is the canonical persistence layer and that migration is complete. An event-driven hub assumes FHIR resources are generated on demand from clinical event streams. These assumptions are not interchangeable. Switching from one to another in production requires a data migration, an architectural rebuild, or both.

Dev had watched a peer institution work through exactly this. Their team had chosen a facade under time pressure, without documenting the rationale for the decision, and discovered two years later that their translation logic could not produce new US Core MustSupport elements without rewriting 40% of their transformation code. The rewrite took eight months; the true cost was eight months of deferred integration projects and clinical workflows waiting for promised capabilities.

Patrick absorbed this. "So the ninety-day remediation was about getting to a position where we could make this decision deliberately?"

"Exactly," Dev said. "Before the incident, we had three options and no time to evaluate them properly. Afterward, we had the same options, documented requirements, tested proofs-of-concept, and the organizational authority to choose."

Sandra looked up. "The compliance clock has not stopped. ONC information-blocking enforcement is not waiting for us."

Dev turned back to the whiteboard and drew the first panel.

Diagram 12.1: Three Architectures, One Hospital: Facade, Hub, and Native Store Side by Side

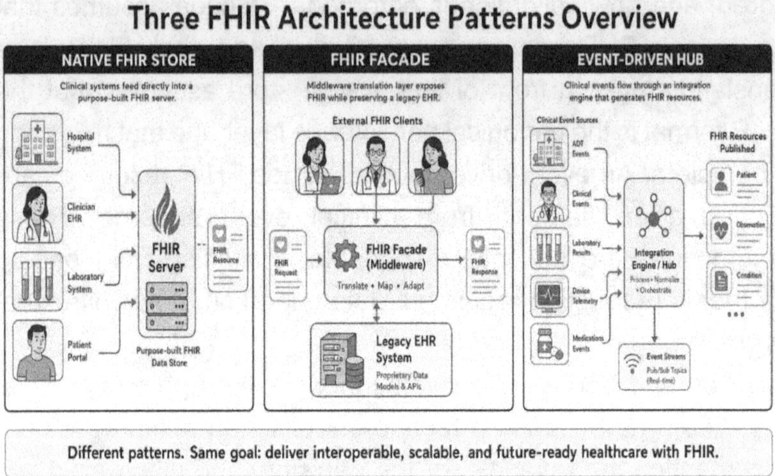

Three FHIR Server Patterns Compared

Maya's question for this section is: what are the three ways a hospital can become FHIR-capable, and how do they differ in what the hospital must build and maintain?

The native FHIR store is the pattern that most closely matches the theoretical ideal of FHIR adoption. A purpose-built server, such as HAPI FHIR Server, Azure API for FHIR, Google Cloud Healthcare API, or AWS HealthLake, becomes the primary persistence layer. Source system data is migrated to the FHIR format before storage, and the server's built-in validation, search, and versioning capabilities operate directly on the stored data without a translation layer.

The native store's advantages are substantial. Conformance validation happens at write time. Search parameter behavior is predictable because the search index is built directly from FHIR resources. When a new US Core version is published, the upgrade path is clear: update the profiles, run the validator, identify non-conformant resources, and remediate. The disadvantage is equally substantial: a native store requires migrating clinical data from whatever format the legacy EHR uses before the system can serve that data. For Riverside, whose EHR had accumulated patient records for over two decades, that migration was not a ninety-day project. It was a multi-year program that required significant data-quality preparation work. The native store was the right long-term destination; it was not the right first step.

Diagram 12.2: How Riverside's Facade Makes a Legacy System Look FHIR-Native — What the Translation Layer Promises the Client and What It Hides About the Source

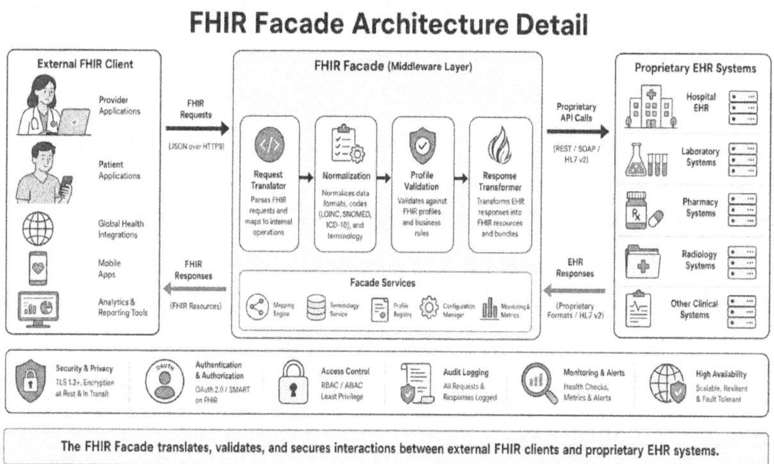

The FHIR facade presents an existing system's data as FHIR resources through a translation layer, without migrating the underlying data. A middleware layer intercepts FHIR API requests, translates them into the legacy EHR's proprietary API calls, transforms the response into conformant FHIR resources, and returns them to the requesting client. The legacy EHR remains the source of truth. The facade's primary advantage is speed of compliance: implementation is bounded by the complexity of translation logic, not by data volume, and legacy workflows are not disrupted. Its primary risk is fidelity: the translation layer must produce resources that are genuinely conformant, not merely structurally valid FHIR that fails to carry the clinical meaning that requesting systems expect.

The event-driven hub most closely resembled the integration engine architecture Dev had been building at Riverside. Source systems emit clinical events: ADT messages, ORU lab results, pharmacy events, scheduling updates. The integration engine receives these events, generates FHIR resources through transformation logic, and routes them to FHIR-capable consumers. At Riverside, extending Mirth Connect to produce FHIR output was a bounded project rather than an open-ended migration program. The challenge is completeness: the hub can only serve queries about data that has passed through the event pipeline. Historical data predating the hub's deployment is invisible to FHIR consumers.

Dev's decision matrix pointed to a hybrid approach: deploy a facade to meet the immediate ONC compliance deadline, run the event-driven hub for new integration work, and plan the native store migration as a three- to five-year program.

Patrick studied the matrix. "The answer is all three, sequenced."

"Sequenced deliberately," Dev said, "with the right pattern in the right role at each stage."

He said it, and then he paused long enough that Patrick noticed. Dev's agreement with the sequenced plan was honest; his concern about it was as well, and he wanted it on the record before the meeting ended. "The risk I am still carrying," he said, "is the facade. We will get it to conformance for the deadline. I am confident about that. What I am less confident about is what happens in year three, when the legacy system has had two more vendor updates, the US Core profile has had a minor revision, and the translation logic has accumulated patches no one wants to refactor. Facades are easy to deploy and hard to retire. We need to write down, today, the conditions under which we will decommission this one, or we will still be running it in year seven." Sandra wrote down: facade decommissioning criteria; owner: Dev; due before go-live. Patrick nodded once. The doubt was logged, and the meeting moved on.

The Facade Pattern in Practice

Maya's question for this section is: if the facade is just a translation layer, why does it take so much work to build correctly?

The work is not in building the facade framework. The work is in making the translation logic produce resources that are genuinely conformant across all the conditions that production data can create, and in maintaining that conformance as both the profile and the underlying legacy system evolve.

Dev had built the first version of Riverside's FHIR facade during the proof-of-concept phase. It took three days to produce a facade that responded to a FHIR Patient read request with a structurally valid resource. It took six more weeks to produce a facade whose responses were conformant with the US Core Patient profile across the full range of patient demographics in Riverside's EHR. The gap between those milestones was not in framework knowledge but in data quality, edge cases, and legacy system behavior.

The first conformance issue concerned patient identifiers. Riverside's EHR stored them across three fields: a medical record number, an enterprise patient identifier assigned during a system consolidation, and a legacy identifier that predated the consolidation. The US Core Patient profile required all identifiers to appear as FHIR Identifier resources with a specific system

URI and use code. Dev's initial translation logic had mapped only the medical record number. A US Core validator flagged the missing enterprise identifier as a MustSupport gap.

The fix was not technically difficult. What made it time-consuming was discovering that many enterprise identifier fields had been populated manually during the consolidation, without format validation at the time of entry. Dev had to build a normalization step that cleaned the values before mapping them. Similar patterns repeated across nearly every data element: dates stored without time zones, name fields mixing given names and suffixes, address fields containing building names where street addresses were expected, race and ethnicity codes from a legacy terminology that predated the OMB categories US Core referenced.

Diagram 12.3: The Six Transformation Steps Between a Legacy Record and a Valid FHIR Response — and the Step Where Dev's Facade Most Often Produces an Invalid Resource

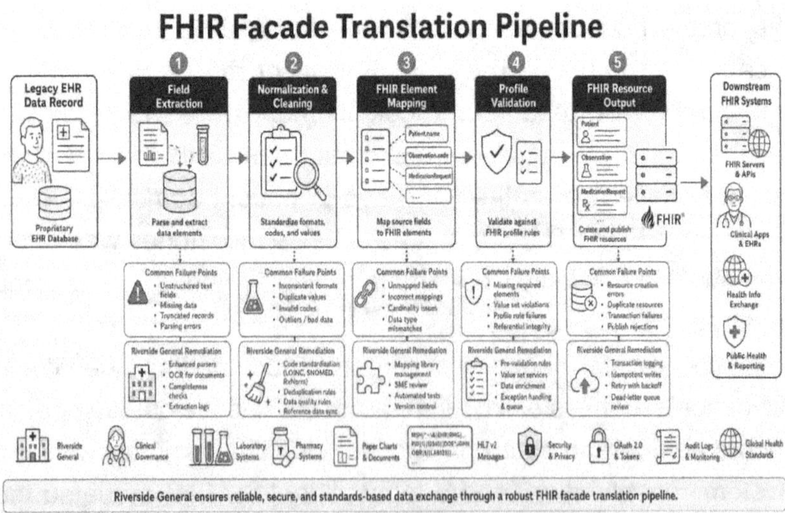

Sandra had been listening with the attentiveness she gave to processes that had compliance implications. "So the facade is also a data quality enforcement layer," she said.

"Correct," Dev said. "And it must be tested against production data, not just synthetic test cases. You cannot know what the legacy data quality problems are until you run the full translation logic against the actual database."

Maya looked up. "What happens if a translation fails in production? Does the patient get incorrect data, or no data?"

"That depends on how you build the error handling," Dev said. "Built correctly, they get an OperationOutcome that explains why the request failed. When built incorrectly, they get either incorrect data or a confusing error. Neither is acceptable in a clinical system." He paused. "Which brings us to the next topic."

OperationOutcome: The FHIR Error Report

Maya's question for this section is: when a FHIR request fails, what does the server actually tell you about what went wrong?

The server returns an OperationOutcome resource. Fluency in reading it separates an experienced FHIR implementer from a developer who has learned the specification but has not yet worked with a production system.

An OperationOutcome carries a structured list of issue elements, each with four fields that Dev reads in a fixed sequence, every time. The first field is severity: fatal means the request could not be processed at all; error means the request failed but was identified and described; warning means the request succeeded with conditions the client should note; information means the server has context to communicate that does not affect success or failure.

The second field is code: a machine-readable category from the FHIR issue-type value set. Common codes included not-found, for absent resources; invalid, for syntactically or semantically invalid elements; required, for missing mandatory elements; value, for elements outside permitted ranges; and processing, for general server-side failures.

The third field is diagnostics: the free-text description of the specific problem. A well-implemented server returned diagnostics specific enough to guide remediation. "The element Patient. identifier[0].system is required but was not present" was actionable. "Validation error" was not. Dev had developed strong opinions about FHIR server quality based on diagnostics quality, because it predicted the quality of the rest of the implementation.

The fourth field is location: a FHIRPath expression pointing to the specific element where the error occurred. In a complex Bundle submission, location values were the difference between a five-minute diagnosis and a forty-five-minute one.

Diagram 12.4: How Dev Reads an OperationOutcome to Find the Actual Error — Why the Issue Array, Severity, and Diagnostics Fields Must Be Read in Sequence

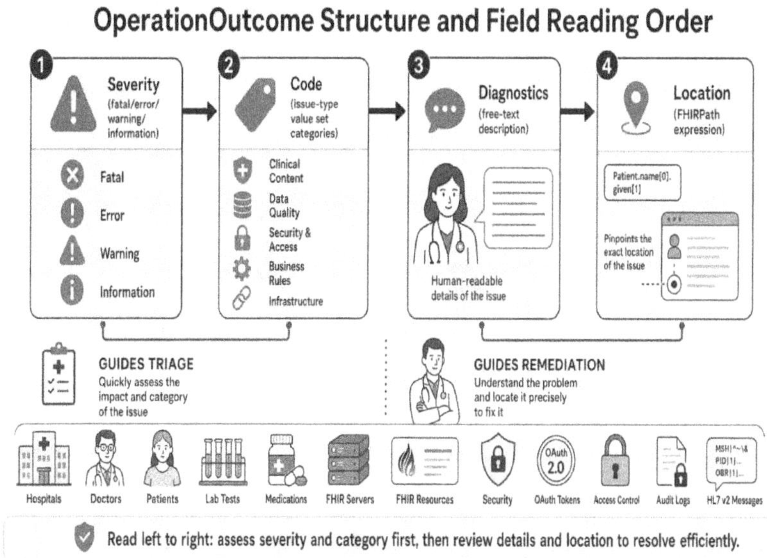

Dev walked Maya through a real OperationOutcome from the facade proof-of-concept. It had been returned when the facade submitted a Patient resource whose birthDate element contained a two-digit year, a value Riverside's legacy EHR had stored for records entered before a Y2K remediation project that had not fully updated historical data. Severity: error. Code: value. Diagnostics: "The value '47' is not a valid date in the format YYYY, YYYY-MM, or YYYY-MM-DD." Location: "Patient.birthDate." Dev had added a normalization step that interpreted two-digit years using a 1935 pivot: years 00-34 mapped to 2000-2034, and years 35-99 mapped to 1935-1999. For patients born before 1935 with two-digit records, the normalization produced a wrong date, and those records were flagged for manual review.

Sandra made a note. "That review workflow needs to be documented as a data quality remediation procedure for audit purposes."

"Already drafted," Dev said. He had learned during the ninety-day sprint that documentation was not a post-implementation task. It was the thing that transformed a proof of concept into a defensible production system.

Idempotency: Safe Operations in Unreliable Networks

Maya's question for this section is: what does it mean for a medical record operation to be safe to repeat, and why does that matter in a hospital?

It matters because networks fail, integration engines retry, and without the right controls, a retry that creates the same patient record again produces a duplicate. Duplicates are not merely data quality problems. They are patient safety problems: a patient whose records are split across two identifiers might receive a medication order that does not account for a documented allergy in the record the prescribing physician did not see. What is less well understood outside the integration engineering community is how often duplicates arise from transport-layer-triggered retries in automated message delivery, even after the original message has already been processed.

FHIR addressed this through conditional operations. The conditional create allowed a sending system to submit a new resource with an implicit instruction: create this resource only if a matching resource does not already exist. The server evaluated the search query and, if it found an existing match, returned that resource rather than creating a new one. The conditional create used the If-None-Exist HTTP header, whose value was a FHIR search query. For a Patient resource, the query searched by the patient's medical record number or enterprise identifier.

Dev showed a short pseudocode block to illustrate the pattern:
FUNCTION submit_patient_if_new(patient_resource, mrn): existing = fhir_GET("/Patient?identifier=MRN|" + mrn) IF existing.total == 0: response = fhir_POST("/Patient", body=patient_resource, header="If-None-Exist: identifier=MRN|" + mrn) RETURN response # 201 Created ELSE: RETURN existing.entry[0].resource # 200 OK, no new record

"The If-None-Exist header makes this idempotent," Dev said. "The server performs the search and conditional check atomically. Even if two requests arrive simultaneously, the server treats them as a single operation producing a single result."

Patrick was direct. "Without that header?"

"A POST always creates a new resource. Run it twice, you get two patients. Run it ten times during a network retry event, you get ten patients."

Diagram 12.5: Why Dev Uses Conditional Creates Instead of Plain POSTs — How Idempotency Prevents Duplicate Resources When the Network Drops Between Retry Attempts

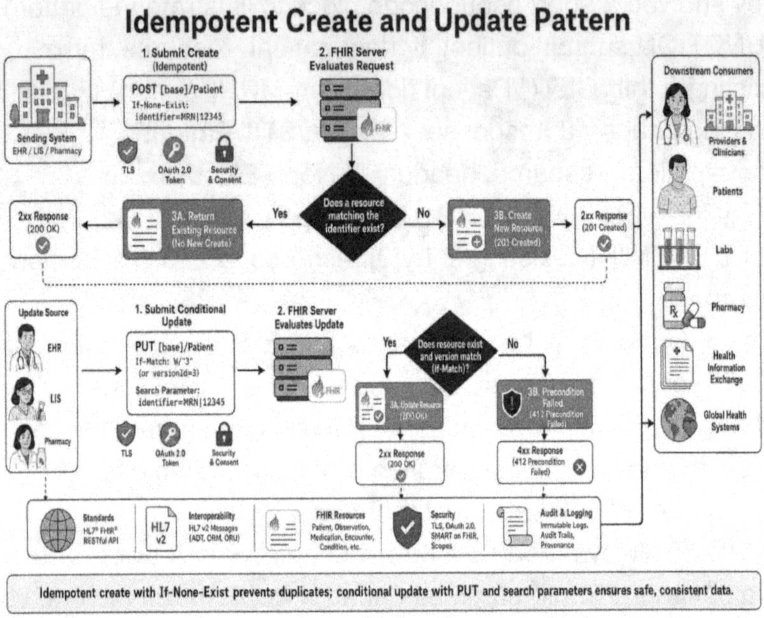

Any operation that an automated system could retry had to be safe to retry. In a healthcare integration environment where network partitions caused sending systems to lose track of whether their last request had succeeded and integration engines automatically retried failed deliveries, the assumption that any individual request would be executed exactly once was not safe. Idempotency was the engineering discipline that ensured the system remained correct even when the assumption was violated.

Resource Versioning and Conflict Resolution

Maya's question for this section is: what happens when two different systems try to update the same patient record at the same time, and how does FHIR prevent one update from erasing the other?

FHIR addresses this through optimistic locking, implemented via HTTP ETags. An ETag is a version identifier the server assigns to each resource version and includes in the HTTP response header. When a client wants to update a resource, it includes the ETag it received from its last read in an If-Match header on its PUT request. The server checks whether the current resource version matches the If-Match value. If they match, the resource has not been modified since the client last read it, and the update proceeds. If they do not match, another client has modified the resource in the intervening time, and the server returns a 409 Conflict response rather than processing the update.

This prevented the write-behind problem. Without ETag-based locking, two clinical systems simultaneously updating the same Patient resource would result in the second update silently overwriting all fields of the first, with no error and no record of the loss.

With ETag-based locking, both systems read the Patient at version 3 and receive ETag "3". Both submit updates with If-Match: "3". The first arrives; the server advances the resource to version 4, returns 200 OK with ETag "4". The second arrives; the server compares version 4 to the requested If-Match of "3" and returns a 409 Conflict. The second system must re-read version 4, merge its intended changes, and resubmit.

Dev showed a short pseudocode sketch of the retry-on-conflict pattern:

FUNCTION update_patient_with_retry(resource_id, new_data, max_retries): FOR attempt IN 1..max_retries: current = fhir_GET("/Patient/" + resource_id) etag = current.headers["ETag"] merged = merge_changes(current.body, new_data) response = fhir_PUT("/Patient/" + resource_id, body=merged, header="If-Match: " + etag) IF response.status == 200: RETURN success

IF response.status == 409: CONTINUE RETURN failure_after_max_retries

The merge step was where the real engineering work lived. Dev had spent two weeks producing a merge strategy document for the eight resource types that multiple clinical systems at Riverside could update concurrently: a Riverside-specific policy defining which system's value prevailed in each conflict scenario.

Diagram 12.6: What Happens When Two Systems Update the Same Resource at the Same Time — How ETag Matching Catches the Conflict Before the Later Write Silently Overwrites Patient Data

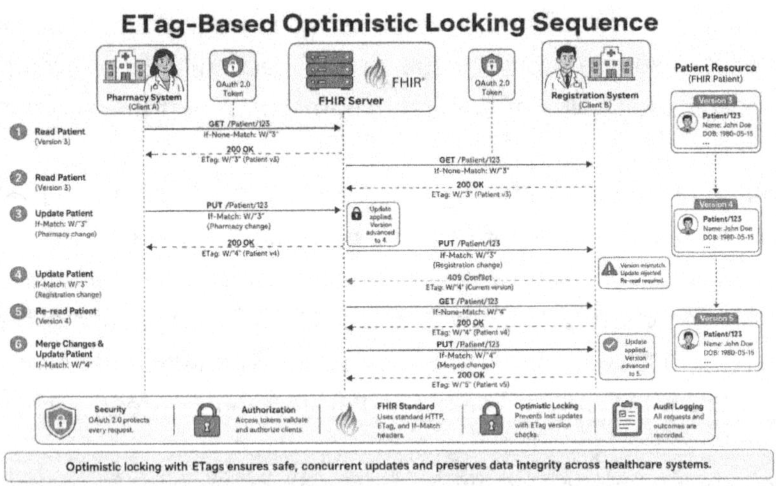

Patrick looked up from the diagram. "Does the FHIR specification tell you how to handle the merge?"

"The specification defines the locking mechanism," Dev said. "The merge strategy is an institutional decision. The specification cannot tell Riverside which system is authoritative for a patient's preferred language field when both the EHR and the patient portal store and update it. That is a data governance decision."

Sandra made a note without looking up. "Data governance decision, owner: clinical informatics and compliance, jointly. I want that documented before we go live."

Provenance: Who Touched This Data

Maya's question for this section is: when a lab result or a medication record appears in a patient's FHIR record, how do we know where it came from and whether we can trust it?

The mechanism FHIR provides is the Provenance resource: structured metadata records that capture, for each clinical resource, where the data originated, when it was recorded, who or what system created it, what transformations it underwent, and who is accountable for its accuracy. In a multi-source FHIR environment, Provenance resources are the difference between a repository that looks comprehensive and one that is actually trustworthy.

At Riverside, data flowing into the FHIR environment would come from at least five distinct origins. The legacy EHR contributed patient demographics, encounter records, clinical notes, and orders. The laboratory information system contributed lab results. The pharmacy system contributed medication administration records. The radiology system contributed imaging reports. The patient portal would contribute patient-reported data. Each source had different levels of clinical authority and different validation standards.

A clinician reviewing a medication list assembled from these sources faced a question the aggregated data could not answer without Provenance records: which entries came directly from the pharmacy system with pharmacist verification, which were transcribed from paper by a registration clerk, which were imported via a v2 message that may have transformed, and which were entered by the patient through a portal that did not validate against the formulary? Without Provenance resources, that question was unanswerable.

The FHIR Provenance resource captured this in a structured format. The target element referenced the resource whose provenance was being recorded. The agent element identified who or what performed the activity, distinguishing between practitioners, organizations, software agents, and integration engines. The entity element described any prior resource from which the target was derived, enabling documentation of transformation chains in which an FHIR resource was generated from a v2 message or a proprietary EHR record.

Diagram 12.7: Provenance Resource Chain for Multi-Source Data

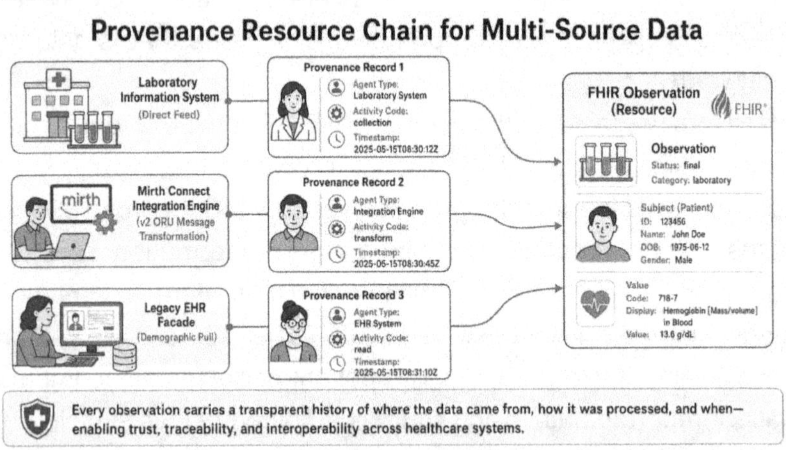

Dev had designed Riverside's Provenance generation strategy during the ninety-day sprint. Resources generated from a v2 message would include a Provenance resource that identifies the sending system, message type, and transformation engine. Resources served by the facade would carry a Provenance resource identifying the legacy system as the agent. Resources submitted directly through the FHIR API would include a

Provenance resource that identifies the responsible practitioner or organization.

Maya looked at the design. "So a physician could look at a medication entry and see that it came through the integration engine from a v2 message, rather than from a direct pharmacy system feed?"

"Yes," Dev said. "Whether the clinical application surfaces that is a vendor UI decision. But the information is in the FHIR data."

"It should surface it," Maya said. "That context affects clinical judgment."

Sandra agreed. "It is also the audit capability that ONC's information-blocking rules assume is present. If you cannot demonstrate where data came from and who is responsible for it, you cannot defend your compliance posture when an enforcement question arises."

The Capability Statement

Maya's question for this section is: before building an integration with a new vendor's FHIR system, is there a way to find out what it actually supports without weeks of trial and error?

Yes. The Capability Statement is that mechanism. Every FHIR server is required to publish a Capability Statement at its /metadata endpoint: a structured FHIR resource, not a marketing document, that declares the server's actual capabilities in machine-readable form. It specifies which resource types the server supports, which RESTful interactions it implements for each resource, which search parameters it honors, which profiles its resources claim to conform to, which terminology services it provides, and which security mechanisms it requires.

Dev had developed a habit of reviewing Capability Statements before any substantive vendor conversation. The Capability

Statement could not be fabricated the way a slide deck could. It reflected what the server actually implemented, and the gaps between what it declared and what a profile required were precisely the gaps that would become integration problems in production.

The fields Dev checked first were the resource-specific capability entries for Patient, Encounter, Observation, Condition, MedicationRequest, DiagnosticReport, DocumentReference, and AllergyIntolerance. For each, he examined three things: the supported interactions list, confirming the server supported read, search, and create rather than only read; the searchParam list, confirming support for the search parameters Riverside's integration logic depended on; and the supportedProfile list, confirming conformance to the specific US Core profiles Riverside's ONC compliance required.

The supportedProfile declaration was the most important and the most frequently misleading entry. A declaration of conformance was a claim, not a guarantee. Verification required running a test suite against the server: submitting test resources, checking validation responses, executing search queries, and comparing results to the profile's requirements. Dev encountered multiple Capability Statements declaring US Core R4 conformance for resources that failed basic validation tests.

Sandra had encountered this from the compliance side. "We accepted a vendor's Capability Statement as evidence of conformance, then discovered during testing that their Patient resource did not carry the race and ethnicity extensions US Core required. Their position was that those extensions were in a future release."

"Which is why we test," Dev said, "not just read."

Diagram 12.8: Capability Statement Evaluation Checklist

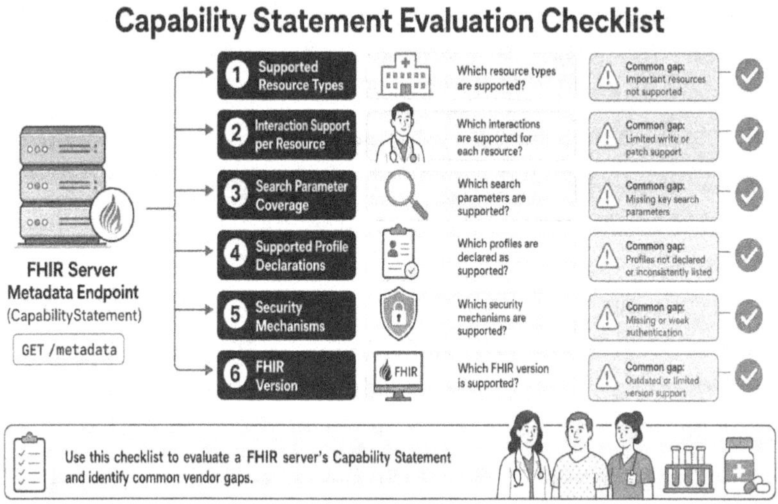

The Capability Statement also served as the simplest possible health check for a FHIR server. In complex integration environments where a server sat behind an API gateway, a load balancer, and a network security layer, the ability to fetch the Capability Statement from the external-facing URL confirmed that the entire network path was open. Monitoring systems at Riverside polled the /metadata endpoint at 15-minute intervals solely to verify reachability. If the poll failed, an alert fired.

Validation at the Gate

Maya drew the last box of the morning at the bottom of the whiteboard, then turned to the group. "I want this one to be the question we close on," she said. Beneath the box she wrote, in deliberate capitals: if bad data enters the FHIR system, why is it so hard to clean it up later, and why is it better to reject it at the start? Sandra was already writing in her notebook. Patrick was watching the whiteboard. Dev set his coffee down.

The answer is in how FHIR data propagates. A non-conformant resource that enters the store does not sit in isolation. The search engine indexes it, returns it in response to queries, and may be the target of references from other resources: a Condition references the Encounter during which it was diagnosed; a MedicationRequest references the Patient for whom the medication was ordered. Every resource that references a non-conformant resource inherits the data quality problem.

The versioning history worsens the compounding. Every update creates a new version, but earlier versions are preserved. A non-conformant resource created at day zero and corrected at day ninety still exists in non-conformant form across its earlier history versions, retrievable by any client querying the history endpoint. Retroactive correction requires either suppressing those history versions or documenting them as superseded.

The correct engineering response is validation at ingestion. Every resource submitted to the store should be validated against the declared profile before persistence. If it fails, the server returns an OperationOutcome with the specific validation errors and rejects the resource. The submitting system receives an actionable error, and the FHIR store never sees the non-conformant data.

Dev implemented this in the Riverside proof-of-concept using the HAPI FHIR Server's built-in validation infrastructure, which applied profile definitions in the StructureDefinition format to incoming resources before persistence. The validation added approximately 40 milliseconds per resource, which is acceptable for Riverside's peak write volume of roughly 3,000 resources per hour. Patrick asked whether that overhead would be a problem at scale. "Not for our write volumes," Dev said. "For institutions writing ten times that volume, it becomes a design consideration. For Riverside, no." The 40-millisecond cost prevented a data quality problem that, once inside the store, would have required a multi-day remediation effort.

Diagram 12.9: Three-Point Validation Architecture

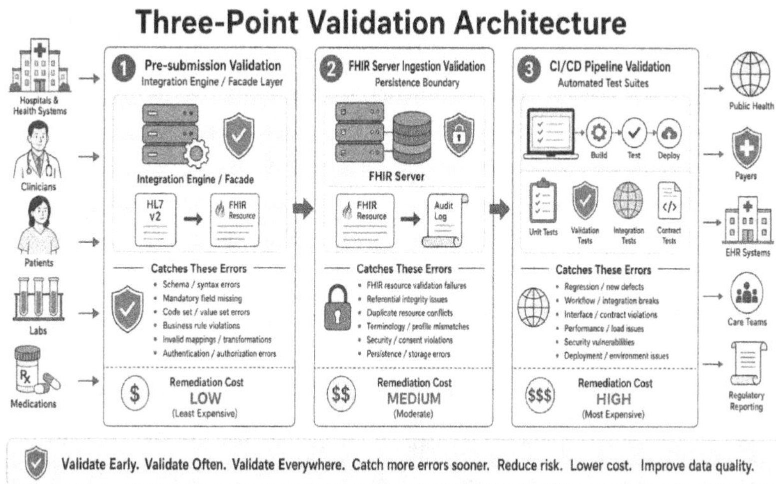

Validation at the server boundary was one of three validation points in Dev's quality assurance framework. The second point was in the production system: the integration engine channel or facade translation layer should validate each resource before submission, catching conformance errors with better diagnostic context at their point of origin. The third point was in the CI/CD pipeline: every code change to a transformation channel should run a validation test suite against production-equivalent test data before deployment.

Sandra asked whether validation rejections were logged for audit purposes. "Yes," Dev said. "The OperationOutcome is the rejection record. The server's audit log captures every rejected submission, retained for the same period as the underlying clinical records."

Maya looked up from her notepad. "If the facade produces a resource that passes validation, does that mean the clinical content is correct, or just the format?"

"The format," Dev said. "Validation confirms structural conformance: required elements present, coded values from the specified value sets, cardinality rules satisfied. A resource can be perfectly conformant and clinically wrong if the source data was wrong."

"So conformance is necessary but not sufficient," Maya said.

"Correct. Conformance is the floor. Clinical accuracy is what we build above it."

Architecture Governance After Go-Live

Maya's question for this section is: once the FHIR system is running, how does the organization ensure it remains conformant as standards continue to change?

Most organizations discovered they had not answered this question until the first time a profile version change was deployed to production. US Core released new versions on a roughly annual cycle, and ONC's certification criteria were updated to reflect those references as new versions were certified. An institution that achieved conformance with one US Core version and then did not update its infrastructure when the next was released was not maintaining conformance. It was accumulating conformance debt at a rate determined by profile update frequency and the complexity of changes between versions.

Dev had built Riverside's FHIR governance framework during the ninety-day sprint. The framework had three components. The first was a profile registry: a maintained list of every US Core profile Riverside's FHIR environment claimed to support, the specific version deployed, and the date each profile was last validated against the live server. The registry made a conformance claim specific and testable: not "we support US Core" but "we support US Core Patient R4 3.1.1, last validated March 12."

The second component was a change monitoring process. Dev had configured automated alerts for new releases from the HL7 FHIR specification, the US Core implementation guide, and the ONC certification criteria. Each alert triggered a review within two weeks: did the new version change any profile Riverside had declared conformance to, and if so, what would the migration effort require? The result was a written assessment delivered to Sandra for compliance documentation and to Patrick for budget awareness.

The third component was a conformance test suite that could be run against the production environment at any time: submitting test resources to the live server, executing search queries, and comparing results to the expected conformance behavior of the declared profiles. Running it quarterly and after any significant system update detected conformance regressions before they became compliance failures.

Sandra had added a fourth component: a vendor conformance attestation requirement. Every FHIR-capable vendor connecting to Riverside's environment was required to provide an annual written attestation of the FHIR and US Core versions they support, along with a Capability Statement for their currently deployed version. Riverside reserved the right to terminate a vendor connection if the attestation revealed a conformance gap the vendor could not remediate within a defined period. The attestation put the compliance obligation on the vendor, not just on Riverside.

Dev found Sandra's addition exactly right. "The governance framework is as important as the architecture," he said. "The architecture is a moment in time. The governance is what makes it last."

Patrick looked at the summary. "Are we ready to make the decision?"

"Yes," Dev said. "Facade for the immediate ONC compliance requirement, event-driven hub for new integration work in the future, native store migration as a funded program in the three-to-five-year capital plan. The decisions are documented, the proofs-of-concept support the recommendation, and the compliance timeline is achievable."

Sandra nodded. Maya looked at her list. One item remained. Dev had identified her three protected clinical workflows during the ninety-day sprint and had designed the facade to leave them untouched. He walked through each one. She checked each box, then set the list down.

"Build it," she said.

Takeaways

Maya's question for this section is: if you had to summarize everything this chapter covered into the decisions that matter most, what would they be?

Architecture decisions made in the design phase are expensive to reverse in production. The choice among a native FHIR store, a FHIR facade, and an event-driven hub should be based on a documented analysis of the organization's actual capabilities: legacy system migration feasibility, conformance profile requirements, regulatory compliance timeline, and ongoing maintenance costs across version upgrade cycles. The hybrid approach- facade for immediate compliance, event-driven hub for new integration work, and native store migration as a funded long-term program- maps the right tool to each stage.

The FHIR facade is the most pragmatic path to immediate compliance for most hospitals, but that pragmatism comes at a price. Facade translation logic must produce genuinely conformant resources, not structurally valid approximations. This requires testing against production data rather than synthetic test cases, and ongoing maintenance as both profile requirements and the underlying legacy system change. The fidelity risk is real, and it is the institution's responsibility to manage it.

Idempotency is a patient safety feature. Conditional create and conditional update operations prevent duplicate record creation from message delivery retries. Every resource creation operation in a production FHIR environment should use conditional create semantics, because the network conditions that cause retries are normal operating conditions, not edge cases.

ETag-based optimistic locking prevents the write-behind problem in multi-system environments. Conflict resolution requires data governance decisions about which system is authoritative for which fields, and those decisions must be made and documented before go-live. The merge strategy document is as much a compliance artifact as it is an engineering one.

Provenance resources are the trust infrastructure of a multi-source FHIR environment. Without them, a clinical user cannot distinguish between a medication entry verified by a pharmacist and one transcribed from paper by a registration clerk. Provenance makes data aggregation trustworthy rather than merely comprehensive, and it is precisely the audit capability that ONC's information-blocking rules assume is in place.

The Capability Statement is the most efficient tool for vendor evaluation. Fetch it before any integration discussion, test conformance claims against it before committing to an integration, and monitor it after go-live through the vendor attestation process. A vendor whose deployed Capability Statement does not match their claimed conformance profile has communicated something important about the reliability of their stated commitments.

Validation at the server boundary is the control that prevents bad data from becoming permanent. A FHIR server that accepts and stores resources without validating them against declared profiles will accumulate non-conformant data, causing search anomalies, clinical decision support failures, and data quality incidents whose retroactive remediation costs orders of magnitude more than the ingestion-time validation overhead. Three-point validation, at the producing system, at the server boundary, and in the CI/CD pipeline, is the complete answer. Conformance ensures the floor is solid; clinical governance is what Riverside builds above it.

Chapter 13 follows Riverside into the full migration planning process: project sequencing, stakeholder management, data quality preparation, and the decision criteria for advancing from the facade to the native store. The governance framework built during the ninety-day sprint will be the difference between a migration that stays on track and one that discovers problems too late to recover without disrupting clinical operations.

14 Planning the Migration

The Board Said Yes

Patrick Osei had spent eleven days on the presentation. He distilled six weeks of internal debate into a single page: the Twenty-Four-Month FHIR Modernization Roadmap for Riverside General Medical Center. Four strategic objectives. Three regulatory compliance obligations. Two vendor categories. One headline budget number. The board approved it in fifteen minutes and moved on to the next agenda item.

What the board did not see was the conversation that had preceded that approval, and the weeks between the architecture planning sessions that Dev Okafor and Dr. Chen had led. The moment Patrick walked into the boardroom, the migration leadership team had worked through the questions the one-page roadmap could not contain: what to do with 217 active integration routes and every Z-segment dependency embedded in them, how to structure a vendor evaluation that protected Riverside against lock-in, what a training program for three distinct audiences actually looked like, and how to write a rollback plan that the team could execute under pressure. The board approved a roadmap. The team still had to build the map.

This chapter is that conversation, reconstructed into a practical migration-planning framework. The planning failures that derail FHIR migrations are consistent enough across organizations that a framework developed from Riverside's experience has wide applicability: scope underestimated because the interface inventory was never completed, timelines collapsed by Z-segment dependencies that appeared only at go-live, clinical staff unable to use a technically successful migration because training had been deferred, and rollback plans written the week before go-live rather than the month before. The organizations that succeed treat migration as a change management project that uses technology, not a technology project that requires some change management.

Patrick carried one half of that equation. He had managed Riverside through regulatory transitions and capital construction projects. He recognized in the FHIR migration the same basic structure: a defined deliverable, a constrained budget, stakeholders whose buy-in had to be earned and maintained, and a schedule that would not survive contact with reality unless it had been built with buffer to absorb the discoveries that planning cannot anticipate. Dev carried the technical half: the architectural judgments, vendor evaluation criteria, phase-sequencing logic, and validation framework. Between them, the framework that emerged from six weeks of structured debate became something more useful than either could have produced alone.

Diagram 13.1: Who Is at the Table When Riverside Plans the R5 Migration

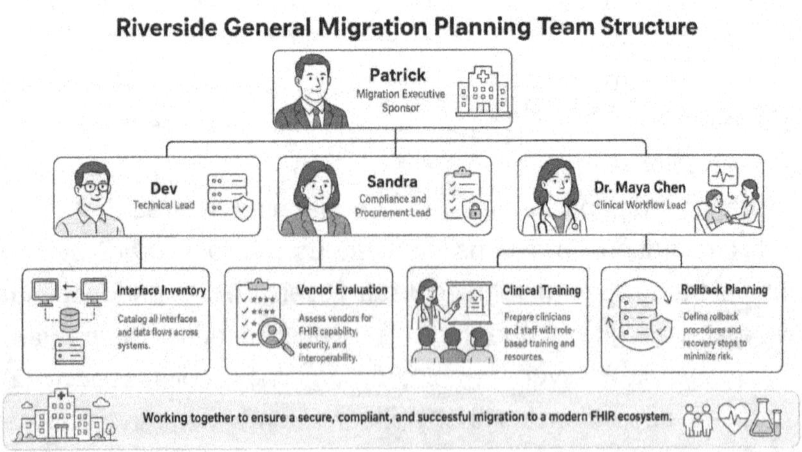

Why Migrations Fail Before They Begin

The projects that derail in health IT migration do not fail because the technology does not work. FHIR servers, transformation engines, SMART authorization frameworks, and validation tools are well-documented, widely deployed, and mature enough for production deployment by any competent health IT team with sufficient time and budget. The failures happen upstream of the technical work.

Dev had seen this pattern before Riverside. In a previous role, a FHIR pilot collapsed not because the FHIR implementation was defective but because the project had never produced a current-state inventory of the interfaces it was replacing. Scope was estimated from memory and from the most-watched dashboards. The post-mortem identified 23 routes that were absent from the migration scope. Eleven had clinical workflow dependencies. Three had patient safety implications requiring immediate rollback. That experience gave Dev a clear understanding of the relationship between planning quality and migration outcome.

Patrick brought a complementary lesson from regulatory transitions at Riverside. The organizations he had watched struggle had capable technical staff and adequate budget. What they lacked was leadership that treated the transition as a change requiring sustained organizational attention. Staff training was scheduled and then postponed. Communication plans were drafted but not executed. Rollback procedures were documented but never distributed to the people who would need them.

These two failure modes, scoping failure and change management failure, share a common root: the misconception that a migration project is primarily a technical project, and that planning work which does not directly produce technology is overhead rather than value. Organizations that succeed in FHIR migrations reject this framing. They invest in current-state documentation because it produces the risk register that drives every subsequent decision. They invest in staff training because an untrained clinical staff generates support volume that consumes the migration team's capacity during hypercare. They write and rehearse rollback plans because a rollback executed in fifteen minutes is categorically different from one improvised over four hours, during which clinical workflows are disrupted.

Diagram 13.2: Common FHIR Migration Failure Modes

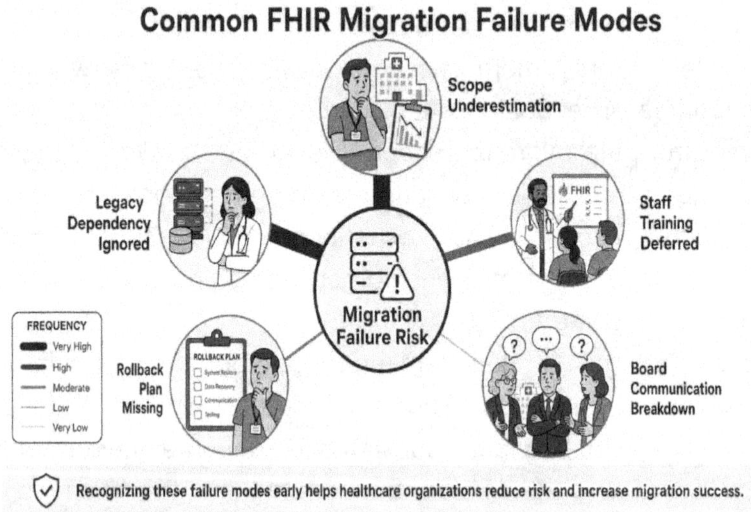

The Current-State Assessment

Dev did not want to do the current-state assessment. He said so directly in the first migration planning session. He had managed Riverside's integration estate for nine years and knew where the interfaces lived, what they carried, and which would require the most attention during migration. Documenting what he already knew felt like time that could be spent building the new system.

Patrick asked Dev a single question: how many interfaces did Dev know about?

Dev thought for a moment and said approximately 180.

Patrick then asked Sandra to pull the integration engine's full route report. It showed 217 active routes. The thirty-seven-route gap was not a failure of Dev's competence; it was a predictable artifact of nine years of incremental growth. Interfaces had been added, modified, and occasionally repurposed in ways not always reflected in the documentation Dev consulted most often. The forgotten routes were the ones that had worked without requiring attention for years, stable enough to disappear into background infrastructure. Those were also the routes most likely to contain undocumented dependencies.

The assessment Dev's team conducted over five weeks became the most-referenced artifact in the migration planning process. For each of the 217 routes, the inventory captured the source system and version, destination system and version, message type, transformation logic, clinical workflow dependency, business owner, error-handling policy, and Z-segment content. Riverside's inventory identified 41 routes carrying Z-segment data; 22 had clinical workflow implications; 8 carried Z-segment content with no direct FHIR representation and no custom extension.

Those eight routes required decisions the technical team could not make alone: What clinical purpose does this data serve? Is it still active or legacy carry-forward? If active, what is the FHIR representation strategy? Who is the accountable business owner? The answers required conversations with clinicians, department heads, and in several cases the original system vendors. A migration plan that had skipped the assessment would have encountered those conversations at go-live, where the context for making good decisions no longer exists.

Diagram 13.3: What Dev's Interface Inventory Must Document for Every Active Route Before the Migration Team Can Assign a Conversion Priority

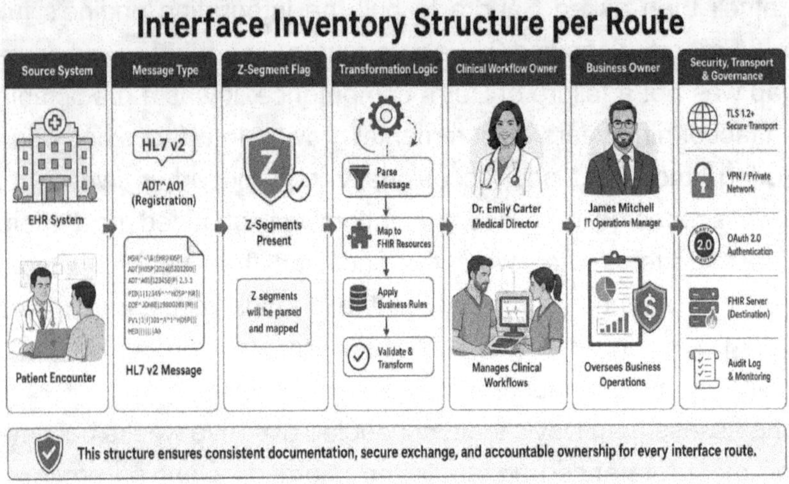

The current-state assessment produced two additional artifacts that significantly shaped the migration plan. The clinical workflow mapping revealed dependencies not visible from routes alone: a medication reconciliation workflow drew on four separate routes (pharmacy dispense, medication order, allergy, and reconciliation confirmation), and migrating any one without the others would have broken the workflow in ways the EHR interface would not have immediately surfaced. The mapping exposed that dependency before go-live, so they could discover it the hard way.

The business owner identification exercise found that a significant fraction of Riverside's routes had no clearly identified current business owner. Routes existed because a department had needed them; accountability had not been maintained as staff turned over. Patrick assigned temporary ownership to the relevant department heads and required each to confirm or transfer ownership as part of the planning process, converting an ambiguity problem into a governance problem he knew how to solve.

The Three-Phase Migration Model

Dev's proposal for the migration model was the bypass road analogy, and it proved useful enough that Patrick used it in every subsequent communication about migration strategy, from the project kickoff to the quarterly board presentations.

When a city needs to reroute traffic from an aging highway via a new bypass, the engineers do not close the old highway until the new one is complete. They build the bypass alongside the old highway and open it to traffic. In contrast, the old highway remains operational, running both simultaneously until the new road has been proven at production volume, and only then will the old highway be closed. If something unexpected occurs in the first weeks of full operation, traffic can be redirected back within minutes. The sequence matters: build new alongside old, prove new at production volume, decommission old route by route.

The three-phase migration model follows the same sequence. Phase One (Parallel Run) builds the FHIR infrastructure alongside the existing HL7 v2 infrastructure without routing any clinical workflows through the FHIR layer. FHIR messages flow in parallel to v2 messages; both layers process the same clinical events. The FHIR output is validated against conformance criteria and compared with the v2 output to verify equivalence, but clinical workflows remain driven exclusively by the v2 layer. Phase One demonstrates that the FHIR infrastructure can reliably and accurately process production-volume clinical data before any clinical workflow depends on it.

Phase One typically runs two to four months at a community hospital of Riverside's size. During that period, the migration team continuously monitors validation results, resolves conformance failures, calibrates transformation logic against edge cases not present in the test data, and builds the operational experience with the FHIR layer needed for confident Phase Two go-lives. The parallel run is also the period during which rollback procedures are built, tested, and rehearsed, because they can be tested in a live parallel environment without clinical consequences.

Phase Two (Coexistence) begins routing specific clinical workflows through the FHIR layer while maintaining v2 for the remaining workflows. During Coexistence, the integration estate operates in a mixed state: some workflows are FHIR-native, while others remain v2-dependent, with the transformation layer managing both simultaneously. This operational complexity is the cost of the risk reduction the phased approach provides. Coexistence at Riverside was planned for approximately 10 months, with a new workflow tier migrating to FHIR every 6 to 8 weeks.

Diagram 13.4: Three-Phase Migration Timeline

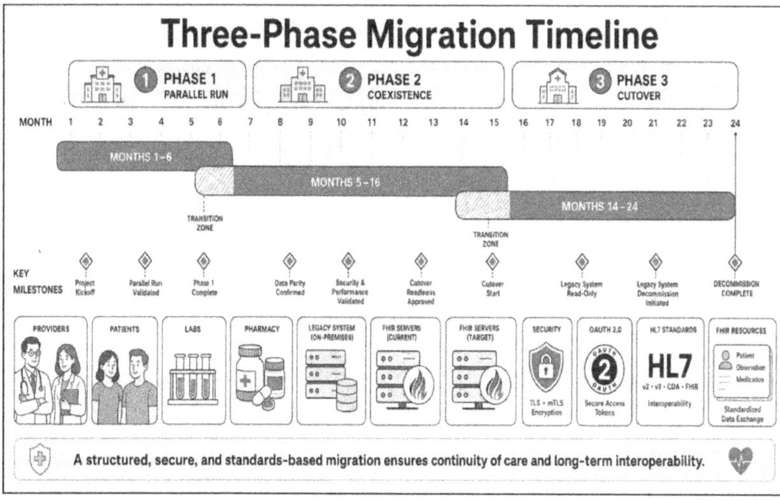

Phase Three (Cutover) transitions the remaining workflows to FHIR and decommissions the corresponding v2 routes. Routes are retired one by one or in small related clusters, never in large batches. Each decommission is preceded by the same go/no-go criteria review that governed Phase Two go-lives and followed by the same hypercare period at maximum team availability. The v2 infrastructure remains intact until the last route has been migrated, validated at production volume, and confirmed stable through its hypercare period.

Patrick's management of the Coexistence phase required a communication discipline the project plan had not originally specified. When clinical staff encountered differences between FHIR-native and v2-dependent workflows running simultaneously, they reported those differences to the help desk in terms that were not always technically precise. A clinician who noticed that a medication list displayed in a slightly different format after the pharmacy workflow migrated to FHIR was not reporting a data inconsistency; they were reporting an unexpected interface change. Patrick instituted a protocol: any change in the visual presentation, timing, or content of clinical data resulting from the migration was documented in a migration change log and communicated to affected departments before the go-live. Clinical staff told in advance that the medication list format would change did not call the help desk. Clinical staff who encountered the change without warning did so, and those calls required investigation time that the protocol avoided.

Use Case Prioritization

Not all use cases carry equal migration risk. The sequencing of use cases is one of the most consequential decisions the migration leadership team makes, and it requires the full team's input rather than a unilateral technical determination.

The risk framework Riverside used assessed each candidate use case against three dimensions: clinical consequence (the patient safety impact of a workflow failure), rollback availability (how quickly and completely the workflow could be returned to v2), and user population scope (how many clinical staff were directly affected and how distributed they were). Applying this framework produced four priority tiers. The first included the Patient Access API and the provider-facing data access portal: external-facing, read-only, low clinical consequence, and compliance-urgent. The second included administrative workflows: appointment scheduling, patient directory synchronization, and referral management. The third included laboratory result delivery and diagnostic imaging notification: high-volume but with established v2 baselines and understood rollback procedures. The fourth, reserved for the final months of Coexistence, included medication order delivery, clinical decision support triggering, and specialty workflows with Z-segment dependencies that required custom FHIR extensions.

Diagram 13.5: Use Case Migration Priority Framework

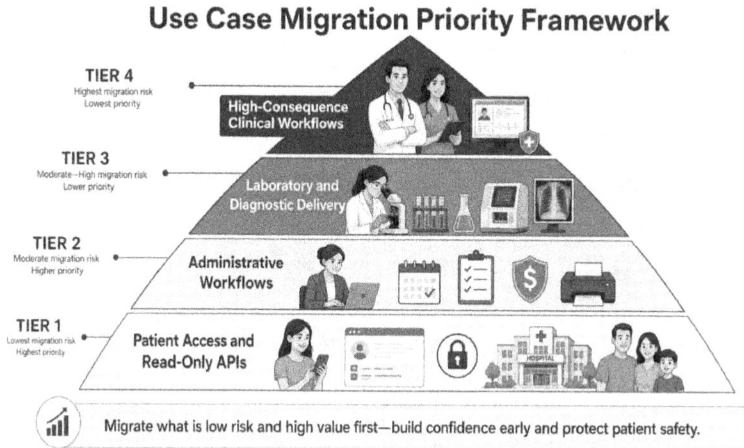

Dr. Chen's contribution identified a risk dimension the technical framework had not weighted. She flagged allergy reconciliation and immunization history delivery, both of which Dev had placed in the second tier, as belonging in the fourth. The technical risk assessment for both was favorable: clean v2 baselines and clear rollback paths. But the failure mode for both was silent incompleteness rather than visible absence. A medication order delivery failure is immediately visible; a clinical staff member who submits an order without receiving confirmation escalates immediately. An allergy reconciliation failure might produce a record that appeared complete but was missing an allergy that had not propagated correctly through the FHIR transformation. Patrick added both workflows to the fourth tier. He introduced a new criterion to the framework: workflows with a failure mode of silent incompleteness should be treated as higher risk regardless of rollback availability. Clinical safety risk in FHIR migrations is not always visible at the integration layer.

Sandra's contribution addressed the compliance calendar's influence on sequencing. The ONC information blocking deadlines required the Patient Access API workflows to migrate in the first tier, not just because they were the lowest risk, but because the compliance calendar mandated it. Sandra also identified a ninth-month window during which the HIPAA security rule audit schedule would consume the migration team's capacity, and the project plan was adjusted to avoid Phase Two go-lives during that window.

Vendor Selection and Procurement

Sandra's vendor evaluation process was structured around five criteria, each addressing a category of risk that the procurement contract needed to mitigate or explicitly accept.

Standards conformance evidence: a vendor claiming FHIR R4 support needed to demonstrate it through verifiable artifacts. A deployed CapabilityStatement accessible at a public test endpoint. Published results from a recognized FHIR conformance testing framework. A documented list of US Core profiles supported. A vendor that could produce polished marketing materials but could not produce a CapabilityStatement at a live endpoint had not yet implemented what it was describing.

Contract portability addressed the long-term risk of vendor selection in FHIR infrastructure. Lock-in provisions are not always present in standard vendor agreements unless explicitly negotiated. Sandra required contractual provisions for data export in standard FHIR bulk data format, API documentation sufficient for independent re-implementation, the absence of proprietary data format requirements for data at rest, and the right to migrate to a different vendor without penalty after the contract term. Vendors that declined to include portability provisions were disqualified.

Regulatory compliance posture: every FHIR infrastructure vendor handling Riverside's patient data was required to provide a Business Associate Agreement that meets Riverside's HIPAA requirements, documentation of ONC certification status relevant to the intended use cases, an information blocking compliance program document, and a security assessment report current within the preceding 12 months. These requirements were stated as conditions of the initial RFP rather than items for negotiation.

Support and maintenance commitments, documented in the service level agreement, specified response times for critical, high, and normal severity issues; FHIR version upgrade timeline commitments covering the anticipated contract term; security patch cadence; and the vendor's process for communicating breaking changes before deployment. Dev translated the operational support requirements the migration would generate into concrete SLA language the procurement contract could enforce.

Diagram 13.6: Vendor Evaluation Five-Criteria Framework

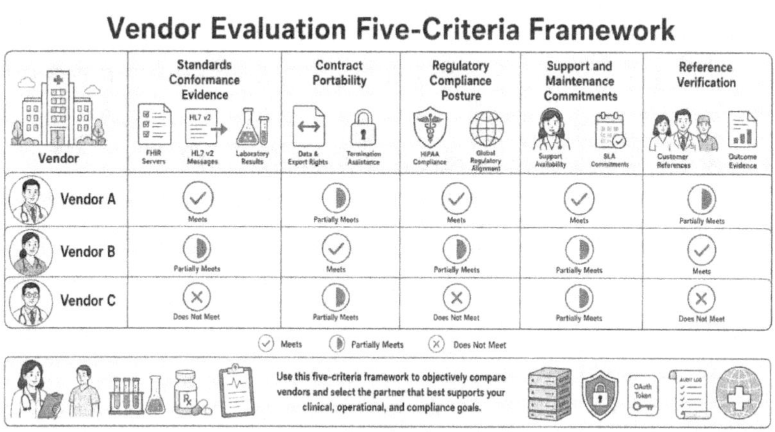

Reference verification required each shortlisted vendor to provide three references: hospitals of comparable size and integration complexity that had completed FHIR migrations using the vendor's platform within the preceding 24 months. Sandra conducted structured interviews to assess the gap between pre-contract representations and actual delivery, vendor support responsiveness during go-live, and what the reference organization would handle differently in hindsight. In two cases, those interviews revealed performance gaps significant enough to eliminate vendors that had otherwise scored well on the first four criteria.

The evaluation took four weeks. The selected vendor had satisfied all five criteria and agreed to portability provisions Sandra judged adequate. Contract negotiation took an additional three weeks. Patrick presented the selection to the board's technology committee as a risk-managed procurement decision, described the five criteria, and explained the portability provisions in terms of the long-term optionality they preserved. The committee approved in a single session.

Staff Training Program Design

The training program for a FHIR migration must address at least three distinct audiences with different knowledge requirements, different learning contexts, and different relationships to the change. A training program designed for only one of those audiences, which is the most common failure mode, will leave the other two underprepared in ways that generate support volume and workflow disruption the hypercare budget cannot absorb.

Dev's technical team needed deep, hands-on FHIR training: the resource model, RESTful interaction patterns, profile validation, SMART on FHIR authorization, bulk data operations, and the specific implementation patterns of the selected platform. Dev structured a ten-week program combining vendor-provided platform training with internally developed exercises built on Riverside's own integration scenarios. Each team member demonstrated competency through a practical assessment before being assigned to Phase Two go-live preparation. The assessment was not graded; its purpose was to identify knowledge gaps while there was still time to address them.

Dr. Chen designed the clinical staff training program around how clinical staff learns most effectively: not through formal classroom instruction but through brief, workflow-integrated demonstrations. The clinical training used six workflow scenarios representing the most common clinical interactions with the systems that would change. Each scenario was delivered as a fifteen-minute demonstration at the unit level, scheduled during shift overlap times, and delivered by department educators rather than IT staff. Clinical staff respond more readily to workflow guidance from colleagues who share their clinical context.

Diagram 13.7: Three-Audience Training Program Structure

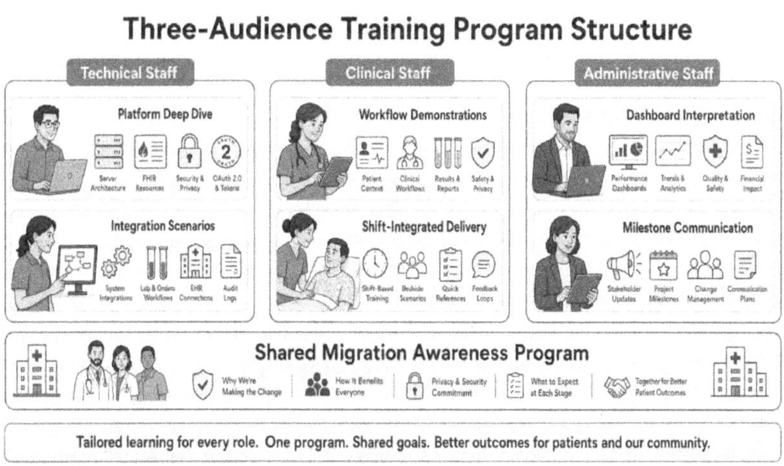

Patrick's administrative staff training addressed a need neither Dev nor Dr. Chen could meet without translation. Department heads, service line managers, and administrative coordinators needed to understand the migration well enough to answer staff questions, interpret monthly progress metrics, and recognize which issues required escalation. Patrick designed a three-hour briefing for all management-tier staff before Phase Two: what FHIR was in non-technical terms, what would change and when, how to read the migration dashboard, and what the rollback plan meant at an operational level.

Training completion was treated as a migration readiness criterion. No Phase Two go-live would be approved unless clinical staff training completion for the affected workflows exceeded eighty-five percent. The threshold reflected the judgment that a fifteen percent training gap could be managed through peer support during hypercare, while a larger gap would overwhelm the hypercare team. The threshold was invoked once: the third-tier go-live was delayed two weeks because laboratory staff training in one unit fell below threshold due to a scheduling conflict with mandatory compliance training. The delay was accepted without controversy because the criteria had been agreed upon before the pressure of go-live arrived.

Go-Live Sequencing and Execution

Go-live sequencing translates the use case prioritization framework from a planning document into a series of specific execution events with defined criteria, defined responsibilities, and defined procedures. The translation from plan to execution is where many migration plans lose fidelity: the prioritization logic agreed on in a planning session three months earlier gets compressed or bypassed by schedule pressure and stakeholder impatience. Patrick's response to this risk was to treat the go/no-go criteria as governance commitments rather than project recommendations, meaning that overriding them required explicit approval from the leadership team rather than a project manager's tactical judgment.

Each Phase Two go-live followed a standardized execution sequence. The go/no-go criteria review was held forty-eight hours before the planned go-live, with Dev, Sandra, Patrick, and the relevant department's clinical or operational lead in attendance. Criteria reviewed included technical validation results from the parallel run, staff training completion rates, vendor support availability confirmation, rollback procedure rehearsal status, and open defect counts by severity. Any criterion not met triggered either a remediation plan with a specific completion date or a recommendation to delay. The criteria distinguished between blocking criteria, which required resolution before go-live could proceed, and monitoring criteria, which were acceptable with documented risk acknowledgment and enhanced hypercare monitoring.

The go-live execution followed a sequence the team had rehearsed during the parallel run. Cutover began at a scheduled time, typically on a Sunday morning, chosen for minimal clinical volume and maximum technical staff availability. The integration engine routing for affected workflows was updated from v2 to FHIR, FHIR validation monitors were activated at heightened alert thresholds, and the hypercare team was positioned at monitoring stations covering the technical layer, clinical operations, and the help desk. The first two hours after cutover were the highest-intensity monitoring period: validation failures, clinical workflow anomalies reported through the help desk, and any system performance degradation that would trigger the rollback decision protocol.

Diagram 13.8: Go-Live Go/No-Go Criteria Decision Flow

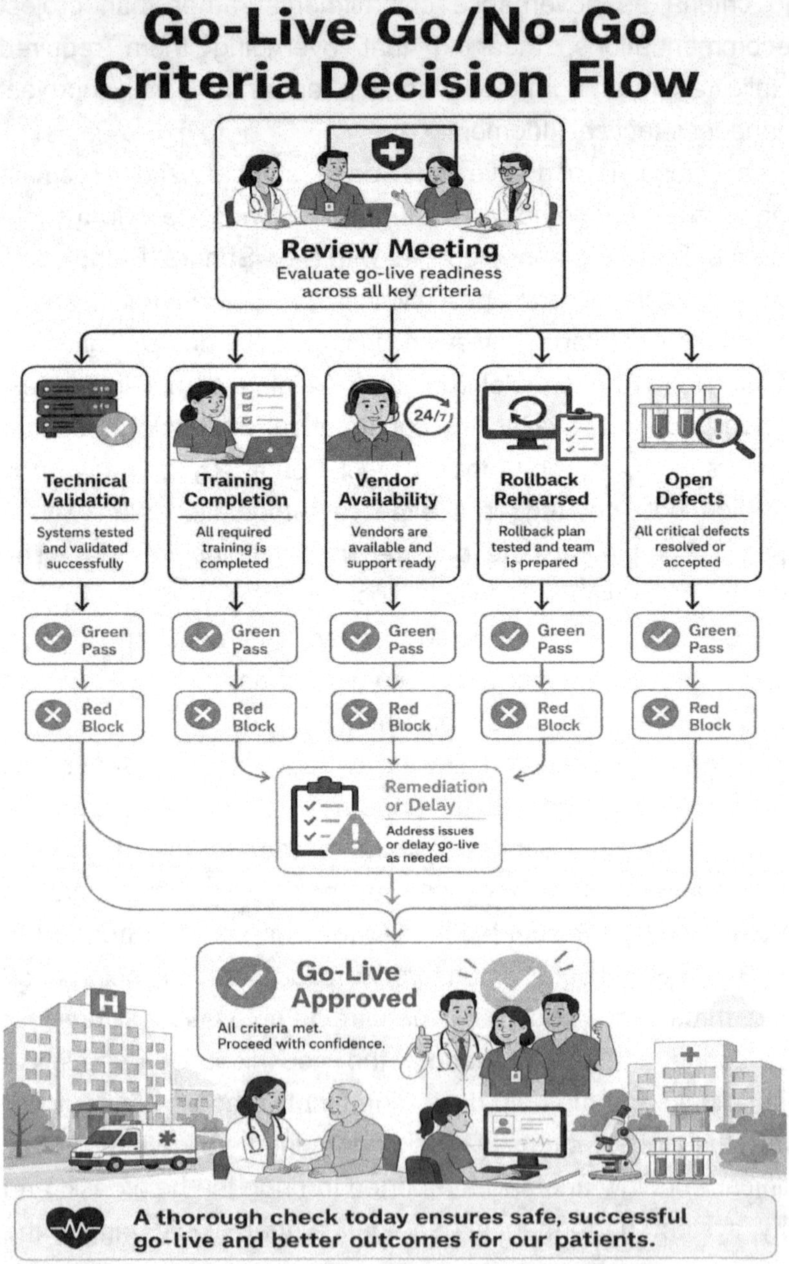

The hypercare period was defined as a maximum of five business days of team availability, followed by a ten-business-day extended monitoring period with reduced but elevated availability, followed by a formal closeout review confirming the workflow was stable and transferring support responsibility to the standard operations team. The definition of maximum team availability was explicit: all team members with workflow-specific expertise reachable within fifteen minutes, on-site coverage during clinical operating hours, and status communicated to Patrick every four hours on the first business day and daily thereafter. That specificity mattered. Vague hypercare commitments tend to be interpreted liberally as go-live novelty fades, and liberal interpretation is the condition under which post-go-live issues become unmanaged incidents.

Sandra's hypercare contribution was the compliance audit trail: an access log capture confirming that new FHIR workflows were logging access events per Riverside's HIPAA policy, a data integrity spot check comparing FHIR resource content against v2 source for a sample of first-day clinical events, and a patient safety issue log confirming zero open items or documenting any items with their resolution timeline. These artifacts were filed in the migration compliance record that Sandra maintained throughout the project.

Rollback Planning and Rehearsal

Dev's insistence on rollback planning generated friction with the project schedule. The initial plan allocated two weeks to rollback documentation; Dev argued that was insufficient for documentation, testing, and rehearsal across all use case tiers. Patrick mediated the rollback plan by reframing it as a risk-management investment. A rollback plan is insurance: its value is invisible when nothing goes wrong, and its cost is very visible when something does. He authorized the additional time and added a budget line for rehearsal logistics.

The decision was vindicated in the third Phase Two go-live. The laboratory result delivery cutover encountered a volume-related performance issue the parallel run had not triggered: Monday morning production volume exceeded test data volume by a factor the capacity planning estimates had underweighted. The rollback to v2 was executed in nineteen minutes by a team that had rehearsed the procedure twice. Laboratory workflows were restored before the issue reached clinical operations, and the FHIR implementation was corrected, retested, and redeployed three weeks later with capacity parameters calibrated against actual production data.

A rollback plan that has been written, tested, and rehearsed answers a specific operational question: if a production incident requires reverting to the v2 configuration after a FHIR go-live, what is the exact sequence of steps, who executes each step, how long does each step take, and what is the total elapsed time from the decision to roll back to the restoration of clinical workflow on v2? The answer should be based on measured execution time from rehearsals rather than estimated time from a planning document.

Diagram 13.9: Rollback Decision and Execution Protocol

Rollback Decision and Execution Protocol

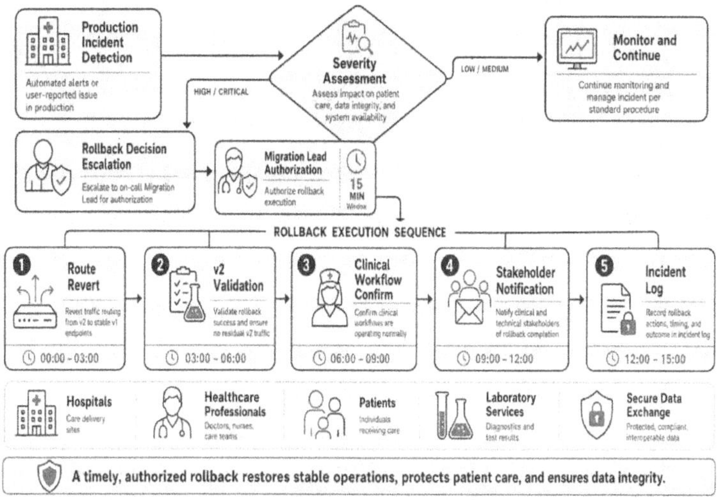

Dev's rehearsal process involved a scheduled half-day exercise during the parallel run phase for each workflow tier: a scripted production incident delivered to the on-call team member, culminating in a documented elapsed time from detection to v2 workflow restoration. The first rehearsal identified two ambiguous steps: the procedure specified verifying that integration engine routing had been updated but did not specify which routing rules to verify or what the verification criteria were. Those ambiguities were clarified before the first go-live. During an actual production rollback, the same ambiguities would have added confusion and delay at the worst possible moment.

The rollback framework also addressed partial rollback, which many organizations omit from planning. When a go-live migrates multiple workflows simultaneously, and a production issue affects only one of them, a full rollback may be more disruptive than a targeted reversal of the affected workflow. Riverside's framework included documented partial rollback procedures at the individual workflow level, tested separately and integrated into the go/no-go criteria for any multi-workflow cutover event.

The Board Communication Framework

Patrick's board presentations followed a five-metric framework built on a single design principle: board members are accountable for governance outcomes, not technical implementation. Presenting implementation-level metrics to a board audience does not simplify technical content; it substitutes the wrong information for the right information, and a board receiving the wrong information cannot perform its governance function.

The five metrics were: regulatory compliance status (a single-sentence assessment of whether Riverside was on track to meet all applicable deadlines); budget variance (actual versus planned spend with a projection to completion); workflows migrated (count and percentage of total scope); open patient safety items (a zero-tolerance standard, meaning every item reported regardless of status); and next milestone (planned date, projected date, and team confidence level). Updated monthly, these five metrics provided complete board-level governance visibility without requiring any board member to understand what a FHIR resource was.

The zero-tolerance standard for patient safety items was the element Patrick was most explicit about. A board that learns about a patient safety issue at its monthly meeting is in a different governance position than one that learns about it months later in a regulatory audit finding. Any item identified during migration activities with potential patient safety implications was reported to Patrick within two hours, and Patrick notified the board chair within twenty-four hours via written summary. This protocol was never triggered during the Riverside migration. Still, its existence was a governance artifact in itself: the board knew the protocol existed, knew it had not been triggered, and that knowledge supported their continued confidence in the migration's progress.

Patrick supplemented the dashboard with a quarterly narrative summary covering what had been learned in the preceding quarter, what decisions had been made and on what basis, the team's current risk assessment, and what the board could do to support migration success. Boards given a specific, concrete answer to that last question are more engaged and supportive than boards that receive progress reports without a clear role.

At the ninth-month mark, schedule compression from the HIPAA audit window and the laboratory go-live delay pushed the Phase Two completion date back by six weeks. Patrick presented the revised timeline with a clear account of the contributing factors, confirmed the revision still met compliance deadlines with three weeks of margin remaining, and recommended accepting the revised schedule rather than compressing the remaining go-lives. The board accepted the recommendation without requiring further justification, reflecting the growing confidence in Patrick's consistently accurate reporting.

Patrick presented the five-metric dashboard at the month-nine board meeting. The room was quiet while he walked through the slides: regulatory compliance on track, budget one point two percent over plan with a credible recovery path, seventy-three percent of workflows migrated, zero open patient safety items, Phase Two completion projected three weeks ahead of the compliance deadline. He had timed the presentation at nine minutes in rehearsal. It ran eleven.

A board member — the chair of the finance committee — asked about the patient-data-request fulfillment rate. It had been ninety-four percent the previous month; the slide showed ninety-one percent this month. Was the number going in the wrong direction? Patrick had expected the question. He explained that the three-point drop reflected a planned expansion of scope: two new data categories had been added to the API the prior week, and early consumer uptake always produced a spike in edge-case requests the system was not yet optimized to fulfill. The rate was expected to return to baseline within sixty days. The chair wrote something down and nodded. The board approved the quarterly update without further questions. Patrick's nine minutes of preparation had taken three weeks to build and had been worth every hour of it.

Governance for a Long Migration

A twenty-four-month migration project creates governance challenges shorter projects do not encounter. Leadership team composition changes, vendor relationships evolve, the regulatory context shifts, and budget assumptions from month one are revised based on discoveries in months six and twelve. Organizations without explicit governance structures end up with migrations that drift: scope quietly adjusted without formal approval, schedule changes accumulating without acknowledgment, and the migration arriving at its completion date with a gap between what was delivered and what was planned that no one can clearly account for.

Dev and Patrick designed a governance structure with four elements: a monthly migration leadership team meeting as the primary decision-making forum (all four leads, standing agenda, and a decision log for every material decision); a quarterly steering committee review for escalation and scope or schedule changes (involving the CEO, CFO, and board technology committee chair); a weekly technical coordination meeting for Dev's operational forum; and a daily fifteen-minute stand-up during active hypercare periods. A formal change control process covered proposed changes beyond defined thresholds: scope changes adding more than five routes, schedule changes moving any compliance-linked milestone by more than two weeks, or budget variances exceeding ten percent of the approved budget for any phase required steering committee approval. The thresholds had been reviewed and accepted by the board chair before the project began.

Measuring Migration Success

Defining migration success before the project begins is both a governance discipline and a practical management tool. An organization that has not defined success criteria before migration starts will measure success by whatever the migration delivered, which is not the same as measuring whether it delivered what the organization needed.

Dev and Patrick defined success criteria at three levels. Technical criteria measured integrity and performance: validation pass rates for FHIR resources, SLA-compliant response times, error rates for clinical data delivery, and the absence of patient safety events attributable to migration-related data integrity failures. These criteria were established as go/no-go conditions for Phase Three final cutover.

Operational criteria measured impact on clinical and administrative operations: migration-related help desk volume as a percentage of total volume, post-migration staff satisfaction survey results, time-to-resolution for migration tickets, and clinical workflows requiring workaround documentation after go-live. These were measured over the six months following the completion of Phase Three.

Strategic criteria measured whether the migration had delivered the capabilities that justified the investment: ONC information blocking compliance, reduction in integration maintenance cost from retiring high-maintenance v2 routes, availability of FHIR-based data access for quality reporting and population health use cases, and the foundation for the next generation of clinical application integration. The board's technology committee reviewed these criteria at twelve and twenty-four months post-completion.

From Plan to Practice

The six weeks Patrick, Dev, Sandra, and Dr. Chen spent building the migration plan before any technical work began were not overhead. They were the most leveraged investment in the entire project, because every decision made during those six weeks influenced the quality and efficiency of the technical work that followed. The interface inventory Dev did not want to produce became the scope document every migration budget line traced back to. The Z-segment analysis, which required three weeks of conversations with vendors and department heads, became the risk register that informed use case prioritization. The vendor evaluation framework Sandra built from first principles became the contractual protection that preserved Riverside's long-term interoperability independence. The rollback procedures Dev fought for in the schedule negotiation became the nineteen-minute recovery that kept the laboratory go-live failure from becoming a clinical operations crisis.

None of those outcomes was inevitable. Each was the consequence of a planning decision made before the pressure of go-live had compressed the team's capacity for deliberate judgment. The organizations that invest in that kind of planning before the pressure begins are the organizations that recover quickly when the unexpected occurs. Every migration encounters the unexpected. The difference between an unexpected event that is managed and one that becomes a crisis is almost always the quality of the planning that preceded it.

The technical work of FHIR migration is well-documented, widely supported, and within the competence of any capable health IT team. The planning work is less well-documented, more frequently deferred, and more consistently correlated with migration outcomes. FHIR migrations fail for planning reasons, not technical reasons. The technology is available, documented, and implementable. The planning failures that derail migrations, including understated scope, ignored legacy dependencies, untrained staff, and unrehearsed rollback plans, are avoidable with the disciplined application of the framework described in this chapter. Organizations that skip planning to accelerate technical work will pay for it at go-live.

The next chapter examines where FHIR is heading: the R5 specification, the emerging role of AI in clinical data integration, and the genomics interoperability frameworks that will define the next generation of health data standards. The migration planning work described here is not the end of Riverside's FHIR journey; it is the foundation on which those capabilities will eventually be built.

15 The Road Ahead

A Talk Three Years in the Making

Patrick Osei had attended the Regional Health IT Leadership Forum three years earlier as an audience member who barely understood the acronyms on the session titles. He sat in the second row of a windowless ballroom, scribbling notes on a hotel notepad, and came away with a list of words to look up later: FHIR, HL7, USCDI, SMART, CDS Hooks, Bulk Data. He did not look up most of them that week because the week after the conference was the week Mrs. Rivera called.

The call from Mrs. Rivera had not felt like the beginning of anything at the time. A patient could not see her lab results in the app she had downloaded. She was entitled to see them. Patrick made the referral, closed the loop, and moved on. What happened next was not moving on. It was two years of audits, migration planning, vendor negotiations, compliance reviews, board presentations, security remediations, staff training, and the slow construction of a health data infrastructure capable of meeting the law's requirements and what patients deserved.

When the conference chair emailed asking whether Patrick would present a session at this year's forum, he said yes. He spent three days writing the talk. The first draft was a technical overview of everything Riverside General had built: accurate, comprehensive, and completely wrong for the audience. The second draft was a compliance summary, equally wrong. The third began with Mrs. Rivera's phone call, told the story of what Riverside built and why, and spent its second half on what was coming next: the FHIR version upgrade cycle, artificial intelligence consuming clinical data at scale, the genomics frontier, the evolving regulatory landscape, and the global spread of FHIR adoption that was transforming interoperability from a US regulatory obligation into an international infrastructure standard.

That third draft was the talk he gave. This chapter is that talk, written as a chapter.

Diagram 14.1: Patrick's Journey from Audience Member to Speaker

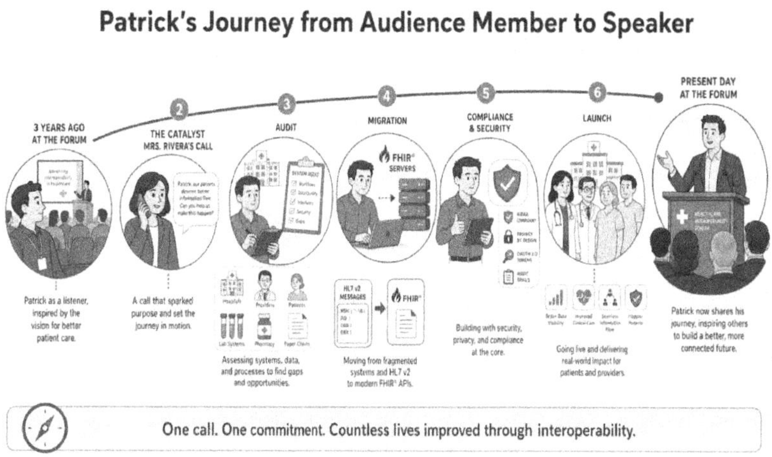

The conference audience Patrick faced was not entirely different from the one he had joined three years before: administrators just beginning to understand what their certified EHR vendor actually delivered, IT directors who had inherited integration environments they had not designed, compliance officers reading information blocking enforcement updates and trying to determine what their organizations needed to do, and a handful of developers attending to find out whether the administrator on the stage had anything useful to say.

Patrick started with Mrs. Rivera. He told the room about the phone call, the gap between what the law required and what Riverside's systems could actually deliver, and the decision to treat that gap as a project rather than a problem to be minimized. He told them the project had been harder than anyone expected and more valuable than anyone had budgeted for. Then he told them what he had learned, and what he believed was coming, and why it mattered for everyone in that room.

FHIR R5: What Changed and Why It Matters

HL7 International published FHIR R5 in March 2023, and its release raised the same set of questions that every major standards release does: Do we need to adopt it immediately? What breaks if we stay on R4? What do we gain if we move? What are our vendors and target implementation guides actually planning to do?

Patrick had asked Dev for a direct read on R5: not the HL7 press release version, not the vendor briefing, but the operational version that an integration engineer who had actually read the specification would give. Dev identified three areas that matter for production teams.

The first is the Subscription framework. In FHIR R4, Subscriptions used a filter-based model that was, in Dev's words, technically functional but operationally unreliable. Filters were difficult to express precisely, notification delivery was inconsistently implemented across servers, and the reliability profile made it unsuitable for high-stakes clinical alerting. R4B introduced a topic-based subscription model that addressed most of those limitations, and R5 makes that model normative. Real-time alerting use cases that Riverside had built on polling could, on a FHIR R5 server, be implemented natively using Subscriptions with a reliability profile that polling cannot match.

The second area is cross-version extensions. Riverside exchanges data with external labs whose FHIR servers run R4, a regional health information exchange upgrading from STU3 to R4, and a payer that has implemented an R4B API for prior authorization. Handling resources from multiple FHIR versions previously required custom mapping logic in the integration layer. R5's cross-version extension framework provides a standardized mechanism for annotating resources with version-specific information, reducing that custom mapping burden.

Diagram 14.2: FHIR Version Evolution and R5 Key Improvements

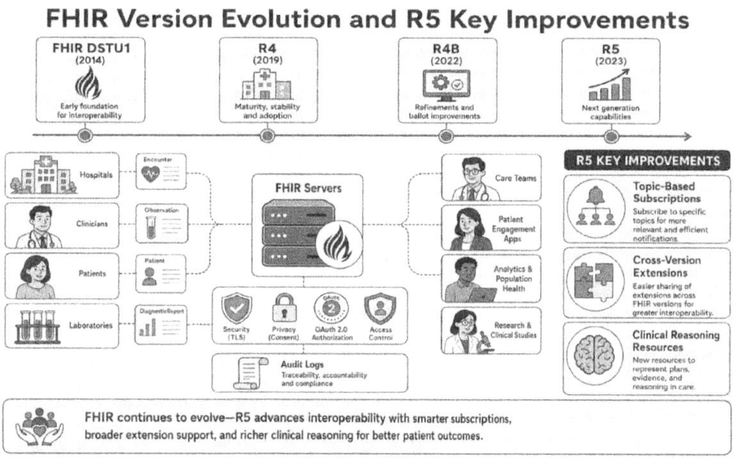

The third area is clinical reasoning. The Measure, MeasureReport, Library, and PlanDefinition resources in R5 are significantly more mature than their R4 counterparts, supporting more sophisticated quality measurement, clinical decision support, and value-based care reporting use cases than R4 can replicate without extensions.

Riverside was not planning an immediate migration to R5. The ONC certification requirements still referenced R4; the US Core remained an R4-based implementation guide; and a FHIR version migration was a multi-year project that the R5 improvements alone could not justify for a mid-sized community hospital. But Riverside was requiring R5 roadmap commitments from new vendors as a standard procurement criterion and architecting new development in ways that would not require structural rework when the R5 transition eventually happened.

Patrick relayed Dev's parting point to the audience: the organizations most surprised by R5 would not be those that had decided to wait on adoption, but those that had not followed the R5 development process and would encounter R5-based implementation guide requirements for the first time in a vendor evaluation. The HL7 ballot process is public, the R5 specification is freely available, and the implementation guides being written against R5 are in the public ballot queue. There is no excuse for being surprised by a standard that has been publicly developed and openly published.

The Regulatory Horizon

The FHIR specification does not exist in a regulatory vacuum, and the environment shaping FHIR adoption in the United States continues to evolve at a pace that requires dedicated attention from every organization with certified health IT obligations. Patrick had asked Sandra to prepare a regulatory horizon briefing, and what Sandra delivered was, characteristically, comprehensive.

The 21st Century Cures Act and the ONC's implementing regulations established the foundational requirements: information blocking prohibitions, the requirement for FHIR-based Patient Access and Provider Access APIs, and the USCDI as the minimum data set for electronic health information exchange. Those foundations were not static. USCDI has progressed through multiple versions since USCDI v1 was incorporated into the ONC certification criteria, and each new version has added data elements that certified EHRs would eventually be required to expose through FHIR APIs.

Sandra's briefing identified four areas requiring ongoing monitoring. First, USCDI version progression: new USCDI versions expand the minimum data set for API access on a regular cycle, and organizations that implemented Patient Access APIs to meet initial USCDI v1 requirements needed processes to maintain conformance as new data elements were added. Second, the Provider Access API requirements under CMS's Interoperability and Patient Access rules, with implementation timelines enforced through CMS audit mechanisms and sanctions for organizations that missed their compliance deadlines.

Diagram 14.3: Regulatory Horizon for US FHIR Implementers

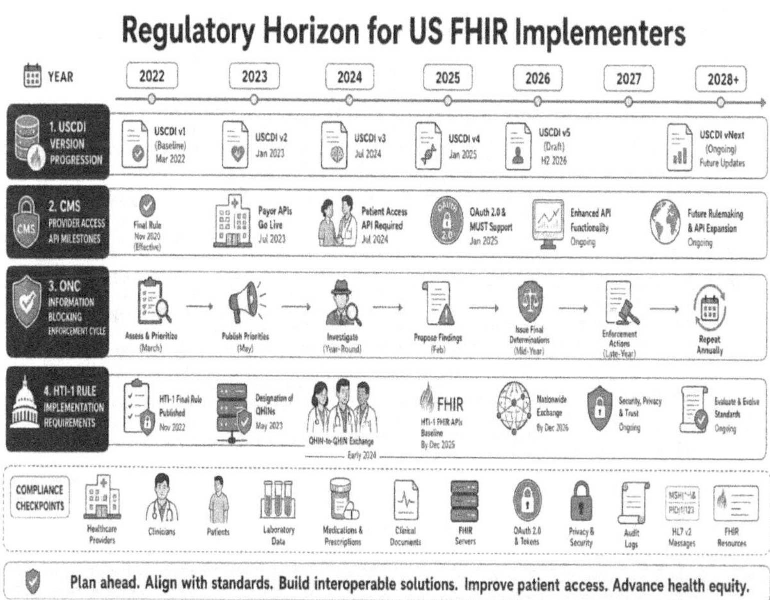

Third, the information blocking enforcement landscape. The Office of Inspector General had begun issuing formal enforcement actions for information blocking violations, and the published enforcement record was establishing the factual patterns that constituted violations: blocking API access for economic reasons, delaying data access in response to portability requests, and implementing technically compliant APIs in ways operationally designed to discourage use. Sandra's message, relayed without softening, was that information-blocking enforcement was real and active, targeting organizations that treated the prohibitions as technical compliance requirements rather than as operational culture requirements.

Fourth, the ONC's Health Data, Technology, and Interoperability rule, commonly called HTI-1, introduced requirements around clinical decision support transparency, algorithmic accountability, and predictive AI in clinical settings. Sandra had flagged HTI-1 as the regulatory development most likely to create unexpected compliance obligations for health systems deploying AI-assisted clinical tools, because its requirements extended beyond the FHIR API layer to the governance frameworks for AI systems that consume FHIR data.

Patrick's message to the conference audience on the regulatory horizon was consistent with what Sandra had delivered to the Riverside board: regulatory compliance in health IT is not a project that ends when the first audit passes. It is an ongoing operational posture that requires dedicated resources, regular monitoring, and processes to translate new requirements into projects before the compliance deadline.

Bulk FHIR and Population Health Intelligence

The individual patient data exchange use case that drove FHIR's initial adoption was necessary but not sufficient. Every conversation Patrick had with peer administrators came back to the same observation: the investment required to build and maintain an FHIR API infrastructure was significant, and if the return was limited to patient-facing access to individual health records, the business case was thin. The business case thickened considerably when the conversation turned to what that same infrastructure enabled for population health analytics. Bulk FHIR, defined by the HL7 FHIR Bulk Data Access specification, is the mechanism by which a FHIR server can export large volumes of resources for a defined patient population in an asynchronous, file-based format. Where the standard FHIR REST API retrieves individual resources one patient and one resource type at a time, Bulk FHIR is designed for population-scale export: all Observation resources for all patients with an active diabetes diagnosis, all MedicationRequest resources for patients on anticoagulant therapy, or all Encounter resources for patients with three or more emergency department visits in the preceding twelve months.

Dev had explained the technical distinction to Patrick using the analogy Patrick found worth repeating at the conference: Bulk FHIR was like a hospital's data infrastructure having a loading dock — where, given a well-specified list of criteria, the system would collect every matching record and have them ready in a standardized format on schedule, rather than requiring individual retrieval one at a time. Population-scale analytics and individual patient queries were complementary capabilities within the same infrastructure, not competing approaches.

Diagram 14.4: Bulk FHIR Architecture for Population Health Export

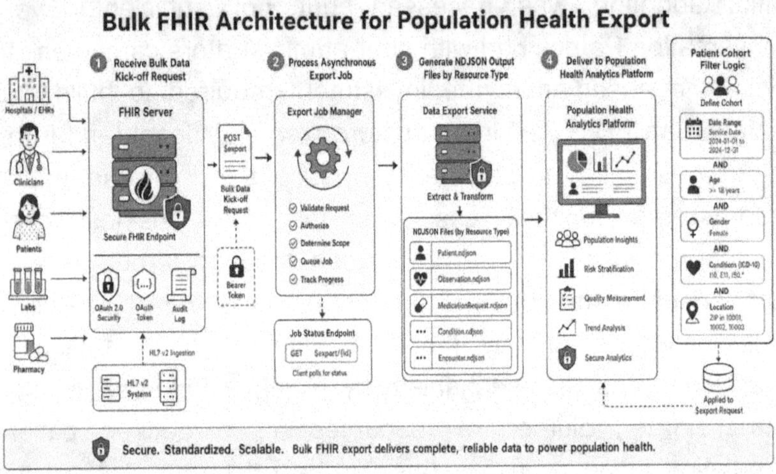

Riverside's population health program had been one of the quieter arguments for the FHIR migration. The migration's initial justification was regulatory compliance. Still, Dev's decision to build a proper FHIR server rather than a thin pass-through proxy meant that Riverside had built an infrastructure capable of supporting Bulk FHIR export for population health analytics as a by-product of compliance work.

The first population health use case Riverside put into production was a diabetic population monitoring program. Dev wrote an automated Bulk FHIR export job that ran nightly, pulling all Observation resources for Riverside's active diabetic population: HbA1c values, fasting glucose measurements, blood pressure readings, and kidney function markers. The export fed a risk stratification model that flagged patients whose glycemic control had deteriorated over the preceding three months and generated outreach tasks for the care management team, and fed a quality measure reporting pipeline that calculated Riverside's performance on the relevant HEDIS diabetes care measures. Neither program required technology that had not existed before FHIR. What FHIR provided was the ability to build them without custom data-extraction projects for each new use case. The Bulk FHIR export format, resource definitions, and clinical terminology bindings were standardized, and a new population health analytics use case could be prototyped in days rather than months. The compliance obligation was the starting gun, not the finish line.

AI Meets FHIR: Clinical Intelligence at Scale

Artificial intelligence had been a topic at the health IT conference three years before, characterized by enthusiasm and imprecision: AI was going to transform healthcare, clinical decision support was going to improve dramatically, and big data combined with machine learning was going to produce insights clinicians could not generate on their own. The AI conversation at the current moment was meaningfully different, not primarily because the underlying technology had advanced. However, it had, but because AI applications were now consuming FHIR data in production clinical environments and the operational questions deferred during the vision phase were unavoidably active.

Dev briefed Patrick on three categories of AI applications that consume FHIR data in clinical settings. The first was clinical documentation assistance. Large language models with access to a patient's FHIR-encoded clinical history, including recent Observation resources, active Condition resources, current MedicationRequest resources, and recent Encounter resources with clinical notes, could generate structured clinical summaries, draft after-visit summaries, and produce pre-populated documentation templates that reduced the clinician documentation burden. The value proposition was clear; the governance questions were the open frontier. Who was responsible for a clinical summary that an LLM generated from FHIR data? What audit trail was required? What consent framework applied when an LLM consumed FHIR data to generate documentation a patient might later read in their own record?

Diagram 14.5: AI Application Categories Consuming FHIR Data

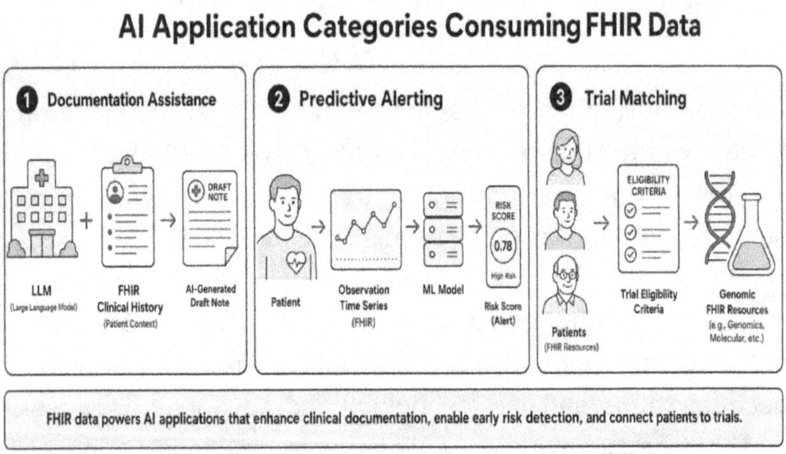

The second category was predictive clinical alerting. Machine learning models trained on FHIR-encoded vital-sign sequences, laboratory result trends, and medication administration records could identify patients whose clinical trajectories were consistent with patterns of deterioration, including sepsis onset, acute kidney injury, and hemodynamic instability. What FHIR enabled was the ability to operationalize more sophisticated models without being tightly coupled to a specific EHR vendor's proprietary data model, reducing the vendor lock-in that had historically made clinical AI deployment a hospital-by-hospital project.

The third category was patient-to-trial matching for clinical research. FHIR-encoded clinical data, when combined with structured genomic data in the MolecularSequence resource format, provided the substrate for automated clinical trial eligibility screening. A patient whose diagnoses were encoded as Condition resources with SNOMED CT codes, whose laboratory values were encoded as Observation resources with LOINC codes, and whose genomic variants were encoded as MolecularSequence resources could be screened against clinical trial eligibility criteria expressed in the same standardized vocabulary, with a precision that manual chart review could not achieve at scale.

Sandra inserted herself into the AI section of Patrick's talk preparation with a point he had not initially planned to include but ultimately decided was the most important thing in the section. AI governance for FHIR-consuming applications was not a future consideration but a present requirement. HTI-1 had established transparency and accountability requirements for clinical decision support tools using predictive AI, including documentation of training data sources, validation population characteristics, and the processes by which model predictions were presented to clinicians. Sandra had reviewed three AI vendor agreements over the preceding six months in which the proposed governance framework was inadequate under HTI-1's requirements, and in each case the gap became apparent only during legal review, after the clinical and IT teams had already committed to deployment.

Patrick's message to the conference was direct: have your compliance officer review the AI governance framework before you sign the contract, not after the implementation is complete. The clinical value of AI-assisted tools is real. The governance obligations are equally real and active today.

The Genomics Frontier

Genomic medicine was, in the view of most clinicians at Riverside General, a frontier that belonged to academic medical centers and large research networks. Dr. Maya Chen had firmly pushed back on that view during the chapter of Riverside's FHIR implementation journey that involved evaluating participation in a regional precision medicine network.

She argued that the frontier was moving. The cost of clinical genomic sequencing had fallen dramatically over the preceding decade, pharmacogenomic testing was becoming standard of care for oncology patients on targeted therapies and expanding into cardiology, psychiatry, and primary care, and the clinical genomics implementation guides developed by the HL7 Clinical Genomics work group were mature enough to support production deployments, not just research pilots.

The FHIR data model for genomics centered on two resources. The MolecularSequence resource encoded the raw genomic finding: the variant, its chromosomal location, the sequencing methodology, and the reference genome against which it was characterized. The Observation resource, with the genomics-specific profiles defined in the HL7 Clinical Genomics implementation guides, encoded the clinical interpretation: whether the variant was pathogenic, likely pathogenic, of uncertain significance, likely benign, or benign, using the classification framework of the American College of Medical Genetics and Genomics.

Diagram 14.6: How a Patient's Genomic Variant Travels From Sequencer to FHIR Resource — and What the Observation Linkage Chain Enables for Precision Prescribing at Riverside

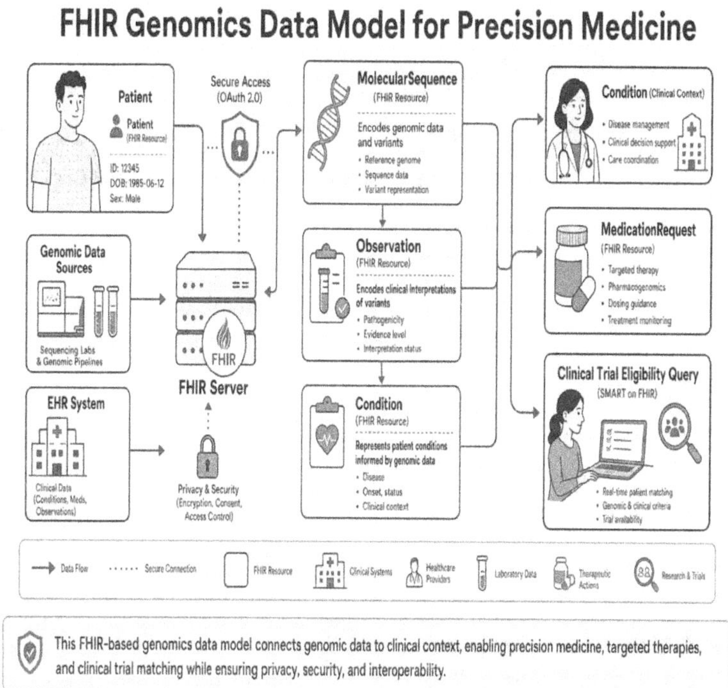

The clinical value of FHIR-encoded genomic data was not primarily in the encoding itself but in what the encoded data could be connected to. A patient whose germline BRCA1 variant was encoded as a MolecularSequence resource, whose breast cancer diagnosis was encoded as a Condition resource with the appropriate SNOMED CT code, and whose current medications were encoded as MedicationRequest resources with RxNorm codes could be queried by a clinical trial matching system, a pharmacogenomic drug interaction checker, and a family health history analysis tool simultaneously through standard FHIR API queries, without custom integration.

Dev had built a prototype genomic data integration for Riverside's oncology program. The oncology EHR module received genomic reports from the reference laboratory as structured HL7 v2 messages with embedded genomic data segments. Dev's integration translated those messages into FHIR MolecularSequence resources and linked Observation resources, wrote them to the FHIR server, and triggered a CDS Hooks call to a pharmacogenomic advisory service that generated drug interaction alerts for clinicians. The prototype took Dev approximately three weeks to build; the most time-consuming part was obtaining genomic data from the reference laboratory in a structured format sufficiently consistent for reliable parsing.

Maya's contribution to Patrick's conference talk was a clinical framing more compelling than his technical description: the goal of precision medicine was not to generate genomic data but to use genomic data to make better treatment decisions for individual patients. FHIR's genomics resources were valuable because they connected genomic findings to the clinical context that gave those findings meaning. A genomic variant without clinical context was a data point. A genomic variant connected to a complete FHIR-encoded clinical record was a clinically actionable finding.

The Global Picture: FHIR Beyond US Borders

Patrick had not originally planned to include an international section in his conference talk, his instinct being that community hospital IT professionals in the mid-Atlantic region would find a discussion of European health data regulation and NHS England's national architecture programs interesting in the abstract but not directly relevant to their operational priorities. Sandra changed his mind, with a point that was not about inspiration from international examples but about practical operational relevance: if any organization in the room was exchanging data with international partners, receiving patients from international healthcare systems, or participating in global research networks, the international FHIR adoption trajectory was directly relevant to what those partnerships would require.

The European Health Data Space regulation, abbreviated as EHDS, entered into force in 2024 and established the legal and technical framework for health data exchange across all 27 EU member states. EHDS required member states to implement national FHIR-based health data exchange infrastructure and to participate in a cross-border patient summary exchange system spanning Europe's national health systems. The EHDS patient summary profile aligned with the International Patient Summary implementation guide, which in turn aligned with FHIR R4, creating structural compatibility with US Core that made cross-border data exchange design conceptually tractable, even when specific data elements and coding requirements differed.

Diagram 14.7: Global FHIR Adoption Landscape

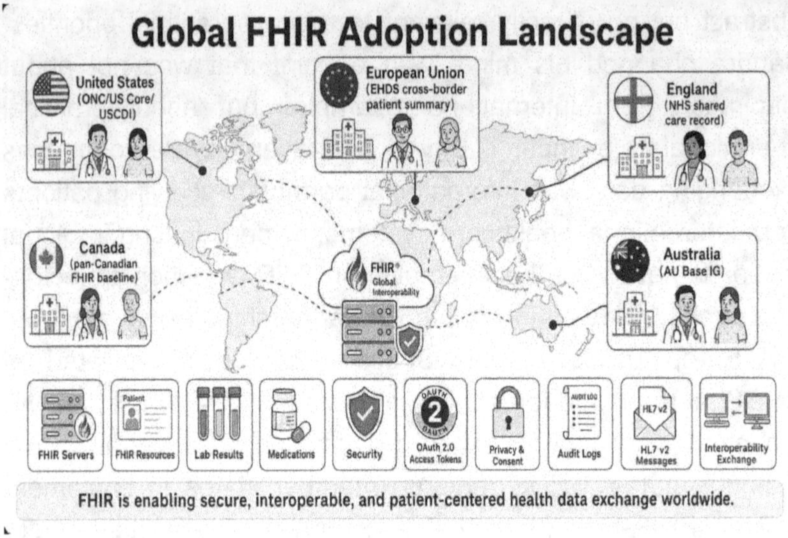

NHS England was deploying FHIR as the foundational API layer to connect previously siloed electronic patient records across hospitals, community health, and social care. The program's federated architecture had individual NHS organizations maintaining their own FHIR servers that exposed standardized APIs in accordance with a national implementation guide, with a national infrastructure layer providing discovery and routing capabilities, allowing clinicians to query across organizational boundaries. England's NHS was one of the world's largest single-payer health systems, and its deployment of FHIR as a national interoperability layer validated the approach at a scale unlike anything seen in the US regulatory environment.

Canada had developed the pan-Canadian FHIR baseline under Canada Health Infoway's auspices, establishing foundational data elements and FHIR profiles for Canadian health data exchange. Not identical to US Core, the baseline reflected differences in Canadian clinical practice, provincial health system structures, and Canadian regulatory requirements, but shared the same FHIR R4 foundation and many of the same clinical terminology systems. Australia published AU Base, the Australian FHIR baseline implementation guide, through the Australian Digital Health Agency, incorporating Australian-specific modifications to reflect local clinical practice patterns and the Medicare Benefits Schedule coding system. The Australian Digital Health Agency was also deploying a national My Health Record system with a FHIR API layer, creating a national patient health summary accessible across Australian healthcare settings.

Patrick's message to the conference audience on the international landscape was modest and actionable: if your organization has any international dimension, understand which FHIR profile your international partners use, determine where it aligns with and diverges from US Core, and document those differences before you need to exchange data.

What Patrick Tells the Conference Audience

The conference session was scheduled for forty-five minutes. Patrick had prepared thirty-five minutes of material, leaving ten for questions. He ran over by fifteen minutes, and the questions continued in the hallway after the session formally ended.

He had started with Mrs. Rivera. He told the room about the phone call, about the gap between what the law required and what the hospital could actually deliver, and about the decision to treat that gap as a project rather than a complaint to be routed and filed. He told them about the two years of work: the audit that revealed how far Riverside's practices were from information blocking compliance, the migration planning Dev led and that Patrick had thought he understood and had not, the security remediation Sandra drove after the risk assessment identified exposures nobody had previously documented, and the Patient Access API launch that ultimately resulted in Mrs. Rivera calling back to say she could see her records and wanted to know how it worked.

That second phone call was the one Patrick had not expected. He had expected, after two years of work and a successful launch, to feel satisfied in the way a completed project produces satisfaction: checklist finished, compliance deadline met, board briefed, auditors signed off. What he had not expected was a patient who was not calling to complain, who wanted to understand why it had worked now when it had not worked before. He spent about twenty minutes on the phone with Mrs. Rivera, explaining in plain language what the Patient Access API was and why it had taken as long as it had. She thanked him and said she hoped it worked that way for everybody.

Diagram 14.8: The Patient at the Center of the FHIR Ecosystem

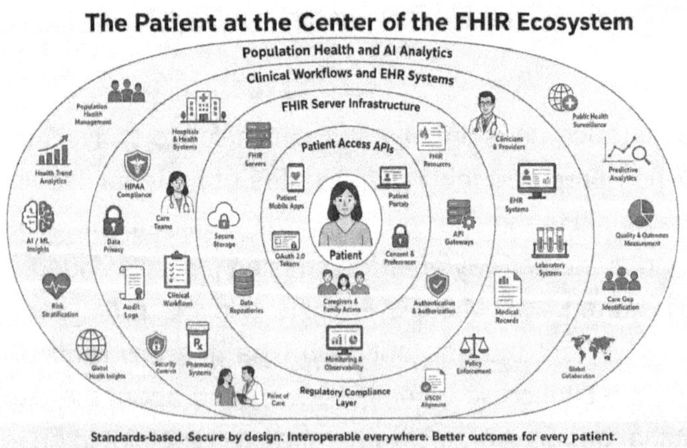

Patrick told the conference audience that Mrs. Rivera's comment about hoping it worked for everybody was the most useful feedback he had received in two years of interoperability work, because it named the thing the work was actually for. Every FHIR resource, every audit log, every integration route, every compliance checklist existed in service of a patient being able to see their own health data in an application of their choosing, and a care team being able to act on a critical lab value before a preventable complication occurred.

He told them about Dev's 217 routes, about the Friday afternoon before the migration cutover when Dev had printed the full route inventory and spread it across the conference room table, and about the careful work that had gone into preserving every route while adding a FHIR layer on top of them. The migration's success was organizational as much as technical: a team that understood why the work mattered, took the time to do it correctly, and refused the false choice between regulatory compliance and clinical continuity.

He told them about Sandra's information-blocking audit, the gap analysis that had identified dozens of practices not compliant with the information-blocking prohibitions, and the remediation requiring changes to clinical workflows, vendor contracts, staff training, and board governance policies. Information blocking compliance was not primarily a technical problem but an organizational culture problem: practices that had prioritized administrative convenience over patient data access, built without deliberate intent, and needed to be rebuilt around a different order of priorities.

And then he told them about the FHIR future he had spent the preceding portion of his talk describing: R5 improvements, Bulk FHIR analytics, AI integration, genomics, global adoption. The work Riverside had done over the past two years was not a completed project. It was a foundation, and the interesting things that could be built on that foundation were only beginning to come into view.

Staying Current Without Being Overwhelmed

The FHIR ecosystem produces documentation at a pace that no single person can match. Between Riverside's initial FHIR implementation planning and the conference where Patrick gave his talk, HL7 International had published FHIR R5, multiple new versions of the US Core, revised implementation guides for SMART on FHIR, and dozens of domain-specific implementation guides. ONC had published the HTI-1 rule, CMS had published updated Interoperability and Patient Access rule requirements, and multiple state-level health data exchange initiatives had published their own FHIR conformance requirements. The volume of information was overwhelming for anyone attempting to track it comprehensively.

Dev's approach to staying current had never been comprehensive tracking. It was a discipline of signal selection: pick a small number of authoritative channels at each layer of the standards stack, accept that everything else would arrive secondhand through those channels, and protect the time required actually to read what comes in. The specific sources matter less than the discipline. Whichever channel is currently authoritative for ballot announcements, regulatory developments, leading-edge implementation work, and community-driven clarification of ambiguous specification points, identify it, subscribe to it, and read it on a published cadence. When a channel goes quiet or a new one displaces it, Dev replaces the line in his subscription list and moves on. The practice is durable; the sources behind it are not. (At the time of this writing, the four channels Dev relied on were the HL7 FHIR announcements mailing list, the ONC blog and Federal Register health IT category, the Argonaut Project's publications, and the

HL7 FHIR Zulip workspace. Any of those may move, fork, or be supplanted by the time a reader picks this book up. The practice of having one channel for each role is what should persist.)

Patrick had added one source category that Dev had not prioritized: the published enforcement record from OIG and ONC enforcement actions. Each action described the specific practices that constitute a violation, making the enforcement record a practitioner's guide to information blocking compliance more efficient than reading the full regulatory guidance.

Diagram 14.9: Staying Current with the FHIR Ecosystem

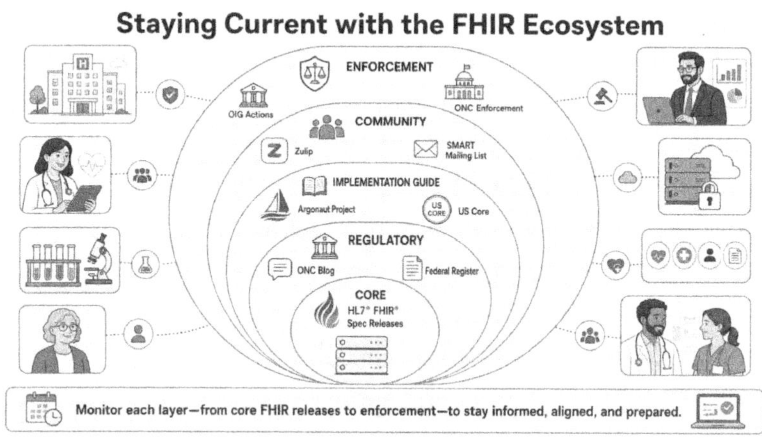

Sandra's staying-current mechanism was rooted in network rather than sources. She had built relationships with compliance officers at peer institutions, with health law attorneys specializing in information blocking and health data privacy, and with ONC staff who participated in public comment processes. The signal she got from those relationships was frequently more current and more actionable than anything in an official channel, because the network transmitted informal consensus, anticipated regulatory developments, and practical implementation questions that had not yet appeared in formal publications.

Maya tracked the clinical literature on health data exchange and clinical decision support, focusing on published outcome studies demonstrating the clinical impact of interoperability investments. A peer-reviewed study demonstrating that structured FHIR-encoded medication data reduced adverse drug events in a care transitions program was more persuasive to a skeptical clinician colleague than any amount of regulatory compliance argument. Patrick's own staying-current practice was, he admitted to the conference audience, the least systematic of the four. He relied on Dev, Sandra, and Maya to filter the technical, regulatory, and clinical information streams, respectively, and to surface the developments that required his attention as an administrator. What he had learned to do well was ask the right questions: when a new regulatory development arrived, ask Sandra what it meant for Riverside's compliance posture; when a new technology capability emerged, ask Dev whether it was mature enough to plan around; and when a clinical impact claim was made, ask Maya whether the evidence supported it. The ensemble approach was the practical answer for administrators who lacked the time or technical background to track the FHIR ecosystem personally.

What Comes After This

Patrick ended his conference talk not with a summary of what he had covered but with a forward-looking observation about the work that remained. Riverside General was not finished. No organization doing serious health data interoperability work was ever finished, and the ones that believed they were finished would be surprised by the next regulatory cycle, the next vendor announcement, or the next clinical use case that required capabilities they had not built.

The FHIR foundation Riverside had constructed was real and valuable. The Patient Access API was live. Mrs. Rivera could see her lab results. The Bulk FHIR export infrastructure was feeding population health analytics that had already measurably improved outcomes for diabetic patients. Dev's integration layer handled FHIR transactions alongside the 217 HL7 v2 routes that had never stopped running, coexisting in a hybrid architecture that was neither elegant nor simple but was operational and maintainable.

What remained was the next layer of the work: the R5 migration decision, the AI governance framework that Sandra and Dev were building for the first AI-assisted clinical tool deployment, the genomics integration that Maya was championing and Dev was prototyping, the regional health information exchange participation that would connect Riverside's FHIR infrastructure to other hospitals and ambulatory practices, and the USCDI version updates requiring ongoing conformance maintenance.

Patrick closed his talk the way he had started it: with Mrs. Rivera. He told the audience that she had called again, not about a problem with her data access but because she had a question about a lab value she had seen in the app and did not understand. She had been connected to a nurse navigator who explained the value, confirmed it was within normal limits, and documented the interaction in the EHR. The interaction had taken twelve minutes. The data that made it possible had taken two years to build.

Twelve minutes for Mrs. Rivera. Two years for Riverside. That ratio, Patrick told the conference audience, was exactly right. The patient experience should be effortless, which meant the institutional work behind it needed to be thorough, careful, and sustained. Every implementation project, every compliance audit, every integration route, every architecture decision existed in service of making a twelve-minute interaction possible for a patient who deserved accurate, timely, accessible healthcare data.

Takeaways

Dev's R5 conviction: the organizations that get blindsided by the R5 transition will not be the ones that decided to wait — they will be the ones that never followed the public ballot process and will encounter R5-based IG requirements for the first time in a vendor evaluation. No new integration work should hard-code R4-specific assumptions. Every new vendor should provide explicit R5 roadmap commitments. At least one team member should be tracking each R5 development area.

Maya's Bulk FHIR conviction: the compliance API and the population health platform are the same infrastructure. An organization that implemented FHIR to meet the ONC Patient Access deadline and has not yet activated Bulk FHIR export has paid for a clinical intelligence capability it is not using. Her criterion for considering the FHIR investment complete: at least one population health program running on Bulk FHIR export data, with outcomes measurable enough to present to the board.

Sandra's AI governance conviction: HTI-1 made AI governance a current compliance requirement, not a future planning exercise. Any AI-assisted clinical tool consuming FHIR data requires documented governance — model training data, validation population, clinical workflow integration, and audit trail — before the vendor contract is signed. If the governance documentation does not exist in the procurement package, the tool does not deploy.

Dev's international FHIR conviction: the differences between US Core and EHDS, NHS England, pan-Canadian, and AU Base implementation guides are not cosmetic — they are regulatory divergences with structural data-mapping consequences. Organizations with international partners need a profile-gap analysis before any cross-border exchange, not after the first batch of validation errors arrives.

Patrick's staying-current conviction: this is a structural problem, not a motivation problem. No one person can track the specification, the regulatory cycle, the IG landscape, and the clinical evidence base simultaneously. Patrick assigned each domain a named owner with a monthly briefing obligation to the team. The organizations that fall behind do so not from lack of intent but from lack of assignment.

The patient is the constant. Every specification version, every regulatory cycle, every architecture pattern, every governance framework in this chapter and in this book exists in service of a patient who deserves accurate, timely, accessible healthcare data. Mrs. Rivera is not an exceptional case study. She is every patient whose care depends on health information flowing accurately and accessibly across the systems that hold it. The organizations that keep her at the center of every interoperability decision will build systems that are not just technically conformant but genuinely worth building, and that is the standard that matters.

Patrick will say, if anyone asks, that the R5 transition is not the part of this work that keeps him up at night; the part that keeps him up is whether the next configuration change deletes a mapping row no one will notice until a patient does. Maya, asked the same question, will answer that the staying-current problem the chapter describes is mostly a workflow problem in disguise: if a new IG is published and no clinician at Riverside reads it, the IG might as well not exist. Dev, by now allergic to the phrase "future-proof," will point out that Bulk FHIR was sitting on the server for a year before anyone asked what it was for. The lesson he draws from that is to assume that the most useful capability in the next version of the standard has already shipped and gone unused. Sandra, last and shortest, will say what she has said since the ransomware incident: the standards are catching up with the threats faster than most healthcare organizations are catching up with the standards, and the gap between those two curves is where the next breach lives. Four opinions. One direction of travel.

16 Closing: Patrick's Hospital, 18 Months Later

There is a new lobby at Riverside General.

Not a dramatic renovation. Patrick did not budget for marble floors or a commissioned installation. What changed is simpler and, in its own way, more significant: the patient check-in kiosks that line the east wall of the entrance now connect directly to the EHR's FHIR endpoint. When a patient arrives and taps the screen, the kiosk pulls a verified medication list. It confirms insurance eligibility in the same round-trip request that was previously handled by three separate systems, two manual reconciliation steps, and a fax machine that broke down often enough that the front desk staff had named it. The fax machine is gone. The medication reconciliation step has not been eliminated; no responsible clinical team eliminates it. It now begins with a verified FHIR MedicationRequest list instead of a handwritten form. That is not a small thing. That is the kind of change that does not appear in a press release but that the nurses notice every morning when they start their shift.

Patrick stood in the lobby the first week the kiosks went live and watched patients use them. Not because he needed to supervise. Because he wanted to see whether eighteen months of planning, negotiation, vendor calls, incident reports, board presentations, regulatory responses, and technical decisions made at two in the morning by people he trusted had produced something a patient could actually use without instructions. Mrs. Rivera, who had been among the first patients to connect a third-party application to her health record through the Patient Access API, used the kiosk without pausing. She tapped through the screens as though she had always done it this way, retrieved her visit summary, and walked toward the elevators. She did not look back at the kiosk. That indifference was the best outcome Patrick could have imagined.

Diagram Close.1: Riverside General Lobby Transformation

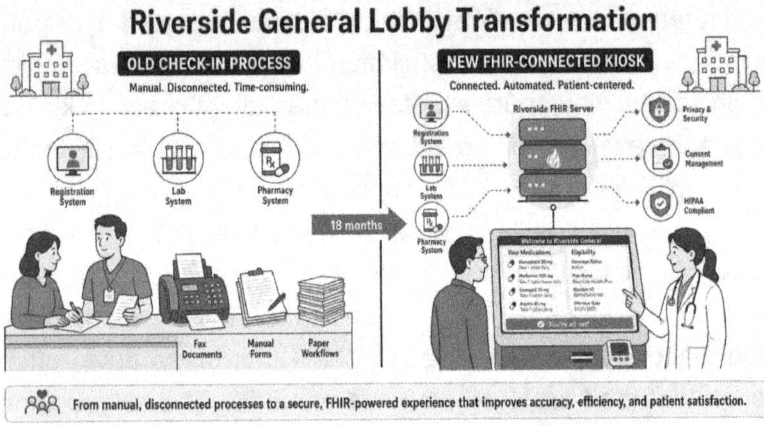

What Changed at Riverside General

The numbers that Patrick tracks on his monthly operational dashboard look different than they did eighteen months ago. The Patient Access API is live and processing an average of 1,400 patient-initiated data requests per week. Portal enrollment has increased by 34 percent since the kiosk deployment, not because the portal itself changed but because the friction between patients and their own data decreased enough that enrollment stopped feeling like a task and started feeling like a convenience. The patient satisfaction scores for the informatics-adjacent categories in the quarterly survey, specifically the items about whether patients can access their records quickly and whether they feel their health information is secure, have both moved above the regional benchmark for the first time since the survey was introduced.

The ONC information blocking complaint that Sandra received eighteen months ago was resolved with a compliance action plan that included the Patient Access API implementation, accelerated lab result release timelines, and a documented information blocking exception review process that Sandra now runs quarterly. The complaint did not result in a formal finding. Patrick reads that sentence carefully whenever he reviews it, because it contains a compressed history of six weeks of Sandra working through evenings and weekends to produce a response that was both technically correct and strategically credible. The absence of a formal finding did not mean the complaint was without consequence. It accelerated the timeline for every improvement the team had already planned, and it gave Patrick a very clear picture of what happens when a healthcare organization is not prepared to explain its data practices to a regulator who has read the statute more carefully than the organization has.

The integration engine, Dev's Mirth Connect instance with its 217 routes, is still running. It will be running for the foreseeable future. But it is now routing FHIR resources alongside HL7 v2 messages. The 23 routes that have been migrated to FHIR-native operation represent the clinical workflows that Dev, Maya, Sandra, and Patrick agreed were the highest-priority migration targets: laboratory result delivery to the patient portal, medication list synchronization between the EHR and the pharmacy system, and the discharge summary exchange interface that connects Riverside General to the regional health information exchange. The 194 routes still running on HL7 v2 are operating correctly and will be migrated according to the migration roadmap, not faster. Patrick had to defend that decision to a board member who believed that if 23 routes could be migrated in eighteen months, then 194 routes could be migrated in eighteen more months. The answer, which Dev helped Patrick construct, was that the 23 routes were chosen because they were the most straightforward candidates for migration. The remaining 194 routes are more complex, more operationally sensitive, and more likely to require vendor cooperation that takes longer than anyone predicts. Rushing a migration to satisfy a board timeline is not a compliance risk. It is a patient safety risk.

Diagram Close.2: FHIR Migration Progress at Riverside General

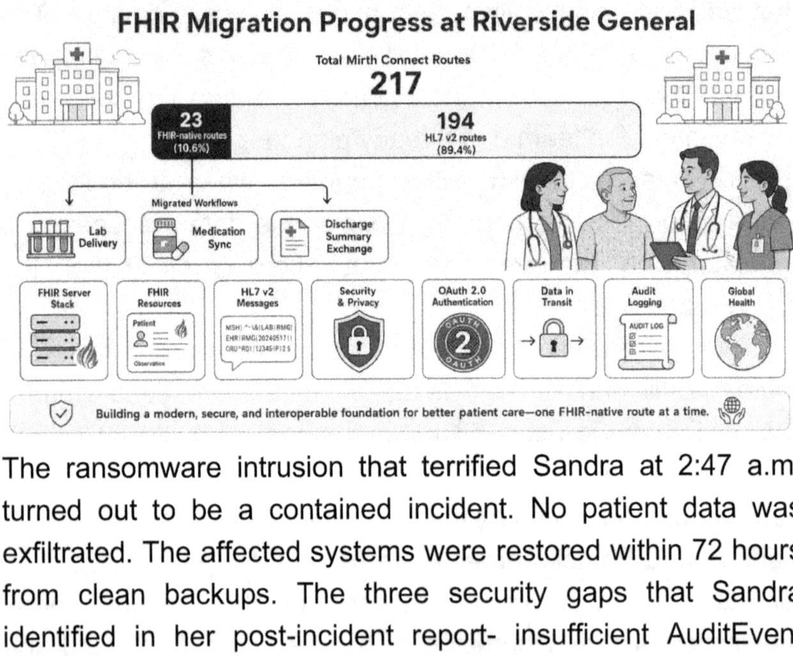

The ransomware intrusion that terrified Sandra at 2:47 a.m. turned out to be a contained incident. No patient data was exfiltrated. The affected systems were restored within 72 hours from clean backups. The three security gaps that Sandra identified in her post-incident report- insufficient AuditEvent logging, overly broad OAuth scopes, and no FHIR Consent resource implementation- were closed within 90 days. The AuditEvent logging configuration now captures every FHIR API interaction in structured format, queryable through the same FHIR server that stores the clinical data. The third-party application scope review that Sandra conducted after the incident resulted in eight applications having their scopes narrowed and two applications being deregistered entirely. The FHIR Consent resource implementation, which took Dev's team six weeks to build and validate, is now the mechanism through which Riverside General documents and enforces patient-directed access restrictions required by HIPAA.

Patrick thinks about that night more than he expected to. Not because it still frightens him, but because it was the clearest demonstration in his professional career that the distance between a contained incident and a catastrophic one is not primarily a function of investment or planning. It is a function of how quickly a team that trusts each other can respond. Sandra called him at 2:49 a.m. and had already identified the probable entry vector. Dev was remote-logged into the integration engine before three o'clock. The incident was contained because the people involved had been working together closely enough, and honestly enough, that they could execute under pressure without waiting for permission. Patrick would fund every technical control Sandra and Dev recommended for the rest of his career at Riverside General, not because the controls are infallible, but because they give competent people something solid to work with when things go wrong at two in the morning.

What FHIR Did Not Solve

It would be dishonest to end this book without saying plainly that FHIR did not solve everything. Healthcare interoperability is a genuinely complex problem, and a standard, however well designed, addresses only the portion of that problem that lives in the specification layer. The areas that fall under organizational behavior, vendor incentives, clinical workflow, and regulatory enforcement are not addressed by the FHIR specification, and Riverside General encountered each of them.

The EHR vendor took 11 months to release an update that exposed the Appointment resource via an FHIR endpoint, despite being contractually committed to that release on a 6-month timeline. Patrick spent more time on vendor management conversations during this migration than on any other administrative task. He learned to treat vendor timelines as optimistic estimates and to build contractual penalties that create actual incentives for delivery. He learned to ask at every vendor meeting what specific technical dependency was blocking the committed deliverable, and to insist on a written answer before leaving the room. These are skills that no FHIR specification teaches.

The clinical data quality problems that Dev discovered during the terminology mapping work in the laboratory information system were not created by the migration. They were revealed by it. When Dev's team began producing LOINC-coded FHIR Observation resources from the laboratory system's output, they discovered that a meaningful fraction of the proprietary codes the laboratory system used for certain specialized tests had no clean LOINC mapping because the laboratory had been using internally defined code values that were never standardized. Those tests required manual review by the laboratory director and a three-month reconciliation process that nobody had budgeted for. FHIR did not create that problem. It made the problem visible, which is the first step toward solving it, but visibility is not the same as resolution. The resolution required human attention, institutional will, and time.

Patient adoption, too, moved more slowly than the optimistic projections suggested. The 34 percent increase in portal enrollment is real and meaningful. It also means that 66 percent of patients who could be using the portal are not. Patrick does not attribute that gap to poor technology or poor communication.

He attributes it to the reality that patients are not waiting for a better patient portal. They are managing their health within the cognitive and time constraints of their actual lives, and a FHIR-connected portal is not always the highest-priority item in those lives. The technology creates the possibility. The adoption is a separate project with its own disciplines, expertise, and timeline.

Diagram Close.3: What FHIR Addressed vs. What Remained

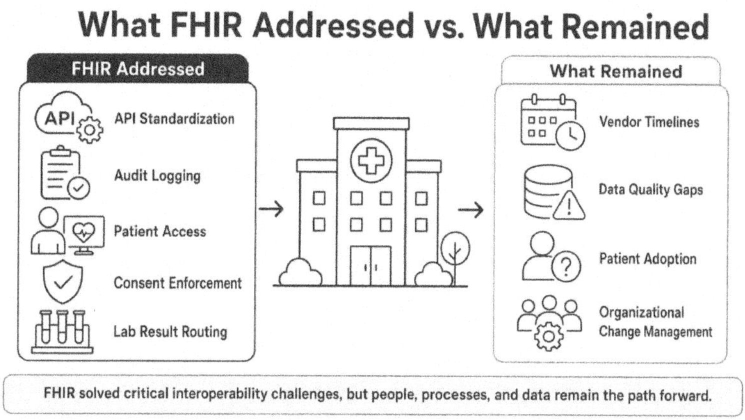

What Changed for Dev

Dev Okafor now leads the technical working group for the regional health information exchange's FHIR integration standards committee. He accepted the role because he recognized, somewhere around month eight of the migration, that the knowledge he had accumulated about v2-to-FHIR mapping, integration engine hybrid architecture, and FHIR validation was not just institutional knowledge for Riverside General. It was community knowledge for every hospital in the region planning a similar migration. He still runs Mirth Connect. He still knows every one of his 217 routes by memory. But he is now teaching the next generation of integration engineers how to build FHIR facades over legacy EHRs, how to write conditional create logic that prevents duplicate patient records, and how to read an OperationOutcome at two in the morning when a production integration is failing. The on-call physician is waiting for a lab result.

Building a FHIR facade over a legacy EHR is not a glamorous exercise. It is, at its core, the work of writing translation logic that takes whatever the source system produces and reformats it into a FHIR resource that passes validation. It requires intimate knowledge of the source system's data model, patient knowledge of the FHIR specification's requirements for each resource type, and the discipline to test against real data rather than synthetic examples. Dev's facades for the laboratory and pharmacy systems took longer to build than any timeline Dev had presented to Patrick, and they passed validation only after iterations no one had anticipated. Dev considers this normal. He tells the engineers he is mentoring that the estimate is always wrong and the code is always harder than the schema makes it look, and that both are true for everyone, not evidence of personal inadequacy.

Dev has strong opinions about FHIR R5's subscription model. He shares them freely, at the standards committee, in the Slack channels he participates in, and occasionally at a volume that startles his team when he is reading a draft specification. His core position is that the topic-based subscription model in R5, while technically superior to the R4 Subscription resource in almost every respect, requires implementation guide authors to be more explicit about event notification semantics than most current IGs are willing to be. He is probably right. The standards committee has not yet resolved the question. Dev will continue to raise it.

What Changed for Maya

Dr. Maya Chen has approved two clinical decision support rules that fire on FHIR Observation resources. The first is a potassium alert rule that triggers when a serum potassium result crosses the critical threshold. The second is a creatinine trend rule that identifies patients whose kidney function is deteriorating across sequential lab results, a pattern that a single result cannot reveal but that a sequence of FHIR Observation resources, queried chronologically through the FHIR server, makes computable in a way it was not before.

Both rules required the laboratory information system to produce LOINC-coded FHIR Observation resources instead of the proprietary-coded results it previously generated. That transition required four months of vendor coordination and three rounds of LOINC mapping validation. Maya attended every validation meeting. Patrick once offered to send a junior member of the clinical informatics team in her place when a scheduling conflict arose. She declined. She did not decline because she does not trust the clinical informatics team. She declined because she understood, by that point, that the mapping decisions made in those meetings were not technical in any separable sense. They were clinical decisions about which code values represent which clinical concepts, and clinical decisions belong in front of a physician.

Maya personally reviews every alert that fires for now, because she does not yet fully trust the rules to catch everything a clinician would. She is right to be careful. The potassium alert rule has fired 34 times in its first three months of operation. Thirty-one of those fires were clinically appropriate. Three were triggered by lab values resulting from hemolysis in the blood draw sample, a pre-analytical error that produces artifactually elevated potassium results. The rule cannot yet distinguish a true critical potassium from a hemolyzed sample. That is a known limitation. Maya has documented it in the clinical decision support governance log, flagged it for the laboratory director, and is working with the vendor on whether a reliability flag can be propagated from the laboratory instrument through the HL7 v2 result message, through the integration engine transformation, and into the FHIR Observation resource as a dataAbsentReason or quality extension. The pathway exists. The vendor has not yet committed to a timeline for implementing it.

She trusts the data more than she did eighteen months ago. She does not yet trust it completely. That is the appropriate posture for a physician who has seen what happens when data fails. It is also, Patrick suspects, the appropriate posture for anyone working in healthcare interoperability, regardless of their role. The data infrastructure Riverside General has built is more reliable, more auditable, and more clinically meaningful than what it replaced. It is not infallible. The work of maintaining that infrastructure, validating its outputs, and continuously improving its accuracy is not a project with an end date. It is a practice, in the same sense that medicine is a practice.

What Changed for Sandra

Sandra Torres passed Riverside General's most recent ONC compliance review with no findings. That is the first time in the hospital's history that an ONC compliance review has concluded with no corrective action items. Sandra attributes the result to three things: the Patient Access API implementation, the AuditEvent logging configuration, and the pre-launch compliance review process that she now conducts for every new data access feature before Patrick approves it for production.

Patrick would add a fourth: Sandra's habit of reading every regulation three times before forming a position, which means that by the time an ONC reviewer asks her a question, she has already considered and documented every interpretation of the rule that the reviewer is likely to raise.

The pre-launch compliance review is Sandra's most important procedural contribution to the way Riverside General builds technology. Before the FHIR migration, new system features underwent security and legal reviews and then went live, often on the same timeline, with compliance as a parallel track that intersected with the technical work only at the end rather than throughout. Sandra changed that. The compliance review now begins when the feature specification begins, not when the feature is ready to deploy. Sandra reviews the proposed data flows, access control design, scope configuration, and audit logging implementation before any code is written. When her review identifies a problem, it costs a conversation and a revision, not a delayed deployment and a compliance remediation. Patrick did not argue with this change when Sandra proposed it. He had watched the alternative play out often enough to know what it costs.

Sandra has been asked to join a federal advisory committee on health data privacy. She accepted the role after consulting with Patrick about whether the time commitment was compatible with her responsibilities at Riverside General. The committee is developing guidance for healthcare organizations navigating the intersection of HIPAA, the Information Blocking Rule, and emerging governance requirements for AI-assisted clinical applications. The last item on that list is the one that occupies most of Sandra's attention outside of work. She has watched the conversation about artificial intelligence and health data shift from a theoretical discussion of future risk to a practical discussion of present obligation, and she is not satisfied with the level of specificity in the current regulatory guidance. She is, characteristically, already three readings ahead of the other committee members. She has submitted two written comments to the committee's public docket. Both received staff acknowledgment. Neither has been incorporated into a draft guidance document yet. Sandra is patient but also persistent, and the combination has served her well in every regulatory process she has participated in.

Diagram Close.4: Sandra's Compliance Evolution at Riverside General

What FHIR Adoption Did and Did Not Solve

The question that Patrick returns to most often, when he has time to think carefully, is what FHIR adoption actually changed at Riverside General versus what it revealed, and whether those are separable categories. The answer he has arrived at after eighteen months is that they are not entirely separable and that the conflation is one of the most common errors that healthcare organizations make when evaluating an interoperability initiative. FHIR adoption at Riverside General produced measurable, meaningful improvements in patient access to health information, the auditability of data flows, the hospital's ability to demonstrate compliance with federal regulations, and the technical foundation for clinical decision support that is actually computable. Those improvements are real. They were not inevitable. They required a substantial investment of time, money, and human attention from four people and dozens of others who do not appear in this book's narrative but who wrote the validation scripts, attended the vendor calls, triaged integration errors at midnight, and updated the technical documentation after every deployment.

What FHIR adoption revealed, rather than solved, is that the underlying data quality in any health information system reflects decades of accumulated decisions about how to configure systems, how to train users, how to map codes, and how to prioritize consistency against speed in the daily operation of a hospital. FHIR makes those decisions auditable and, therefore, improvable. But the improvement requires institutional commitment to quality that goes beyond any individual standard. Riverside General's laboratory data quality is better today than it was eighteen months ago because Maya and the laboratory director attended the mapping validation meetings. It is not better because FHIR required it.

The organizational dimension of interoperability, the patterns of trust, communication, and shared accountability that make technical integration clinically meaningful, is the dimension that standards bodies cannot specify and implementation guides cannot mandate. Patrick learned, over eighteen months, that the most important architectural decisions he made were not about

FHIR profiles or API gateway configurations. They were about who was in the room when a difficult decision needed to be made, how much authority he gave those people to act without waiting for his approval, and how he handled the moments when the technical reality did not match the plan he had approved. Those decisions shaped the outcome at Riverside General more than any specification document could have.

FHIR gave Riverside General a common language. The team gave that language meaning. Both are necessary. Neither is sufficient without the other.

What You Now Carry

You have arrived at the end of this book with something that most people who encounter the words FHIR and HL7 do not have: an understanding of what those words actually mean, why they exist, how they work in practice, and what it genuinely costs to make them work for real patients in real hospitals. That cost is measured in time, organizational effort, technical investment, and human attention. It is not reducible to a budget line or a project plan. It is the accumulated cost of every decision made correctly, every vendor negotiation conducted patiently, every compliance review conducted before a feature launched rather than after, and every 2 a.m. incident response that held together because the team had built the trust to move fast without breaking more than they fixed.

You understand that HL7 v2 is not a failure but a foundation. It has been carrying healthcare data reliably for decades, and it will continue to carry it for decades more in most of the hospitals you will encounter in your career. You understand that FHIR is not a destination but a direction: a set of specifications that creates the conditions for interoperability without guaranteeing it, and that requires continuous investment in implementation quality to deliver on its design promises. You understand that clinical terminology is not bureaucratic overhead but the semantic layer without which technical connectivity is clinically meaningless, and that an FHIR Observation resource with a proprietary code value is not interoperable, regardless of the format in which it travels.

You understand that regulations exist because patients have been harmed by the absence of the behaviors those regulations require, and that compliance with the Information Blocking Rule and the HIPAA Security Rule is not a tax on innovation but a floor beneath which patient trust cannot be maintained. You understand that security in a FHIR environment is not a configuration setting but a continuous practice, and that the AuditEvent resource, the OAuth scope, and the FHIR Consent resource are not abstract technical constructs but the specific mechanisms through which a patient's expectation of privacy is either honored or violated at the moment of each API call.

Take what you have learned here into whatever role you occupy in the healthcare data ecosystem. If you are an administrator, ask your vendors to demonstrate a Capability Statement before you sign a contract, and ask your compliance officer to explain the information blocking exceptions before your next ONC review. If you are a developer, read the implementation guide before the FHIR specification, because the IG will tell you what the specification requires you to implement, and the specification will explain why. If you are a compliance officer, start the compliance review at the beginning of the feature specification, not at the end, and invest in your relationship with the technical team until they understand why you are asking the questions you are asking. If you are a clinician, stay in the mapping validation meetings. The code values are clinical decisions. They belong to you.

Riverside General is not a finished story. The migration roadmap has 194 routes remaining. The EHR vendor is six months behind on the Appointment resource. The patient adoption numbers have room to grow. Sandra's committee has not yet published the AI governance guidance that the healthcare data ecosystem needs. Dev's R5 subscription argument has not yet persuaded the standards committee. Maya's creatinine trend rule is still being refined. Patrick is already in preliminary conversations with the regional health information exchange about whether Riverside General can serve as a reference implementation site for the next generation of FHIR-based care coordination workflows. This conversation will require another board presentation, another technical assessment, another compliance review, and probably another incident that no one has yet imagined.

The work continues. It always continues. That is not a warning. It is the condition under which meaningful work in healthcare happens, and it is the condition that makes the people who do it, including you, more valuable than any system they manage.

You are, in every meaningful sense, one of them.

17 Appendix A: Glossary of Key Terms

The following 50 terms are defined in plain English for readers with varying levels of technical background. Each definition is written to stand alone without requiring knowledge of adjacent terms. Where a term has a specific technical meaning in the FHIR or HL7 context that differs from everyday usage, that distinction is noted.

A

ACK / NACK (Acknowledgment / Negative Acknowledgment) The guaranteed-delivery handshake mechanism used in HL7 v2 messaging. A sending system holds a message in its outbound queue until it receives either an ACK, confirming successful receipt and processing, or a NACK, indicating that the message arrived but could not be processed. Without this mechanism, messages can vanish silently in transit with no record of the loss.

ADT (Admit, Discharge, Transfer): The HL7 v2 message type that notifies all connected systems whenever a patient's registration status changes. An A01 event fires when a patient is admitted; A03 fires on discharge; A08 fires on a demographic update. ADT messages are the heartbeat of a hospital integration environment, triggering record creation, billing account opening, and bed management in ancillary systems.

AllergyIntolerance (FHIR Resource): A FHIR resource that represents a patient's allergy or adverse reaction to a specific substance, whether a medication, a food, or an environmental agent. It captures the allergen identity, the clinical severity, the type of reaction observed, and the verification status of the allergy, making it the structured foundation for medication safety decision support in FHIR-connected workflows.

AuditEvent (FHIR Resource): A FHIR resource that records a single access interaction with a FHIR server, including who acted, what resource was accessed, when the access occurred, and whether it succeeded or failed. AuditEvent resources are the structured, queryable audit trail that HIPAA's audit control requirement demands, and they are the first place a compliance officer turns when investigating a suspected unauthorized disclosure.

B

Binding Strength: The level of enforcement applied to a ValueSet on a coded element in a FHIR profile. The four levels are: required (only codes from the specified ValueSet are valid, with no exceptions), extensible (local codes may be used if no suitable code exists in the ValueSet), preferred (the ValueSet is recommended but not enforced), and example (the ValueSet is purely illustrative). Misunderstanding binding strength is one of the most common sources of interoperability failures in production FHIR environments.

Bundle (FHIR Resource): A FHIR container resource that groups multiple individual resources into a single package for exchange, storage, or atomic processing. Bundle types serve distinct purposes: a transaction Bundle allows multiple creates and updates to succeed or fail together as a single unit; a document Bundle wraps a complete clinical record; a searchset Bundle returns results from a FHIR search operation. Nearly every FHIR API response involving more than one resource arrives wrapped in a Bundle.

Bulk FHIR (Bulk Data Access): The FHIR Bulk Data Access specification, which defines an asynchronous server-side export mechanism for retrieving large volumes of FHIR data without blocking the server. Rather than requesting individual resources, a bulk export operation requests all resources of specified types for a defined population; the server processes the request asynchronously and delivers the results as NDJSON files. Bulk FHIR is the appropriate pattern for population health analytics, quality reporting, and payer-to-payer data exchange.

C

CapabilityStatement (FHIR Resource) A FHIR resource, retrievable at the /metadata endpoint of any FHIR server, that declares which resource types the server supports, which operations it implements, which search parameters it honors, and which profiles its resources conform to. The CapabilityStatement is the server's formal self-description and the most informative document available when evaluating a vendor's actual conformance posture. It is typically the first thing an integration engineer requests from a new FHIR partner.

CDA (Clinical Document Architecture): An HL7 standard for the structure and semantics of clinical documents, including discharge summaries, progress notes, referral letters, and consultation reports. CDA documents are XML-based and follow the HL7 v3 Reference Information Model. The federal Meaningful Use incentive program required the Consolidated CDA (C-CDA) variant and remains widely exchanged between EHR systems for transitions of care today.

CodeableConcept (FHIR Data Type) A FHIR data type that carries both a machine-readable code drawn from a standard terminology system and a human-readable text description of the same concept. CodeableConcept is the primary bridge between FHIR resources and clinical terminologies. It enables a system that does not recognize a specific code to display meaningful text while preserving semantic utility across heterogeneous environments.

CodeSystem (FHIR Resource) A FHIR resource that defines a set of codes and their meanings, representing the authoritative definition of a terminology vocabulary within the FHIR ecosystem. CodeSystem resources describe terminologies such as LOINC, SNOMED CT, or a locally defined vocabulary. ValueSets draw their membership from CodeSystems, selecting curated subsets of codes appropriate for specific clinical contexts.

Consent (FHIR Resource) A FHIR resource that represents a patient's documented agreement to or refusal of specific uses of their health information, expressing that decision in machine-executable form. Consent resources encode the scope of permission, the time period during which it applies, the parties to whom it grants or denies access, and the specific data or actions it addresses. In FHIR-based access control architectures, Consent resources serve as the structured policy layer governing patient-directed data sharing.

Coverage (FHIR Resource) A FHIR resource representing a patient's health insurance coverage, including the plan details, the payer, the subscriber identity, the policy and group numbers, and the effective coverage period. Coverage is a foundational resource in financial FHIR workflows, referenced by Claim and ExplanationOfBenefit resources to link clinical services to the insurance arrangement under which reimbursement is sought.

CPT (Current Procedural Terminology) The AMA-published code system for physician procedures, services, and diagnostic tests submitted to payers for reimbursement. CPT codes are required for claims processing throughout the United States. They are subject to a commercial licensing requirement: any organization that transmits or displays CPT codes in a health IT product, API, or electronic transaction must hold an active license from the American Medical Association.

D

DiagnosticReport (FHIR Resource) A FHIR resource that represents the findings and interpretation of a diagnostic study, including laboratory panel reports, radiology interpretations, and pathology reports. A DiagnosticReport links to the individual Observation resources containing the discrete result values, carries the performer identity, and can include a human-readable narrative or a PDF attachment of the formatted report as the clinician received it.

E

EHR (Electronic Health Record) A digital system that collects, stores, and manages patient health information across clinical encounters, replacing paper-based records with structured, queryable data. EHRs are the primary source systems for FHIR data in most healthcare organizations. In the United States, EHRs deployed in clinical settings are subject to ONC certification requirements that mandate FHIR API support for patient data access.

Encounter (FHIR Resource) A FHIR resource that represents a patient's interaction with healthcare services, whether an inpatient admission, an outpatient visit, an emergency department presentation, or a telehealth session. The Encounter resource carries the visit class, the care team, the service provider organization, the location, and the period during which care was delivered. It is the central context that ties most clinical resources to a specific episode of care.

EVN (Event Type Segment) The HL7 v2 segment that carries information about the triggering event that caused an ADT message to be generated, including the event type code, the date and time the event was recorded, and the operator who triggered it. EVN provides the temporal and causal context that downstream systems need to process ADT events correctly, distinguishing between a same-day admission correction and a genuine discharge.

F

FHIR (Fast Healthcare Interoperability Resources) is a modern standard for exchanging healthcare data, built on the same web technologies that power everyday websites and applications: HTTP, JSON, REST, and OAuth. Published by HL7 International, FHIR Release 4 (R4) is the version mandated by United States federal regulations for certified health IT products. Pronounced "fire," FHIR is now the primary vehicle for patient data access, payer data exchange, and clinical integration in US health IT.

FHIR R4 FHIR Release 4, published by HL7 International in January 2019 as the first normative (production-stable) release of the FHIR standard. FHIR R4 is the version mandated by ONC certification requirements and the basis for the US Core Implementation Guide, making it the operative FHIR standard for US healthcare interoperability. Its normative status means that the core resource definitions it contains are subject to formal change management processes and will not be broken without a formal deprecation process.

FHIR R5 FHIR Release 5, published by HL7 International in March 2023, introduces significant improvements including a redesigned topic-based Subscription framework, cross-version extension support, and matured clinical reasoning resources. R5 is the current complete FHIR release and the target version for forward-looking implementation guide development, but US regulatory requirements and ONC certification criteria continue to reference R4 as of 2026.

H

HCPCS (Healthcare Common Procedure Coding System) A CMS-maintained extension of CPT that adds codes for services, supplies, and equipment not covered by CPT, including durable medical equipment, ambulance services, and certain outpatient drugs. A US government agency publishes HCPCS codes, and they are freely available from CMS without a licensing fee, in contrast to the licensed CPT codes they supplement.

HIE (Health Information Exchange) An organization or technical infrastructure that enables the sharing of clinical data among healthcare providers, patients, and payers across organizational and geographic boundaries. HIEs serve as regional or national intermediaries that aggregate patient data from multiple source systems and route it to requesting organizations, reducing the need for point-to-point integrations between every pair of trading partners.

HIPAA (Health Insurance Portability and Accountability Act) The federal law enacted in 1996 that establishes privacy and security protections for protected health information (PHI). HIPAA's Privacy Rule governs permissible uses and disclosures of PHI; the Security Rule establishes technical, physical, and administrative safeguards for electronic PHI. HIPAA compliance is a baseline obligation for any healthcare organization that deploys an FHIR API that handles patient data.

HL7 (Health Level Seven International) The international standards organization that develops and maintains healthcare data exchange standards, including both the HL7 v2 messaging standard that underlies most current hospital integrations and the FHIR standard that governs modern API-based exchange.

The "seven" refers to the seventh layer of the OSI networking model (the application layer), where clinical data exchange operates.

HL7 v2 Health Level Seven version 2, a pipe-delimited text-based messaging standard first published in 1987 and still the most widely deployed clinical messaging standard in the United States. HL7 v2 messages carry patient admissions, laboratory results, medication orders, and clinical documents between hospital systems via the MLLP transport protocol. Most hospitals run thousands of HL7 v2 messages per day through their integration engines, and v2 will remain in active use for many years alongside FHIR.

I

ICD-10-CM International Classification of Diseases, Tenth Revision, Clinical Modification: the US-specific adaptation of the WHO's disease classification system, containing approximately 70,000 diagnosis codes used for billing, epidemiology, public health surveillance, and administrative reporting. ICD-10-CM codes are required on all insurance claims submitted in the United States and are published annually by CMS and the National Center for Health Statistics as a freely available public-domain resource.

ICD-11 The eleventh revision of the International Classification of Diseases, published by the WHO in 2019 with significant improvements in clinical concept coverage, ontological structure, and support for multiple-condition coding. ICD-11 is in international use and underpins the European Health Data Space's diagnostic coding requirements, but as of 2026, the United States has not mandated a transition from ICD-10-CM to ICD-11.

Implementation Guide (IG) A published package that bundles FHIR profiles, extensions, value sets, code systems, terminology bindings, and explanatory narrative into a complete specification for a defined use case or national context. Implementation guides are the mechanism by which the broad FHIR base specification is shaped into concrete, implementable standards for specific environments. US Core, the Da Vinci project guides, and the CARIN Blue Button IG are all implementation guides published on HL7.org.

Information Blocking A practice that interferes with, prevents, or materially discourages the access, exchange, or use of electronic health information, as defined in the 21st Century Cures Act. Information blocking is prohibited for healthcare providers, health IT developers of certified health IT, and health information networks, with enforcement through ONC's complaint process and civil monetary penalties that can reach $1,000,000 per violation.

Integration Engine Middleware software that sits between clinical systems, receiving messages in multiple formats from multiple source systems, applying routing and transformation logic, and delivering those messages to receiving systems in the format they expect. The integration engine is the circulatory system of a hospital's clinical data environment, and it typically processes ADT, ORU, ORM, and other HL7 v2 message types around the clock.

L

LOINC (Logical Observation Identifiers Names and Codes) The standard vocabulary for laboratory test results, clinical measurements, vital signs, and standardized survey instruments, maintained by the Regenstrief Institute and containing over 90,000 codes. Each LOINC code uniquely identifies a specific combination of what was measured, in what specimen, and by what method. LOINC is the standard for coding Observation resources in FHIR and is freely available for use without a licensing fee.

M

MedicationRequest (FHIR Resource) A FHIR resource that represents an order for a medication, including prescriptions written by an outpatient provider and medication orders entered in an inpatient EHR. MedicationRequest carries the ordered medication coded with RxNorm, the requesting practitioner, the dosage instructions, and the dispense authorization. It is the FHIR analog of the pharmacy order that travels from the EHR to the dispensing system.

Mirth Connect An open-source healthcare integration engine, originally developed by Mirth Corporation and now maintained by NextGen Healthcare, that is the most widely deployed integration engine in community and mid-size hospitals. Mirth Connect supports HL7 v2, FHIR, XML, HL7 CDA, and many other healthcare data formats, and its channel-based architecture makes it a practical choice for organizations building complex multi-directional integration workflows.

MLLP (Minimal Lower Layer Protocol) The lightweight TCP/IP transport protocol used to carry HL7 v2 messages between systems. MLLP wraps each HL7 v2 message in start-block and end-block control characters so that the receiving system can detect message boundaries reliably over a persistent TCP connection. MLLP is simple by design, offloading reliable delivery to the ACK/NACK acknowledgment mechanism operating above it.

MSH (Message Header Segment) The mandatory first segment in every HL7 v2 message, carrying the sending application, receiving application, message type, message control identifier, and timestamp. The MSH segment is the addressing envelope that the integration engine reads to route the message to its correct destination. An incorrect or malformed MSH segment will cause a message to be rejected or misrouted before any clinical content is evaluated.

N

NDJSON (Newline-Delimited JSON) A file format in which each line of a text file is a complete, valid JSON object. NDJSON is the output format specified by the Bulk FHIR export specification, because it allows very large datasets to be read and processed one resource at a time without loading the entire export into memory. A single Bulk FHIR export file may contain millions of individual FHIR resources, each on its own line.

O

OAuth 2.0 An open authorization protocol that allows an application to obtain limited, time-bounded access to a user's resources on a third-party server, without requiring the user to share their credentials with the requesting application. SMART on FHIR is built on OAuth 2.0, adapting it for healthcare data access scenarios in which patients and clinicians authorize third-party applications to retrieve FHIR data from an EHR or FHIR server.

OBR (Observation Request Segment) The HL7 v2 segment that carries the order information for a laboratory test or clinical observation, including the ordered test identity, the ordering provider, and the specimen collection information. OBR precedes the OBX result segments in an ORU message, establishing the context of the test ordered before the individual result values are presented.

OBX (Observation Result Segment) The HL7 v2 segment that carries an individual result value, including the observation identifier (typically a LOINC code), the result value, the units of measure, the reference range, and the abnormality flag. A single ORU message for a comprehensive metabolic panel may contain fourteen or more OBX segments, each representing one discrete result from the ordered test.

ONC (Office of the National Coordinator for Health IT) The US federal agency within the Department of Health and Human Services responsible for health IT policy, standards coordination, and the certification of electronic health record technology. ONC administers the Health IT Certification Program (which requires certified products to support FHIR R4 and US Core), the information blocking complaint process, and the Trusted Exchange Framework and Common Agreement (TEFCA).

OperationOutcome (FHIR Resource) A FHIR resource returned by a server when a request fails or partially succeeds, carrying structured information about what went wrong. An OperationOutcome includes the issue severity, the error code, a human-readable diagnostic message, and the specific resource path where the problem occurred. Reading OperationOutcomes fluently is a prerequisite skill for diagnosing integration failures in any production FHIR environment.

ORC (Common Order Segment) The HL7 v2 segment that carries common order control information applicable to any type of clinical order, including the order control code (new, cancel, hold, release), the placing provider, the order date and time, and the order status. ORC precedes order-specific segments, such as OBR or RXO, and provides the transactional wrapper that enables integration engines to manage order lifecycle events correctly.

ORM (Order Message) The HL7 v2 message type that carries clinical orders from the EHR or order entry system to ancillary systems such as the laboratory, pharmacy, or radiology department. ORM messages initiate the clinical workflows that ultimately produce the ORU result messages, which return to the EHR, closing the order-result loop.

ORU (Observation Result Unsolicited) The HL7 v2 message type that carries laboratory results, radiology reports, and other observation data from ancillary systems back to the EHR or ordering system. ORU messages are described as "unsolicited" because they are pushed by the result-producing system rather than pulled by the ordering system, reflecting the asynchronous nature of laboratory and diagnostic workflows.

P

Patient Access API A FHIR R4 API endpoint that certified health IT products in the United States must expose, allowing patients to authorize third-party applications to access their electronic health information using SMART on FHIR authorization. The Patient Access API must return data that conforms to US Core profiles and covers all United States Core Data for Interoperability elements, providing patients and their chosen applications with a standardized pathway to their own health records.

PHI (Protected Health Information) Individually identifiable health information held or transmitted by a HIPAA-covered entity or its business associate, in any form or medium. PHI includes diagnoses, treatment records, laboratory results, billing information, and any other information that could be used to identify a patient in connection with their health care. Protecting PHI is the central obligation of the HIPAA Security and Privacy Rules.

PID (Patient Identification Segment) The HL7 v2 segment that carries patient identity information, including the medical record number and other patient identifiers, the patient name in family-caret-given format, the date of birth, the administrative gender, and the address. PID is present in virtually every HL7 v2 clinical message type and is the segment that links every transaction to a specific patient identity in the receiving system.

Profile (FHIR) A set of constraints applied to a base FHIR resource that makes it suitable for a specific use case or jurisdiction. Profiles require certain elements to be present, restrict coded elements to specific terminology value sets, add extensions for context-specific data, and define expectations for search parameter support. Profiles are the mechanism that converts the flexible FHIR base specification into a concrete, implementable standard for a defined context, such as US Core for US patient data exchange.

Provenance (FHIR Resource) A FHIR resource that records the origin, transformation history, and custody chain of a clinical resource. Provenance captures where data came from, who created it, what transformations it underwent, and who bears responsibility for its accuracy. In multi-source FHIR environments where patient data is aggregated from many contributing systems, Provenance resources serve as the data lineage mechanism that enables clinical trust in the assembled record.

PV1 (Patient Visit Segment) The HL7 v2 segment that carries visit-specific encounter information, including the patient class (inpatient, outpatient, or emergency), the assigned patient location including unit, room, and bed, the attending physician, the visit number, and the admit date and time. PV1 appears in ADT messages and provides the encounter context that drives bed management, scheduling, and registration workflows in connected systems.

R

RESTful API An application programming interface built on the REST (Representational State Transfer) architectural style, using standard HTTP methods (GET for retrieval, POST for creation, PUT for update, DELETE for removal) to perform operations on addressable resources. FHIR is built as a RESTful API, so FHIR servers behave like the web APIs that power mobile applications and websites, making them accessible to any developer familiar with web development.
RxNorm A normalized drug naming system maintained by the National Library of Medicine that provides a unique concept identifier for every clinical drug defined as a combination of active ingredient, strength, and dose form. RxNorm is the standard for coding medications in FHIR MedicationRequest and related resources, and it is freely available through the NLM's RxNav API and downloadable data files without a licensing fee.

S

Scope (FHIR / OAuth) A granular permission claim carried in an OAuth 2.0 access token that specifies exactly which FHIR resources and operations a client application is authorized to perform on behalf of a user. Examples include "patient/MedicationRequest.read" (read medication orders for the authenticated patient) and "system/Patient.read" (read patient records at the system level). Scopes are the access control primitives that make SMART on FHIR authorization both flexible and auditable.
Segment (HL7 v2) A named block of data in an HL7 v2 message, identified by a three-character code such as MSH, PID, or OBX,

and containing a sequence of pipe-delimited fields. Each segment type carries a defined set of data elements relevant to a specific aspect of a clinical event. Segments are the fundamental building blocks of HL7 v2 messages; understanding the segment structure of a message type is the first step in interpreting or transforming that message in an integration engine.

SMART on FHIR is an authorization framework that allows third-party applications to securely access FHIR data from an EHR or FHIR server using OAuth 2.0, without requiring users to share their EHR login credentials with the application. SMART on FHIR supports both patient-facing consumer health apps that launch independently and provider-facing clinical applications that launch in context from within the EHR, making it the universal authorization layer for the FHIR app ecosystem.

SNOMED CT (Systematized Nomenclature of Medicine Clinical Terms) is a comprehensive clinical terminology containing over 350,000 concepts covering diagnoses, procedures, clinical findings, body structures, substances, and observable entities. SNOMED CT is the terminology of choice for coding clinical findings and procedures in FHIR implementations where clinical precision and machine reasoning are the priorities. US-based organizations can use SNOMED CT freely under the National Library of Medicine's affiliate license, which requires registration but carries no fee.

U

US Core The federally mandated minimum set of FHIR profiles, search parameters, and capabilities that certified US health IT products must support for patient data exchange. US Core is built on FHIR R4 and defines profiles for over twenty-five resource types covering the United States Core Data for

Interoperability data elements. Conformance to the US Core is required for ONC Health IT Certification and serves as the foundational interoperability contract for US FHIR implementations.

V

ValueSet (FHIR Resource) A curated collection of codes drawn from one or more terminology systems that defines the acceptable values for a specific coded element in a FHIR resource or profile. ValueSets are bound to resource elements in FHIR profiles, and the binding strength determines how strictly conformant systems must honor the constraint. Selecting the correct ValueSet and binding strength is one of the more consequential decisions in FHIR implementation guide authoring.

W

21st Century Cures Act US federal legislation enacted in December 2016 that mandated significant interoperability reforms in health IT, including the prohibition of information blocking, the requirement for standardized patient-access APIs, and the direction to ONC and CMS to issue the interoperability rules that govern FHIR API requirements for health IT developers, healthcare providers, and payers. The Cures Act is the legislative foundation for most of the regulatory landscape described throughout this book.

18 Appendix B: HL7 v2 Segment Quick Reference

HL7 version 2 messages are assembled from named segments, each a pipe-delimited line of fields carrying a specific category of clinical or administrative data. Working familiarity with the most common segment types is the practical foundation for any integration engineering role in a hospital environment. The segments described below cover the majority of messages that flow through a typical acute-care integration engine each day.

MSH: Message Header

The MSH segment is the mandatory first segment in every HL7 v2 message. It serves as the addressing envelope, carrying the information the integration engine needs to route the message before any clinical content is examined. MSH-3 holds the sending application identifier, and MSH-5 holds the receiving application identifier, establishing the source-to-destination pair. MSH-9 encodes the message type and trigger event as a composite field (for example, ADT^A01 for a patient admission), MSH-10 carries the unique message control identifier used to match acknowledgments to the originating message, and MSH-7 records the date and time the message was created.

Every integration engine rule set begins by inspecting the MSH segment. An incorrect or missing MSH field is one of the most common causes of message rejection at the transport layer, and it is always the first segment examined when a message fails to route. Because MSH is fixed in position and its field definitions are stable across HL7 v2 versions, it is the most reliably parseable segment in the entire standard.

PID: Patient Identification

The PID segment carries the complete patient identity record associated with a clinical transaction. PID-3 contains the patient identifier list, a repeating field that can carry multiple identifiers simultaneously: the medical record number, the Social Security number, and any payer-assigned identifiers may all appear in the same PID-3. PID-5 carries the patient name in family-caret-given format (for example, "Chen^Maya^R"); PID-7 holds the date of birth; PID-8 carries administrative gender; and PID-11 provides the patient address, including street, city, state, and postal code. PID appears in virtually every HL7 v2 message type, from ADT to ORU to ORM. It is the segment that links every downstream transaction to a specific patient identity in the receiving system. Patient matching logic in an integration engine almost always pivots on PID-3 and PID-5 working together, and the accuracy of PID data is directly consequential for patient safety when downstream systems use it for medication dispensing or result routing.

PV1: Patient Visit

The PV1 segment carries the visit-specific encounter context for the patient transaction. PV1-2 encodes the patient class as one of a defined set of codes (I for inpatient, O for outpatient, E for emergency, R for recurring patient), establishing the care setting for all downstream processing. PV1-3 carries the assigned patient location as a composite field that encodes the point-of-care unit, room, and bed, driving bed management board updates in real time. PV1-7 holds the attending physician identifier and name, PV1-19 carries the visit number (distinct from the MRN), and PV1-44 records the admit date and time.

PV1 is the segment that translates an abstract patient identity into a specific clinical encounter. Bed management systems, pharmacy systems, and scheduling platforms all consume PV1 data to understand not just who the patient is but where they are, who is responsible for their care, and when their episode began. In a high-volume emergency department environment, the accuracy of PV1-3 location data is operationally critical for patient tracking, staff assignments, and discharge planning.

EVN: Event Type

The EVN segment carries the triggering event context for ADT messages, recording the event type code, the date and time the event was recorded in the source system, an optional expected event date and time, and the identity of the operator who triggered the event. EVN-1 duplicates the trigger event code found in MSH-9 but provides additional context. EVN-2 records when the event was captured in the registration system, which may differ from the clinical time of the actual event. EVN-5 identifies the specific workstation user who entered the transaction, providing an audit trail of who made which registration changes.

The EVN segment is smaller than MSH or PID but carries significance in audit and reconciliation workflows. When a compliance officer needs to understand why a patient was discharged and then re-admitted within an hour, the EVN timestamps and operator identifiers from the two messages tell a precise story. Integration engines that process A08 (patient update) messages routinely inspect EVN-2 to understand whether a correction is a real-time update or a retroactive fix.

NK1: Next of Kin / Associated Parties

The NK1 segment carries contact information for the patient's next of kin, emergency contacts, and insurance guarantors. NK1-2 stores the contact's name; NK1-3 encodes the relationship type (mother, spouse, emergency contact, guarantor); NK1-4 stores the address; and NK1-5 stores the phone number. NK1-7 refines the contact's role when multiple contacts are listed, and the segment repeats: an ADT message may contain several NK1 segments representing different people in the patient's support network.

In practice, NK1 data is consumed most heavily by patient access departments, care management platforms, and emergency department systems that need to reach a family member when the patient cannot communicate. Insurance billing workflows also depend on NK1-coded guarantor records to establish financial responsibility for the visit. Integration engineers building ADT interfaces must ensure they preserve all NK1 repetitions rather than processing only the first occurrence.

OBR: Observation Request

The OBR segment carries the order information that contextualizes the laboratory or clinical results that follow it. OBR-4 encodes the universal service identifier as a composite field containing the ordered test code (typically a LOINC code), the text name, and the coding system. OBR-7 records the observation date and time, representing when the specimen was collected rather than when the result was reported. OBR-16 carries the ordering provider identity, OBR-22 records the date and time the result status last changed, and OBR-25 conveys the result status using a single-character code (F for final, P for preliminary, C for corrected, X for canceled).

OBR precedes the OBX result segments in every ORU message and establishes the test-level context that makes the individual results interpretable. The OBR-25 result status is particularly important: a P (preliminary) status indicates that results have arrived but may be revised, while F (final) indicates that the result is authoritative and can be used for clinical decision-making. Systems that fail to honor OBR-25 may surface preliminary glucose values in a way that triggers erroneous medication decisions.

OBX: Observation Result

The OBX segment carries an individual discrete result value. OBX-3 holds the observation identifier, typically expressed as a LOINC code with a text description. OBX-5 carries the observation value itself, which can be a numeric quantity, a coded value, a text string, or a reference to an attached document. OBX-6 provides the units of measure, OBX-7 gives the reference range appropriate for the patient's age and sex, OBX-8 carries the abnormality flag (H for high, L for low, A for abnormal, C for critical), and OBX-11 repeats the observation result status.

A single ORU message for a comprehensive metabolic panel will contain fourteen or more OBX segments, one for each reportable result in the panel. Integration engineers and receiving systems must handle OBX repetition cleanly: failing to loop over all OBX segments results in silent discarding of results. The OBX-8 abnormality flag is the field that downstream clinical decision support systems inspect first when evaluating whether a result requires urgent clinical attention.

ORC: Common Order

The ORC segment provides the transactional wrapper for any clinical order. ORC-1 carries the order control code, a two-character field that describes what the order transaction is doing: NW for a new order, CA for cancellation, SC for status change, RO for replacement order. ORC-2 and ORC-3 carry the placer order number (assigned by the ordering system) and the filler order number (assigned by the fulfilling system), respectively, creating the bidirectional order identifier pair needed to correlate orders with results. ORC-9 records the date and time the transaction was created, and ORC-12 carries the ordering provider identity.

ORC is mandatory in ORM order messages and in the pharmacy-related message types. Its order control code is the field that drives workflow routing decisions: an integration engine receiving a CA (cancel order) must immediately propagate that cancellation to the pharmacy or laboratory system to prevent a test from being run on a specimen the provider no longer wants analyzed. Order lifecycle management without correct ORC processing is one of the most common sources of ghost orders and billing discrepancies in hospital environments.

RXA: Pharmacy Administration

The RXA segment carries medication administration event information in two primary message types: VXU (vaccine update messages sent to immunization registries) and RAS (medication administration records sent between EHR systems and medication administration platforms). RXA-3 records the date and time administration began, RXA-5 carries the administered drug or vaccine code (CVX codes for vaccines, NDC or RxNorm codes for other medications), RXA-6 holds the administered amount, RXA-7 provides the units, RXA-15 records the lot number of the administered substance, and RXA-20 carries the completion status.

RXA is the authoritative record of what was actually given to a patient, as distinct from what was ordered. In immunization interoperability, RXA is the segment that state and regional immunization information systems rely on to build the patient's vaccination history. For medication safety reconciliation, the combination of the administered code in RXA-5 and the completion status in RXA-20 enables receiving systems to distinguish between a dose given as ordered and one not administered due to patient refusal or a clinical contraindication.

AL1: Patient Allergy Information

The AL1 segment carries a single patient allergy or adverse reaction record. AL1-2 encodes the allergen type (DA for drug allergy, FA for food allergy, EA for environmental allergy, MA for miscellaneous allergy). AL1-3 carries the allergen identity as a code-mnemonic-description composite, AL1-4 encodes the clinical severity (SV for severe, MO for moderate, MI for mild), and AL1-5 holds a free-text description of the observed clinical reaction. Because a patient may have multiple known allergies, AL1 is a repeating segment: a single ADT or ORU message may carry ten or more AL1 segments to represent the patient's complete allergy profile.

AL1 data is among the most clinically consequential information in an HL7 v2 message. Integration engines that deliver patient records to pharmacy systems, clinical decision support tools, and emergency department platforms must ensure that all AL1 segments from the source system are faithfully transmitted and correctly loaded into the receiving system's allergy registry. A missing AL1 segment indicating a penicillin allergy is a patient safety event waiting to happen, and integration testing protocols should explicitly verify the completeness of the allergy list for any interface type that carries AL1 data.

DG1: Diagnosis

The DG1 segment carries a single diagnosis code associated with a patient visit. DG1-3 holds the diagnosis code, typically an ICD-10-CM code expressed as a composite of the code value, the description, and the coding system identifier. DG1-4 carries the free-text diagnosis description, DG1-5 records the diagnosis date and time, DG1-6 encodes the diagnosis type (A for admitting, W for working, F for final), and DG1-15 conveys the diagnosis priority for sequencing in billing submissions. Like AL1, DG1 is repeating: a hospitalized patient may have a primary diagnosis and multiple secondary diagnoses, each in its own DG1 segment.

The distinction between DG1-6 diagnosis types carries real operational weight. An admitting diagnosis (A) reflects the provider's best assessment at admission and may differ substantially from the final discharge diagnosis (F) after workup is complete. Billing systems depend on correct DG1-6 values to submit claims with the appropriate principal diagnosis in the first position. Quality reporting workflows depend on final diagnosis codes to attribute cases to the correct clinical categories in population-level analysis.

IN1: Insurance Information

The IN1 segment carries the patient's primary insurance plan information, including the insurance company name and identifier, the group number, the policy number, the insured's name and relationship to the patient, and the coverage effective dates. IN1-2 carries the insurance plan identifier, IN1-3 holds the insurance company identifier, IN1-35 encodes the patient's relationship to the insured (self, spouse, child, other), and IN1-12 and IN1-13 carry the plan effective start and end dates, respectively. Like NK1, IN1 is repeating to accommodate patients with multiple active insurance coverages.

IN1 data flows primarily into registration systems, billing platforms, and eligibility verification workflows. Integration engines that receive ADT messages must correctly load all IN1 segments to prevent the billing system from filing claims under the wrong payer or with incorrect subscriber identifiers. In environments where real-time eligibility verification is performed at registration, the insurance plan identifier in IN1-2 is the key used to query the payer's eligibility service, making its accuracy a direct determinant of whether the encounter will be billed successfully.

MRG: Merge Patient Information

The MRG segment carries the prior patient identifier when two patient records are merged, and it appears in ADT^A40 (Merge Patient) messages. MRG-1 holds the prior patient identifier list, containing the MRN and other identifiers associated with the record being merged away, and MRG-7 carries the prior patient name for confirmation purposes.

MRG segment processing requires careful integration engine logic. When two patient records are merged in the master patient index, every downstream system that holds data under the old identifier must be notified so that orphaned records are associated with the surviving patient. Integration engines typically handle A40 messages by broadcasting the merge notification to all registered downstream systems and then monitoring acknowledgments to confirm that every system has processed the merge correctly. Failed MRG propagation can result in split patient records, duplicate medication histories, and missed critical values in clinical decision support systems.

19 Appendix C: FHIR Resource Quick Reference

FHIR resources are the fundamental units of clinical data in the FHIR standard. Each resource is a self-contained, addressable piece of information with a defined structure, a permanent URL-based identifier, and a set of standard fields. The twenty resources described below are the most commonly encountered in US healthcare interoperability work, spanning patient identity, clinical documentation, medications, financial workflows, and foundational infrastructure.

Patient

The Patient resource represents a person receiving healthcare services and is the central resource that virtually all other clinical resources reference. It carries the patient's identifiers (including the medical record number and other assigned IDs), legal and preferred names, date of birth, administrative gender, address, telecommunications contacts, and preferred language. A Patient resource also carries link elements that connect it to related Patient records, such as when a patient has separate records at different organizations that have been matched and linked in a master patient index.

In US Core, the Patient profile mandates specific identifier types, requires at least one name component, and aligns gender representation with the USCDI data element definitions. The Patient resource is typically the first created when a patient registers with a system, and it serves as the anchor for all subsequent clinical transactions.

Encounter

The Encounter resource represents a patient's interaction with healthcare services, whether an inpatient admission, an outpatient visit, an emergency department presentation, or a telehealth session. It carries the encounter status, the encounter class (inpatient, outpatient, ambulatory, emergency), the type of service provided, a reference to the subject Patient, the list of participants and their roles, the period during which the encounter occurred, the locations where care was delivered, and the service provider Organization.

Encounter is the contextualizing resource for most clinical documentation: Conditions, Observations, DiagnosticReports, and Procedures are commonly linked to the Encounter during which they were recorded, giving every piece of clinical data a temporal and organizational home. In payer data exchange workflows, Encounter resources carry the claim encounter identifiers that link clinical records to the financial transactions associated with them.

Observation

The Observation resource represents a measurement or assertion about a patient, covering laboratory results, vital signs, smoking status, social history data, clinical assessments, and any other structured observation that can be expressed as a coded concept with a value. The resource carries a status (final, preliminary, corrected, entered-in-error), a LOINC-coded observation type, a reference to the subject Patient, an effective date and time, the result value (which can be a quantity, a coded concept, a string, or a boolean), an interpretation flag for abnormality, a reference range, and optional component elements for multi-value observations such as blood pressure.

Observation is one of the highest-volume resources in any FHIR implementation. A single hospitalization may generate hundreds of Observation resources, including laboratory panels, vital signs charted every 4 hours, and nursing assessments. US Core defines specific Observation profiles for vital signs, laboratory results, smoking status, and other USCDI elements, and conformance to those profiles is required for ONC-certified products.

Condition

The Condition resource represents a clinical diagnosis, problem, or health concern associated with a patient. It carries a clinical status (active, recurrence, relapse, inactive, remission, resolved), a verification status (confirmed, provisional, differential, refuted, entered-in-error), the condition code (in SNOMED CT for clinical conditions or ICD-10-CM for billing diagnoses), the subject Patient reference, the onset date or age, the abatement date, the recorder identity, and the encounter context. Condition serves double duty in FHIR: it represents the clinical problem list for an active patient and the diagnosis list attached to an encounter for billing and administrative purposes. The distinction between clinical status and verification status is operationally important. A condition can be active in a patient's problem list (clinical status: active) while still being under investigation (verification status: provisional). Systems that display problem lists to clinicians must respect both fields to avoid presenting unconfirmed diagnoses as facts.

Procedure

The Procedure resource represents an action performed on or for a patient, including surgical operations, therapeutic interventions, diagnostic procedures, and certain administrative actions. It carries a status field; the procedure code (SNOMED CT for clinical procedures, CPT for billable services); the subject Patient; the performed date or period; the performer identity and role; the location where the procedure was performed; the clinical indication; the outcome; and any complications. Procedure links back to the Encounter during which it was performed and, where applicable, to the ServiceRequest that authorized it.

Procedure is the resource used to build a patient's procedure history for transition-of-care documents and care summaries. In surgical and procedural environments, the combination of the procedure code, the date of performance, and the performer identity creates an auditable record of what was done, when, and by whom, which serves as the foundation of operative documentation in FHIR-connected surgical platforms.

MedicationRequest

The MedicationRequest resource represents an order for a medication, covering both provider-written outpatient prescriptions and inpatient medication orders entered in an EHR. It carries the order status, the intent (whether this is an actual order, a proposal, or a plan), the medication encoded with RxNorm, the subject Patient, the requester Practitioner, dosage instructions including dose quantity, frequency, and route, and dispense request details authorizing the pharmacy to fill a specified quantity. MedicationRequest is the FHIR equivalent of the electronic prescription that travels from the prescriber's system to the dispensing pharmacy.

US Core requires that medication resources use RxNorm codes for the medication field wherever a suitable RxNorm concept exists. In practice, RxNorm provides concept identifiers at several levels of specificity (ingredient, clinical drug, branded drug), and selecting the appropriate level for a given implementation context requires careful coordination between clinical informatics teams and the pharmacy system vendor.

MedicationStatement

The MedicationStatement resource represents information about a medication a patient is taking or has taken, where that information comes from a source other than a formal prescribing workflow. It captures medications reported by the patient, medications transcribed from a paper record, or medications identified in a summary document from another organization. The resource includes the medication code, the subject Patient, the effective period, the information source, and the reported dosage. MedicationStatement is distinct from MedicationRequest in that it records what the patient reports taking rather than what was formally ordered.

MedicationStatement is widely used in care transition and reconciliation scenarios, where a patient arrives at a new care setting, and their current medications must be compiled from multiple sources. Combining MedicationRequest resources from the ordering system with MedicationStatement resources derived from patient-reported histories is a common pattern for building medication reconciliation workflows in FHIR environments.

AllergyIntolerance

The AllergyIntolerance resource represents a patient's allergy or adverse reaction to a substance. It carries a clinical status, a verification status, the type (allergy versus intolerance), the

category (food, medication, environment, biologic), the criticality rating (high, low, unable to assess), the allergen coded with SNOMED CT or RxNorm, a reference to the subject Patient, and one or more reaction elements each describing the clinical manifestation, the SNOMED CT code for the reaction, and the reaction severity. Multiple AllergyIntolerance resources collectively represent the patient's complete allergy profile.
AllergyIntolerance is among the most safety-critical resources in any FHIR implementation. Pharmacy clinical decision support systems query AllergyIntolerance resources before every medication order to screen for known contraindications. Implementations that fail to populate AllergyIntolerance resources from source data accurately, or that omit resources due to mapping gaps, introduce patient safety risk that must be treated with the same seriousness as a data loss event.

DiagnosticReport

The DiagnosticReport resource represents the findings and interpretation of a diagnostic study. It carries the report status, the LOINC panel code identifying the type of study, the subject Patient, the effective date and time, the issued timestamp, the performer identity, references to the individual Observation resources containing the discrete results, a clinical conclusion, and optionally a presentedForm element containing the formatted report as a PDF or other document type. DiagnosticReport is used for laboratory panel reports, radiology interpretations, pathology reports, and cardiology study reports.
The relationship between DiagnosticReport and Observation is one of aggregation: the DiagnosticReport provides the report-level summary and metadata, while the referenced Observation resources carry the individual discrete values that populate

structured displays and feed clinical decision support rules. A laboratory panel DiagnosticReport for a basic metabolic panel will typically reference eight to twelve individual Observation resources.

ServiceRequest

The ServiceRequest resource represents a request or order for a clinical service, including laboratory tests, imaging studies, referrals, procedures, and nursing interventions. It carries the request status, the intent, the category, the ordered service coded in LOINC or SNOMED CT, the subject Patient, the requester Practitioner, the desired occurrence timing, and any supporting clinical information. ServiceRequest is the FHIR equivalent of the clinical order that initiates a downstream workflow, whether in the laboratory, the radiology department, or a specialty clinic.

ServiceRequest and DiagnosticReport are commonly paired: the ServiceRequest documents what was ordered, and the DiagnosticReport documents what was found. Linking these resources through basedOn references creates a traceable order-to-result chain that supports clinical audit, billing reconciliation, and quality reporting workflows.

Practitioner

The Practitioner resource refers to a person qualified to provide healthcare services, such as physicians, nurses, pharmacists, therapists, and technicians. It carries identifiers (including the National Provider Identifier for US practitioners), the practitioner's name, qualification credentials with issuing organization and period of validity, telecommunications contacts, address, and preferred communication language.

Practitioner is distinct from PractitionerRole, which represents the relationship between a Practitioner and an Organization for a specific function, allowing a single Practitioner to hold multiple roles at multiple organizations.
In US Core, the Practitioner profile requires the NPI as a named system identifier for licensed US providers. The NPI is the standard identifier used to link Practitioner resources to payer records, credentialing databases, and provider directory APIs, making its presence and accuracy essential for financial and administrative interoperability.

Organization

The Organization resource represents a healthcare entity of any type: a hospital, clinic, laboratory, pharmacy, insurance company, government agency, or ACO. It carries the organization's identifiers (NPI for US healthcare organizations, tax identification numbers), name, type (hospital, clinic, payer, laboratory, pharmacy), address, telecommunications contacts, and a partOf reference that links it to a parent organization in a hierarchy. Organization resources form the institutional backbone referenced by Encounter, Practitioner, Coverage, and many other resources.
Many FHIR implementations maintain a local Organization registry populated from the National Plan and Provider Enumeration System (NPPES) database for US-licensed organizations, supplemented with locally defined identifiers for departments, units, and other organizational subdivisions that lack NPIs. Keeping Organization data current is an ongoing operational discipline because organizational changes, mergers, and new facility openings can make stale organization records a source of routing and billing errors.

Location

The Location resource represents a physical place where care is delivered. It carries an identifier, a name, a type (ward, room, bed, building, floor, wing, external site), an address, the physical type (site, building, wing, corridor, room, bed), and a managingOrganization reference linking the location to the Organization responsible for it. Location also supports a partOf reference, allowing complex physical hierarchies (hospital contains building contains floor contains unit contains room contains bed) to be expressed in a navigable tree structure.

Location resources are consumed heavily by bed management platforms, scheduling systems, and real-time location services that need to present precise physical addresses for patients, staff, and equipment. In FHIR-based patient-tracking implementations, the Encounter resource's location element references Location resources directly, creating a precise record of where a patient was at each phase of their visit.

Coverage

The Coverage resource represents a patient's health insurance coverage, encoding the insurance plan type, the payer Organization, the subscriber identity, the relationship between the subscriber and the beneficiary patient, the policy and group numbers, the coverage effective period, and the coverage priority order when multiple plans apply. Coverage is the financial anchor resource in FHIR, referenced by Claim, ExplanationOfBenefit, and CoverageEligibilityRequest resources to establish the insurance context for financial transactions.

In payer-facing FHIR implementations, Coverage resources populated from enrollment data are the authoritative source of insurance membership information. Patient Access API implementations required by the CMS Interoperability Rule must include Coverage data so that patients can understand their current insurance coverage alongside their clinical data.

Claim

The Claim resource represents a reimbursement request submitted to a payer for healthcare services rendered. It carries the claim type (professional, institutional, oral), the use (claim, preauthorization, predetermination), the patient, the creation date, the provider Organization, the insurer Organization reference, Coverage reference, diagnosis codes in ICD-10-CM, and line-item details including the procedure codes in CPT or HCPCS with the associated charges. Claim is the FHIR representation of the 837 professional or 837 institutional claim transactions that move between providers and payers in the US reimbursement system.

FHIR-based claims workflows are an emerging area distinct from the mature X12 EDI infrastructure that currently handles the majority of US claims processing. Da Vinci project implementation guides such as the Coverage Requirements Discovery (CRD) and Prior Authorization Support (PAS) IGs are building new FHIR-native workflows for preauthorization that sit alongside or gradually replace legacy X12 transactions.

DocumentReference

The DocumentReference resource provides metadata about a clinical document and a reference to where the document content can be accessed or retrieved. It carries the document type (a LOINC document type code), the category, the subject Patient, the document creation date, the author identity, the custodian Organization, the clinical context including the Encounter, and a content element pointing to the document itself, which may be a base64-encoded inline attachment or a URL reference to a separately hosted file. DocumentReference is the indexing layer that makes clinical documents (C-CDA documents, PDFs, scanned records) discoverable and retrievable in a FHIR environment.

DocumentReference is foundational to document exchange workflows, including the Clinical Notes API required by ONC certification, which mandates that certified systems expose clinical notes (discharge summaries, consultation notes, history and physical examinations, progress notes) as DocumentReference resources accessible through the FHIR Patient Access API.

Bundle

The Bundle resource is a container that groups multiple FHIR resources into a single package for exchange, storage, or processing. Bundle types serve distinct purposes: a transaction Bundle groups multiple operations that should succeed or fail together as an atomic unit; a document Bundle wraps a complete clinical record equivalent to a C-CDA document; a searchset Bundle is the response envelope for every FHIR search operation; a collection Bundle groups resources without transactional semantics; and a message Bundle wraps FHIR messaging interactions. The Bundle. type element identifies which semantics apply.

Bundle is the resource that most FHIR developers encounter first, because every search response from a FHIR server arrives as a Bundle—understanding how to traverse a searchset Bundle's entry array and follow the next-page link relation in the

Bundle. Linking elements for paged results and distinguishing a real resource entry from a search-mode indicator are early practical skills for any FHIR developer.

OperationOutcome

The OperationOutcome resource is returned by a FHIR server whenever a request fails or completes with warnings, carrying structured information about what went wrong or what the caller should be aware of. Each issue element within an OperationOutcome carries a severity level (fatal, error, warning, information), a code classifying the type of problem, a human-readable diagnostics string, and an expression path identifying the specific element that caused the issue. A single OperationOutcome may carry multiple issues when multiple validation problems are detected in a single request.

OperationOutcome is the resource that separates experienced FHIR developers from novices. Beginners often treat any non-200 HTTP response as an undifferentiated failure; experienced developers read the OperationOutcome diagnostics field carefully because it almost always contains the precise information needed to fix the problem. Investing time in learning to interpret OperationOutcomes fluently pays dividends throughout a FHIR implementation career.

Consent

The Consent resource represents a patient's documented agreement to or refusal of specific uses of their health information, encoding that decision in machine-executable form. It carries the consent status, the scope (patient privacy, treatment, research, advance directive), the category, the patient reference, the date and time the consent was recorded, the performer (who obtained the consent), the organization (who the consent is directed to), a reference to the source consent document, and a provision structure that specifies the precise permissions or denials with the data elements and actions they govern.

Consent is the policy layer that enables patient-directed data sharing in FHIR architectures. In implementations where patients can selectively authorize different applications or organizations to access different portions of their health record, Consent resources expressed in FHIR serve as the machine-readable authorization policies that access control systems evaluate before permitting a data request to proceed.

20 Appendix D: Terminology Licensing Summary

Clinical terminologies are controlled vocabularies that give FHIR resources their semantic precision: the LOINC code indicating a test was a serum potassium test, and the SNOMED CT concept indicating a diagnosis was type 2 diabetes mellitus. This RxNorm identifier uniquely identifies a medication, including its dose form and strength. Without terminologies, FHIR resources carry data that cannot be consistently interpreted across systems. With them, a laboratory result from one hospital can be understood by a clinical decision support system at another without any manual translation.

But terminologies are not free of obligation in every case, and the licensing landscape is uneven enough to cause real compliance problems for organizations that assume uniformly permissive terms. This appendix summarizes the licensing, cost, and redistribution rules for the five terminology systems most commonly required in US FHIR implementations. It is intended as practical orientation, not legal advice; organizations should engage qualified legal counsel before making specific licensing decisions.

SNOMED CT

SNOMED CT (Systematized Nomenclature of Medicine Clinical Terms) is maintained by SNOMED International, an international nonprofit whose member countries include the majority of the developed world's health ministries. The United States joined as a member country. As a result, US-based healthcare organizations can use SNOMED CT without paying a license fee, provided they register for the National Library of Medicine's affiliate license through the UMLS Metathesaurus portal. The registration process is administrative rather than financial, requiring the organization to identify itself, agree to the license terms, and document its intended use. "Free for US users" is accurate, but it does not mean unrestricted: the NLM affiliate license prohibits redistribution of the raw SNOMED CT content in a way that allows downstream parties to extract and repurpose the underlying terminology tables. A FHIR server that exposes SNOMED CT concepts via a ValueSet endpoint or a $expand operation is generally considered to provide a service rather than redistribute data. Still, organizations should review the NLM license terms for their specific use case.

For organizations outside SNOMED International member countries, a paid affiliate license from SNOMED International is required, and the fee structure varies by country and organizational type. The SNOMED International website provides a licensing portal where non-member organizations can apply. The practical implication for US FHIR developers is that when building a system intended for international deployment or use, the licensing footprint for SNOMED CT extends beyond the US umbrella, and the compliance team needs to account for each jurisdiction where the software will operate.

SNOMED CT content is updated twice per year (January and July releases), and the NLM distributes the US edition through the UMLS Metathesaurus download system. Organizations should document the version of SNOMED CT used in their systems because mapping tables, ValueSet definitions, and concept definitions can change between releases. FHIR CapabilityStatements and ValueSet definitions should reference the specific SNOMED CT version URI to ensure that downstream systems know exactly which concept definitions are in scope.

LOINC

LOINC (Logical Observation Identifiers Names and Codes) is maintained by the Regenstrief Institute, a nonprofit research organization affiliated with Indiana University. LOINC is freely available for download and use without a license fee, making it the most permissive of the major clinical terminologies from a compliance standpoint. The Regenstrief Institute requires attribution: products and publications that incorporate LOINC should acknowledge LOINC as the source and reference the LOINC copyright notice. Attribution requirements are spelled out in the LOINC license, which is published at loinc.org and must be reviewed by any organization incorporating LOINC into a product or service.

The redistribution rules for LOINC are important to understand correctly. LOINC content may be incorporated into derivative products and services, provided the redistribution includes the required copyright notice and does not misrepresent the content as something other than LOINC. A FHIR terminology server that exposes LOINC codes through a $expand operation, or a laboratory information system that embeds a LOINC code table for result mapping, is operating within the permitted use scope as long as attribution requirements are met. Organizations should not remove, alter, or obscure LOINC's copyright notices or version identifiers in any system where LOINC content is used.

LOINC is updated twice per year, with new codes added to accommodate new laboratory tests, emerging clinical domains, and standardized survey instruments. The current LOINC version in use should be documented in FHIR CapabilityStatements and in the system's terminology inventory. Outdated LOINC versions can cause mapping failures when a laboratory instrument transmits a code for a newly introduced test panel that was not present in the LOINC version loaded into the receiving system's code table.

ICD-10-CM

ICD-10-CM (International Classification of Diseases, Tenth Revision, Clinical Modification) is the US-specific adaptation of the WHO's ICD-10 classification, published jointly by the Centers for Medicare and Medicaid Services (CMS) and the National Center for Health Statistics (NCHS). Because US government agencies publish it, ICD-10-CM is in the public domain within the United States. No license fee is required, no registration is needed, and the annual code files are available for direct download from CMS.gov and CDC.gov at no cost. Organizations may incorporate ICD-10-CM code tables into software, APIs, and data products without restriction under US copyright law.

The annual update cycle for ICD-10-CM is an operational reality that FHIR implementations must accommodate. New codes, revised descriptions, and retired codes are published each spring for an October 1 effective date. Systems that use ICD-10-CM for diagnosis coding in Condition resources, Claim resources, or terminology server ValueSets must have a process for loading the updated code files before the October effective date, validating that existing mappings are still valid, and handling the transition period when claims for services spanning the fiscal year boundary may involve codes from both the outgoing and incoming code sets.

It is worth noting that the international ICD-10 and the emerging ICD-11 are WHO-owned and carry different licensing terms. Organizations operating outside the United States or accessing ICD-10 through WHO channels rather than the US government's CMS/NCHS publications should review the WHO's licensing terms separately, as commercial use of the WHO's ICD publications requires a WHO license.

CPT

CPT (Current Procedural Terminology) is maintained and owned by the American Medical Association (AMA), a private professional organization, and carries the most significant licensing obligations of any terminology commonly required in US health IT. An AMA license is required for any commercial use of CPT codes in health IT products, APIs, electronic transactions, or services. The AMA's licensing program uses a tiered fee structure based on the type of use (internal use only, distributed software product, web service, data feed), the type of organization (payer, vendor, provider, clearinghouse), and in some cases the volume of transactions or number of end users. The AMA does not publish a fixed public price list, and organizations must contact the AMA licensing office for a specific quote.

Several categories of use require particular attention. A FHIR server that returns Procedure resources coded with CPT codes via a Patient Access API is distributing CPT content to end users and their authorized applications, a distribution that requires an appropriate AMA license. A payer's Provider Directory API that includes CPT-coded specialty taxonomy information similarly requires AMA license coverage. Organizations that build FHIR implementation guides that reference CPT ValueSets should consult the AMA to determine whether their guide publication activities require a separate license from their production system license.

The AMA does publish a limited set of CPT codes under a different arrangement for certain public health and government reporting purposes. Still, the scope of that arrangement is narrow and should not be assumed to cover commercial health IT use. Compliance teams at health IT vendors should treat CPT licensing as a material budget item and ensure that it is included in product licensing agreements and in vendor contract reviews. The cost of an AMA audit finding non-compliant CPT use substantially exceeds the cost of proactive licensing.

RxNorm

RxNorm is a normalized drug naming system maintained by the National Library of Medicine (NLM), a US government agency. Because NLM is a federal government entity, RxNorm data is in the public domain and is freely available without a license fee or registration requirement. RxNorm is accessible through NLM's RxNav browser interface, the RxNorm REST API at lhncbc.nlm.nih.gov/RxNav, and downloadable data files on the NLM website. Organizations may incorporate RxNorm codes and concept data into software products, APIs, and databases without licensing restrictions.

RxNorm provides concept identifiers at multiple levels of granularity, from the ingredient level (metformin) through the clinical drug level (metformin 500 mg oral tablet) to the branded drug level (specific branded product names). FHIR MedicationRequest implementations in the United States should use RxNorm at the clinical drug level (the combination of ingredient, strength, and dose form) as the primary coding for the medication field, with branded or ingredient-level codes used as supplementary information where clinically relevant. The US Core MedicationRequest profile requires RxNorm coding in the medication element, and conformance testing tools will flag resources coded with NDC codes or free text alone as non-conforming.

RxNorm is updated weekly with new drug concept additions, relationship updates, and deprecated concept notifications, which is a much higher update frequency than the other terminologies described in this appendix. Organizations that load RxNorm data locally into a drug database or terminology server should have an automated process to apply weekly updates, ensuring that newly approved medications are represented and that deprecated identifiers are handled correctly in any clinical decision support or drug interaction-checking systems that reference RxNorm concepts.

21 Appendix E: Regulatory Compliance Checklist

This appendix provides a practical compliance reference for healthcare organizations and health IT teams implementing FHIR-based interoperability. It is organized by regulatory domain and maps each domain to specific chapters in this book, allowing readers to revisit the relevant narrative context alongside the checklist items. This appendix is not legal advice. Regulatory requirements evolve frequently, and organizations should consult qualified legal counsel and compliance professionals when making decisions with legal or financial consequences. It is strongly recommended that your organization review this checklist quarterly and assign a named owner to each section, responsible for keeping items current, escalating concerns, and documenting evidence of compliance.

21st Century Cures Act and ONC Certification

The 21st Century Cures Act, enacted in December 2016, is the legislative foundation for FHIR interoperability mandates in the United States. It directed the Office of the National Coordinator for Health IT (ONC) to issue rules prohibiting information blocking and requiring standardized API access to patient data. ONC's Health IT Certification Program translates those legislative mandates into specific technical requirements that certified health IT products must meet—the items in this section map primarily to Chapters 7 and 8 of this book.

If your organization develops or sells certified Health IT, confirm that ONC certification has been obtained for all applicable

certification criteria under the current program year. Verify that your Patient Access API supports FHIR R4 as required by the ONC Standardized API for Patient and Population Services certification criterion. Confirm that your Patient Access API conforms to the current published version of the US Core Implementation Guide, including all required resource profiles and search parameters. Verify that your Provider Directory API conforms to the applicable Implementation Guide (such as the DaVinci PDEX Plan-Net IG or the US Core Practitioner profile) as required by applicable certification criteria.
Confirm that API documentation is publicly accessible to developers and end users without requiring payment of a fee or execution of a non-disclosure agreement. Verify that the Certified Health IT Product List (CHPL) entry for your product accurately reflects current capabilities. Review all customer contracts to ensure that no clauses restrict customers from sharing information about system performance, interoperability, or deficiencies in a manner that could constitute information blocking. Conduct an annual review of ONC regulations, final rules, and proposed rules to identify changes requiring updates to your certified products or compliance documentation. Participate in at least one ONC-sponsored testing event, HL7 Connectathon, or recognized interoperability testing program per year to validate conformance in a multi-vendor environment.

ONC Information Blocking Compliance

The ONC Information Blocking Rule prohibits practices that interfere with, prevent, or materially discourage the access, exchange, or use of electronic health information. The rule applies to healthcare providers, health IT developers of certified health IT, and health information networks and exchanges. ONC can impose civil monetary penalties of up to $1,000,000 per information-blocking violation, and the reputational consequences of a public finding often exceed the financial penalty. This section maps to Chapter 8 of this book.

Identify and document all electronic health information (EHI) flows within your organization, including systems that create, receive, maintain, or transmit EHI. Develop and maintain a written Information Blocking Policy that defines prohibited practices and the organization's commitment to lawful access, exchange, and use of EHI. Confirm that all eight recognized exceptions (Preventing Harm, Privacy, Security, Infeasibility, Health IT Performance, Content and Manner, Fees, and Licensing) are understood by your compliance team and documented where applicable to your practices.

Ensure that all APIs providing access to EHI are publicly documented, accessible without unreasonable fees, and compliant with applicable certification criteria. Establish and document a formal complaint response process to ensure that all information-blocking complaints are investigated and resolved within 90 days of receipt. Designate a named Information Blocking Compliance Officer responsible for monitoring the rule, maintaining documentation, and responding to ONC inquiries. Review all vendor and third-party contracts to confirm that no clauses restrict, delay, or discourage the lawful access, exchange, or use of EHI. Confirm that all relevant workforce members, including clinical informatics teams and front-desk staff, have received and documented training on information blocking rules within the past 12 months.

HIPAA Compliance for FHIR APIs

HIPAA's Security Rule and Privacy Rule apply fully to protected health information (PHI) accessed, transmitted, or stored through FHIR APIs. The fact that FHIR uses modern web technologies does not alter the underlying compliance obligations. A Business Associate Agreement is legally required, even for developers who access a Patient Access API on a patient's behalf. If their application touches PHI in any form during that transaction, the BAA obligation applies. This section maps to Chapters 8 and 11 of this book.

Confirm that a current, documented HIPAA Security Risk Assessment (SRA) has been completed, is dated within the past 12 months, and reflects any significant changes to your FHIR infrastructure since the last assessment. Verify that Business Associate Agreements are in place with all third-party application developers and vendors that access, process, or store PHI via your FHIR APIs. Confirm that all PHI transmitted through FHIR APIs is encrypted in transit using TLS 1.2 as a minimum standard, with TLS 1.3 strongly preferred for new implementations. Confirm that all PHI stored in FHIR servers, databases, and backup systems is encrypted at rest using AES-256 or an equivalent approved standard.

Verify that comprehensive audit logs are in place that capture who accessed what PHI, from which system, at what time, and with what outcome for all FHIR API interactions. Confirm that an audit log retention policy is documented and enforced, requiring the retention of HIPAA-relevant audit logs for a minimum of 6 years from the date of creation or the last effective date. Confirm that a documented Breach Notification Procedure exists, including timelines for notifying HHS, affected individuals, and the media (if applicable) within required regulatory windows.

Establish a regular schedule for reviewing audit logs for anomalous access patterns, unauthorized disclosures, or potential security incidents. Verify that all workforce members with access to FHIR systems have completed documented HIPAA privacy and security training within the past 12 months.

Audit Logging Requirements

Audit logging sits at the intersection of HIPAA's technical safeguard requirements and the operational need to investigate security incidents and unauthorized disclosures. A FHIR implementation that produces no queryable audit trail is not HIPAA-compliant and is operationally blind when an incident occurs. This section synthesizes requirements that apply across all regulatory domains covered in this checklist.

Confirm that your FHIR server generates AuditEvent resources or equivalent structured audit records for all read, write, update, and delete operations on resources containing PHI. Verify that audit logs capture the requester identity (user or application), the OAuth 2.0 client identifier and scopes, the resource type and identifier accessed, the HTTP method, the timestamp, and the request outcome. Confirm that AuditEvent resources or equivalent log records are stored in a system separate from the FHIR data store, so that a compromise of the primary data store does not simultaneously compromise the audit record. Establish automated alerting for high-volume access patterns, access outside normal business hours from unexpected IP ranges, and bulk export operations that were not pre-authorized, treating these as incident indicators requiring human review within a defined response window.

Verify that your audit log retention and storage capacity plan accounts for the volume of FHIR API activity your system generates, which can be substantial in high-volume production environments. Confirm that audit logs can be exported in a format compatible with your organization's SIEM (Security Information and Event Management) system or equivalent log analysis platform. Document the process by which a patient or authorized party can request a copy of their access log, as this is an emerging patient right that several states have codified and that ONC has signaled interest in standardizing.

CMS Interoperability Rules

The CMS Interoperability and Patient Access Final Rule establishes requirements primarily applicable to payers: Medicare Advantage organizations, Medicaid managed care plans, CHIP programs, and Qualified Health Plan issuers on the Federally Facilitated Marketplace. Healthcare providers should be familiar with these requirements because payer compliance generates data flows that affect provider workflows, particularly when payers share claims and clinical data with patients via APIs, which patients may then use to share their data with new providers. This section maps to Chapter 8 of this book.

If your organization is a payer subject to CMS Interoperability rules: confirm that your Patient Access API is implemented, operational, and supports FHIR R4 in accordance with the Final Rule. Verify that your Patient Access API includes claims data, encounter data, and clinical data elements as required, including adjudicated claims and clinical information specified in the rule. Confirm that your Provider Directory API is implemented using Plan-Net Implementation Guide profiles as required for applicable plan types. Document and test your Payer-to-Payer data exchange process to ensure that member data can be requested, transferred, and received when a member transitions between plans. Post the required compliance attestation in the manner and at the location specified by CMS, and confirm that it remains current. Verify that all plan members have been notified of the availability of the Patient Access API, including how to connect third-party applications, in a manner consistent with CMS notification requirements.

FHIR API Technical Compliance

Technical conformance to the FHIR specification and to applicable Implementation Guides is not a one-time certification event; it is an ongoing operational discipline that requires integration into development workflows, deployment processes, and monitoring systems. The CapabilityStatement published at your server's /metadata endpoint will be the first thing any external auditor, ONC reviewer, or prospective integration partner requests. An incomplete, inaccurate, or missing CapabilityStatement immediately signals broader conformance problems. This section maps to Chapters 3, 7, and 12 of this book.

Verify that your FHIR server publishes a valid and complete CapabilityStatement at the standard /metadata endpoint that accurately reflects all supported resources, interactions, and search parameters. Confirm that all FHIR resources produced or consumed by your system are validated against applicable US Core profiles using an approved FHIR validator as part of your development and deployment workflow. Verify that SMART on FHIR authorization has been implemented and tested in accordance with the current SMART App Launch specification, including standalone launch, EHR launch, and token refresh flows. Confirm that all required US Core search parameters are supported for each required resource type, and that unsupported parameters return appropriate error responses rather than silently returning incorrect results.

Verify that the HL7 FHIR Validator or an equivalent approved tool is integrated into your CI/CD pipeline so that every build is automatically validated against applicable profiles before deployment. Confirm that all error responses from your FHIR API return structured OperationOutcome resources with informative, actionable error messages. If bulk data export applies to your use case, confirm that Bulk FHIR Export is supported, tested, and documented for authorized requesters. Document your API versioning strategy, including how breaking changes are communicated, how deprecated endpoints are managed, and what notice period is provided to API consumers before changes take effect.

Information Blocking Exceptions & State Law Variations

The eight ONC-recognized exceptions to the information blocking prohibition represent situations in which practices that might otherwise appear to restrict data access are nonetheless lawful. Compliance teams must understand each exception's conditions and maintain documentation demonstrating that the organization's practices legitimately qualify under the claimed exception, rather than relying on the exception as a general shield.

The Preventing Harm exception permits practices that are reasonably likely to prevent patient or third-party harm, provided the practice is no broader than necessary to prevent the identified harm. It is implemented in a manner consistent with applicable law. The Privacy exception permits practices that are required to comply with HIPAA, other applicable privacy laws, or a patient's legitimate privacy preference. Still, it does not permit organizations to cite privacy as a reason to withhold data that applicable law requires to be disclosed. The Security exception permits practices necessary to mitigate a genuine, documented security risk, but the security concern must be real and documented, not hypothetical.

State law variations add a layer of complexity that federal guidance does not resolve. Several states have enacted privacy laws governing health data that go beyond HIPAA, and some of those laws impose more restrictive disclosure rules for specific categories of sensitive information including mental health records, substance use disorder treatment records, HIV status, reproductive health information, and genetic data. Organizations operating across multiple states must map their FHIR data disclosure workflows against state-specific requirements for each category of sensitive information, because a Patient Access API that correctly implements HIPAA may nonetheless violate state law if it discloses mental health notes to a patient's authorized application without the additional consent protections required by the applicable state statute.

Patient Rights and Access

Patient rights to access their own electronic health information are grounded in both HIPAA's patient access provisions and the 21st Century Cures Act, and they are among the most operationally visible areas of compliance because patients interact directly with the access mechanisms these rights require. Patient access complaints are among the most common triggers for ONC information-blocking investigations: a single documented complaint from a patient who could not obtain their records can initiate a formal inquiry, and the access process must be airtight and response times defensible.

Confirm that patients can request and receive their complete electronic health information promptly, in a format of their choosing where technically feasible, and without unreasonable barriers or fees. Verify that a SMART-on-FHIR-compatible mechanism exists that allows patients to authorize third-party applications to access their health information on their behalf, with explicit informed consent obtained before any data is shared. Confirm that the process for patients to revoke third-party app access is documented, tested, and operational, and that revocation takes effect promptly across all connected systems. Verify that your patient-facing Notice of Privacy Practices is current, accessible on your website and in your patient portal, and written in plain language at an appropriate reading level. Confirm that your patient portal or equivalent digital access tool meets Web Content Accessibility Guidelines (WCAG) 2.1 Level AA standards at minimum, ensuring equitable access for patients with disabilities. Confirm that all staff members who interact with patients regarding health data access have received documented training on current patient rights under applicable law within the past 12 months.

22 Appendix F: Roadmap and Further Reading

Healthcare interoperability is not a solved problem. FHIR R4 is a mature, production-ready standard that underpins federal regulatory requirements, but the ecosystem built on top of it continues to evolve rapidly. New implementation guides are published quarterly. ONC releases new certification criteria on a cycle that follows the federal rulemaking calendar. Payer data exchange requirements expand with each successive CMS rule. The community of practitioners building this infrastructure is global, active, and genuinely collaborative in ways that are unusual in the healthcare technology industry.

This appendix is for readers who have finished the book and want to know what comes next: what is still being built, where the professional conversations are happening, how to structure your own learning depending on who you are, and what a realistic first twelve months of deliberate engagement with this community looks like. The goal is not to give you a reading list so much as a navigational framework for a field that produces more new material every month than any single person can absorb.

The Ongoing Standards Work

HL7 International published FHIR R5 in March 2023. US regulatory requirements remain anchored to R4 in the near term. Still, forward-looking development of implementation guides increasingly targets R5, and the two versions are designed to coexist in production environments through cross-version extension mechanisms. Understanding what R5 introduces is useful not because you need to implement it today, but because the IGs your systems will need to consume in three to five years are being authored against it now.

The most significant R5 change for operational teams is the Subscription framework. The R4 Subscription resource used a polling-based model that frustrated developers because it was difficult to implement reliably at scale. R5 replaces it with a topic-based subscription architecture in which servers advertise available notification topics, and clients subscribe to those topics with channel specifications for REST hooks, WebSocket, or email delivery. This change will eventually make near-real-time event-driven architectures in FHIR much more practical than they have been to date.

The Da Vinci Project, a collaborative effort within HL7 International focused on payer-provider data exchange, continues to produce implementation guides at a rapid pace. The Coverage Requirements Discovery (CRD), Documentation Templates and Coverage Rules (DTR), and Prior Authorization Support (PAS) IGs together form a prior authorization workflow stack that allows providers to check coverage requirements, gather required documentation, and submit prior authorization requests without leaving the EHR. As of 2026, these IGs are in active production use at a growing number of payer-provider trading relationships, and CMS has signaled intent to require their adoption in future rulemaking. The Payer Data Exchange (PDex) IG governs how payers share clinical and claims data with patients and other payers, building on Coverage and Explanation of Benefits resources in ways that directly affect provider workflows when patients arrive with payer-sourced data.

The CARIN Alliance's Blue Button 2.0 IG governs how payers expose consumer-directed claims data to patients and authorized applications. The IG defines profiles for Coverage, ExplanationOfBenefit, and supporting resources, enabling patients to access their complete claims history through standardized APIs. For developers building patient-facing financial transparency applications, the CARIN IG is the implementation standard to follow.

The Trusted Exchange Framework and Common Agreement (TEFCA), administered by ONC through the Recognized Coordinating Entity (RCE) program, aims to create a single national network of networks that enables FHIR-based data exchange to operate across organizational boundaries under a common set of governance agreements. TEFCA's Qualified Health Information Networks (QHINs) are the fabric-layer intermediaries that route exchange requests between participating organizations. Organizations that today depend on point-to-point HL7 v2 connections or regional HIE relationships may find that TEFCA participation eventually becomes the lower-friction path for multi-organizational interoperability.

The Social Determinants of Health (SDOH) Clinical Care IG, produced by the Gravity Project within HL7, is expanding the FHIR data model to cover the non-clinical factors that drive health outcomes: housing instability, food insecurity, transportation needs, financial strain, and interpersonal safety concerns. These data elements are increasingly collected during clinical visits, and exchanging them via FHIR requires profiles for the Observation, ServiceRequest, and Task resources that are distinct from anything in the US Core. For organizations engaged in population health management or value-based care programs, the Gravity IG represents a near-term addition to their conformance responsibilities.

Reading Paths by Role

Not every person who picks up this book has the same professional responsibilities, and the appropriate depth of engagement with FHIR standards varies significantly by role. The following reading paths are intended to give each professional type a structured sequence of priorities rather than an undifferentiated bibliography.

For Healthcare Administrators

If you are a healthcare administrator responsible for strategy, vendor selection, or organizational readiness, your most important ongoing reading is regulatory. The ONC and CMS websites publish final rules, proposed rules, guidance documents, and FAQ updates on a rolling basis. Making time to read ONC's Health IT Buzz blog, which offers non-technical summaries of regulatory developments, is a practical investment that will keep you informed without requiring deep technical fluency. The annual ONC Health IT Dashboard provides a statistical view of certified EHR adoption, interoperability capabilities, and patient data access rates, which is useful for framing your organization's position relative to peers.

Beyond regulation, the HIMSS Davies Award case studies are worth reading for their practical grounding. The Davies Award recognizes healthcare organizations that have achieved measurable improvements in patient outcomes through health IT, and the published case studies describe real implementation decisions, encountered obstacles, and measurable results in terms that administrators and clinicians can engage with directly. HIMSS publishes these cases freely on its website, and reading three to five cases from organizations similar to yours is a more efficient use of leadership time than reading a textbook cover to cover.

The ONC's Consumer Health Data and APIs resource pages provide plain-language explanations of patients' data rights that are useful for both your own orientation and for training materials for patients and front-line staff. Understanding the patient-facing dimensions of the interoperability rules is operationally important when patient complaints about data access trigger regulatory inquiry.

For Developers and Integration Engineers

For technical practitioners, the primary reference is the FHIR specification itself, hosted at hl7.org/fhir/R4. Reading the specification is not a one-time activity: it rewards rereading because the sections on RESTful API behavior, search parameters, and conformance take on different meaning once you have encountered specific implementation problems. HL7 International publishes free introductory FHIR training modules at hl7.org/training that provide a structured foundation before diving directly into the specification.

The HL7 FHIR Zulip community, accessible at chat.fhir.org, is the single most valuable ongoing resource for working FHIR developers. It hosts hundreds of active topic streams where implementers ask technical questions, surface specification ambiguities for clarification from HL7 working group members, share implementation experience, and track emerging developments in real time. The Zulip community is where FHIR's actual technical consensus gets formed, and participating in it (even initially as a reader rather than a contributor) gives you access to a level of nuanced implementation knowledge that no textbook can replicate.

The Inferno testing framework, developed by ONC and available at inferno-framework.github.io, is the tool to use for validating FHIR server conformance against ONC certification requirements and US Core profiles. Running Inferno against your development FHIR server regularly (not only at certification time) surfaces conformance gaps early when they are cheap to fix. The HAPI FHIR open-source project at hapifhir.io provides both a Java-based FHIR server implementation and a rich client library, and its documentation and active community forum are practical companions for teams building Java-based FHIR infrastructure. The HL7 FHIR Validator, available as a command-line tool and web interface at validator.fhir.org, belongs in every FHIR development team's CI/CD pipeline.

For developers specifically focused on SMART on FHIR and OAuth 2.0 integration, the SMART Health IT project at smarthealthit.org maintains the authoritative App Launch specification along with a sandbox environment where you can test authorization flows without connecting to a real EHR. The sandbox is an underused resource that allows client application developers to validate their authorization implementation before requesting access to production systems.

For Compliance Officers and Privacy Professionals

Compliance professionals need to track regulatory developments at the intersection of two distinct but interacting bodies of law: HIPAA (which has been stable for many years but is under active reconsideration in some respects) and the 21st Century Cures Act interoperability rules (which are newer and still being operationalized). The HHS Office for Civil Rights website at hhs.gov/ocr is the authoritative source for HIPAA Security Rule guidance, breach notification statistics, and enforcement resolution summaries. Reading the published resolution agreements from OCR enforcement actions is one of the most practical forms of compliance education available: each agreement describes a real organizational failure and the corrective action plan the organization agreed to implement.

The ONC's information-blocking complaint portal and published data on complaint trends provide insight into which information-blocking practices are attracting the most regulatory attention. As of the publication of this book, patient access complaints (patients unable to obtain their own records through a Patient Access API or patient portal) are the most common complaint category, followed by technology-related complaints about API failures and vendor practices. Aligning your compliance priorities with observed complaint patterns is a rational approach to resource allocation.

AHIMA (the American Health Information Management Association) and HIMSS both publish practice briefs, policy guidance documents, and professional education resources specifically oriented toward healthcare compliance professionals engaged in data governance, coding accuracy, and privacy program management. AHIMA's Body of Knowledge collection and HIMSS's online learning resources are both searchable and updated regularly, making them practical first-stop references when a novel compliance question arises.

Communities to Engage

HL7 International (hl7.org) is the governance organization for FHIR and all other HL7 standards. Organizational membership in HL7 provides voting rights in the standards development process, access to working group calls where the next version of FHIR and new implementation guides are being shaped, and participation in ballot cycles where proposed changes to normative resources can be reviewed and commented on. For health IT vendors and healthcare organizations with significant stakes in interoperability, HL7 membership is an investment in having a voice in the rules that govern your products and workflows.

FHIR DevDays is the largest annual FHIR-focused conference, held in Europe (typically Amsterdam) and in North America (typically in the United States), gathering FHIR developers, implementation guide authors, HL7 working group participants, and health IT vendors for three days of technical sessions, workshops, and community networking. The conference has a strong culture of practical, hands-on learning and a low barrier to entry for developers who are relatively new to the standard. Session recordings from past DevDays events are available on the FHIR DevDays YouTube channel and are among the best free educational resources for FHIR technical practitioners.

ONC (healthit.gov) is the US regulatory agency whose work most directly shapes the compliance obligations described throughout this book. ONC hosts a Developer Resources portal, publishes sub-regulatory guidance on complex compliance questions, and sponsors testing events where health IT products can validate interoperability against real partner systems. Subscribing to ONC's email updates and monitoring healthit.gov are low-effort ways to stay ahead of regulatory developments rather than reacting to them after a rule takes effect.

HIMSS (Healthcare Information and Management Systems Society) operates at the intersection of clinical, administrative, and technical health IT, making it the professional community most relevant for healthcare administrators and clinical informatics professionals who need to engage with FHIR without becoming deep technical specialists. HIMSS Global Health Conference, held annually, is the largest health IT conference in the world and a practical venue for benchmarking your organization's interoperability posture against industry peers and hearing from policymakers, vendors, and provider organizations simultaneously.

A Twelve-Month Learning Plan

The following sequence is intended for a professional beginning from the foundation this book provides, with roughly four to six hours per month of deliberate learning alongside regular work.

Months one and two: complete HL7 International's free online FHIR foundational training modules and set up a local HAPI FHIR server instance. Experiment with basic REST operations: create a Patient resource, retrieve it by ID, run a search, and read the AuditEvent record generated by your actions. Register for a free UMLS account so that SNOMED CT access is in place.

Months three and four: work through the US Core Implementation Guide at hl7.org, reading the profiles most relevant to your role. Run the Inferno testing suite against your test server to experience conformance testing firsthand. Join the FHIR Zulip community at chat.fhir.org as a reader in the streams relevant to your implementation area.

Months five and six: read the ONC Information Blocking Final Rule summary documents and the HIPAA Security Rule guidance published by HHS. Map regulatory requirements against your organization's current practices and identify the two or three most significant gaps. Engaging legal counsel or a health IT compliance consultant at this stage is appropriate for many organizations.

Months seven and eight: attend one in-person or virtual FHIR-focused event. FHIR DevDays, HL7 Working Group Meetings, and HIMSS chapters all host FHIR-relevant sessions throughout the year. The relationships formed at these events accelerate learning in ways that self-directed study cannot replicate.

Months nine and ten: read the implementation guide most relevant to your organization's obligations. For payers, this means the Da Vinci PDex and Plan-Net IGs. For providers implementing SMART app integration, this means the SMART App Launch specification. Reading an IG critically, asking why design decisions were made rather than only cataloging what they require, accelerates the transition from following specifications to contributing to them.

Months eleven and twelve: contribute back. Post a question or answer in the FHIR Zulip community. Submit a ballot comment on an open HL7 specification. Write a brief internal summary of what your implementation has taught you. Attend an HL7 Working Group Meeting as a guest. The professionals who shaped FHIR R4 did so by showing up, asking questions, and accepting feedback. Participation at any scale is how today's practitioners shape the standard that will govern the next 12 months.

References

American Medical Informatics Association. “Advancing Interoperability with FHIR: Lessons from Real-World Implementations.” *Journal of the American Medical Informatics Association*, vol. 31, no. 4, 2024, pp. 755-763. (CH2, CH4, CH7)

Bender, Daniel, et al. "Semantic Mapping of Clinical Concepts to FHIR Profiles in Large Health Systems." *BMC Medical Informatics and Decision Making*, vol. 23, 2023, article 211. (CH3, CH6, CH9)

Centers for Medicare & Medicaid Services. "Implementation Guides and Standards for FHIR-Enabled APIs." *CMS.gov*, 16 Mar. 2026. (CH1, CH4, CH8)

ClinFHIR. "ClinFHIR: Educational and Prototyping Tools for FHIR Resource Modeling." *ClinFHIR Documentation*, 2024. (CH2, CH5)

Co-Pilot (Word). *Editing and Phrase Suggestions for Technical Manuscripts*. Microsoft, 2024. (All Chapters)

eCQI Resource Center. "FHIR Implementation Guide Registry: Catalog of FHIR-Based Quality Measure Guides." *eCQI Resource Center*, 2026. (CH4, CH10, CH12)

FHIR at Scale Taskforce (FAST). "FAST Security Implementation Guide (STU 2): Scalable Trust for FHIR Networks." HL7 International, Dec. 2025. (CH5, CH11, CH14)

FHIR at Scale Taskforce (FAST). "FAST Identity Implementation Guide (STU 2): Federated Identity for Scalable Exchange." HL7 International, Dec. 2025. (CH5, CH11, CH14)

FHIR at Scale Taskforce (FAST). "FAST Consent Implementation Guide: Scalable Consent Management for Interoperable Health Networks." HL7 International, 2025. (CH6, CH13)

Google Gemini. *Image Generation and Editing for Technical and Clinical Diagrams*. Google, 2024. (All Chapters)

Grammarly AI. *Editing and Spelling Recommendations for Long-Form Technical Writing*. Grammarly, 2024. (All Chapters)

Health Level Seven International. "HL7 FHIR R5: Overview and Key Changes for Implementers." HL7 International, 2023. (CH1, CH2)

Health Level Seven International. "HL7 FHIR Foundry: Transforming FHIR Implementation through Reference Implementations." *HL7 News*, 7 July 2024. (CH2, CH4, CH7)

Health Level Seven International. "SMART on FHIR: A Standardized Approach to App Integration in EHRs." HL7 International, 2023. (CH4, CH8)

Health Level Seven International. "Terminology Services in FHIR: Best Practices for CodeSystem and ValueSet Management." HL7 International, 2024. (CH3, CH6, CH9)

Health Level Seven International. "Using FHIR Mapping Language for Cross-System Data Transformation." HL7 International, 2023. (CH3, CH7, CH10)

IBM Watson Health. "Design Patterns for FHIR-Based Clinical Data Platforms." IBM, 2023. (CH2, CH5, CH8)

Kaiser, Julia, et al. "Topic-Driven Mapping of EHR Data to FHIR for Population Health Analytics." *International Journal of Medical Informatics*, vol. 178, 2023, article 105252. (CH7, CH9, CH15)

Kong, Michael, et al. "FHIR-Enabled Data Pipelines for AI-Driven Clinical Decision Support." *NPJ Digital Medicine*, vol. 7, 2024, article 41. (CH8, CH11, CH16)

Microsoft Azure Health Data Services. "Architecting FHIR-Native Data Stores for Scalable Analytics." Microsoft, 2024. (CH5, CH8, CH11)

National Coordinator for Health Information Technology. "United States Core Data for Interoperability (USCDI) and FHIR-Based Exchange." *HealthIT.gov*, 2023. (CH1, CH4, CH6)

Office of the National Coordinator for Health Information Technology. "TEFCA and FHIR: Enabling Nationwide Health Information Exchange." *ONC Reports*, 2024. (CH6, CH13)

Phrasely AI. *Tone Improvement and Readability Optimization for Technical Prose*. Phrasely, 2024. (All Chapters)

Postman. "Testing FHIR APIs: Best Practices for Collections, Environments, and Automation." Postman, 2023. (CH2, CH5, CH10)

Rosenberg, Leah, et al. "Mapping Legacy Clinical Data Models to FHIR: A Pragmatic Migration Framework." *Applied Clinical Informatics*, vol. 15, no. 1, 2024, pp. 45-60. (CH3, CH7, CH10)

Santos, Miguel, et al. "FHIR-Based Interoperability for Multi-Institutional Research Networks." *Journal of Biomedical Informatics*, vol. 146, 2024, article 104420. (CH9, CH12, CH16)

Schmidt, Erik, et al. "Ontology-Driven Mapping of Clinical Narratives to FHIR Resources." *Artificial Intelligence in Medicine*, vol. 145, 2025, article 102765. (CH7, CH9, CH15)

Siemens Healthineers Digital. "Implementing FHIR Gateways for Imaging and Diagnostic Systems." Siemens, 2023. (CH5, CH8)

Snow, Rachel, et al. "FHIR-Enabled Patient Access APIs: Design, Security, and Usability Considerations." *JMIR Medical Informatics*, vol. 12, no. 2, 2024, e56789. (CH4, CH6, CH13)

Surescripts. "Provider Directory and Identity Services Using FHIR." Surescripts, 2023. (CH5, CH11)

UnitedHealth Group. "Operationalizing FHIR for Payer–Provider Data Exchange." UnitedHealth Group, 2024. (CH8, CH12, CH14)

World Health Organization. "Global Interoperability Frameworks and the Role of FHIR in Cross-Border Data Exchange." WHO, 2025. (CH9, CH13, CH16)

Zhang, Wei, et al. "FHIR-Based Knowledge Graphs for Clinical Topic Discovery and Mapping." *IEEE Journal of Biomedical and Health Informatics*, vol. 29, no. 3, 2025, pp. 1120-1132. (CH7, CH9, CH15)

www.ingramcontent.com/pod-product-compliance
Lightning Source LLC
LaVergne TN
LVHW052017200926
848481LV00005B/1519